Operator Theory
Advances and Applications
Vol. 87

Editor
I. Gohberg

Operator Theory
Advances and Applications
Vol. 67

Editor
I. Gohberg

Riemann's Boundary Problem with Infinite Index

N.V. Govorov

Introduction, Appendix and Editing
by I.V. Ostrovskii

Translation from the Russian
by Yu.I. Lyubarskii

Springer Basel AG

Editor

I.V. Ostrovskii
Institute for Low Temperature Physics & Engineering
47, Lenin pr.
Kharkov, 310064
Ukraine

Originally published in 1986 under the title *Kraevaya zadacha Rimana s beskonechnym indeksom* by Nauka, Moskva.

A CIP catalogue record for this book is available from the Library of Congress, Washington D.C., USA

Deutsche Bibliothek Cataloging-in-Publication Data
Govorov, Nikolaj V.:
Riemann's boundary problem with infinite index / N. V.
Govorov. Introd., app. and ed. by I. V. Ostrovskii. Transl.
from the Russ. by Yu. I. Lyubarskii. - Basel ; Boston ; Berlin :
Birkhäuser, 1994
(Operator theory ; Vol. 67)
Einheitssacht.: Kraevaja zadaca Rimana s bezkonecnym indeksom
<dt.>
ISBN 978-3-0348-9655-9 ISBN 978-3-0348-8506-5 (eBook)
DOI 10.1007/978-3-0348-8506-5
NE: GT

Originally published by Birkhäuser Verlag in 1994
Softcover reprint of the hardcover 1st edition 1994

Printed on acid-free paper produced from chlorine-free pulp
Cover design: Heinz Hiltbrunner, Basel

ISBN 978-3-0348-9655-9

CONTENTS

ON THE BOOK AND ITS AUTHOR

The present book treats the following two topics from complex analysis: (1) asymptotic properties of functions analytic within an angle; and (2) Riemann's boundary problems. Respectively, the book falls into two parts consisting of Chapters 1 to 3 and Chapters 4 to 6. Part I prepares the needed analytic tools for Part II, though, of course, it is interesting as it is.

To be more precise, Part I is devoted to constructing a theory of analytic functions of the completely regular growth within an angle. This theory can be regarded as a further development of the well-known theory of entire functions of the completely regular growth due to Levin and Phlüger. A principal difference comes from the fact that the asymptotic behavior at infinity of a function analytic in an angle does not depend on the distribution of zeros as completely as that of an entire function, for it is affected by boundary values. There was no tool for taking these effects into consideration until N.V. Govorov invented it. Thereafter it became possible to develop a theory as complete as that for entire functions of the completely regular growth. A complete account of this theory is being published here for the first time: only the main results were announced before, Refs. [8–10].

In Part II a theory of Riemann's boundary problems with infinite index of a power rate is constructed. This theory can be regarded as an extension of the theory of boundary problems with finite index, developed by F.D. Gakhov. A new principal step is made: the role of polynomials, to within which the general solution of the problem was obtained in the case of a finite index, passes to entire transcendent functions, with rigid limitations on their growth and decrease. Singling out these limitations and constructing all entire functions which obey them is a very subtle analytic problem. Its solution is based on the theory of the completely regular growth within an angle (developed in Part I). As a result, a theory as complete as that for a finite index has been obtained. A complete account of this theory is published for the first time as well: only fragments were published earlier (Refs. [3–7,11,13,16,17]).

One of the most important applications of the theory developed in Part II is N.V. Govorov's solution [14] of the well-known Paley problem on the growth of entire functions of finite order. This result is described in the Appendix written by me.

The author of this book, N.V. Govorov, was born on 5 December 1928, in *stanitsa* (cossacks' settlement) Kazanskaya of the Krasnodar region; his parents were teachers. From 1948 to 1957 he worked as a teacher in his native settlement, in 1950 he graduated — as an extramural student — from Groznyi Teachers College and in 1957 from Rostov University. He taught mathematics in Novocherkask Polytechnic Institute and its branch in the town of Shachty. That was when his mathematical talent blossomed and he obtained the main results given in the present monograph. In 1969 N.V. Govorov received the degree of Doctor of Mathematics and the academic rank of a Professor. From 1970 until his tragic death on 24 April 1981, N.V. Govorov worked as Head of the Department of Mathematical Analysis of Kuban' University actively engaged in preparing new courses and teaching young mathematicians. His original mathematical talent, vivid reactions, kindness bordering on self-sacrifice made him highly respected by everybody who knew him.

In preparing this book for publication I was given substantial assistance by E.D. Fainberg and A.I. Heifiz, while V.M. Govorova took a significant part of the technical work with the manuscript. Professor C. Prather contributed substantial assistance in preparing the English translation of the book.

I.V. Ostrovskii.

PREFACE

The classic statement of the Riemann boundary problem consists in finding a function $\Phi(z)$ which is analytic and bounded in two domains D^+ and D^-, with a common boundary — a smooth closed contour L admitting a continuous extension onto L both from D^+ and D^- and satisfying on L the boundary condition

$$\Phi^+(t) = G(t)\Phi^-(t) + g(t).$$

Here $G(t)$ and $g(t)$ are considered to be known functions satisfying the Hölder condition on L, with $G(t)$ non-vanishing and $\Phi^+(t)$, $\Phi^-(t)$ being the continuations of $\Phi(z)$ onto L from D^+, D^- respectively. When $g(t) \equiv 0$, the problem is called *homogeneous*, otherwise — *non-homogeneous.*

A complete and efficient solution to this problem was obtained by F.D. Gakhov [1] in 1936. The main characteristic of the problem is the index

$$\kappa = \mathrm{Ind}_L G(t)$$

defined as the increment of $\arg G(t)$ along L divided by 2π. If $\kappa \geq 0$, then the homogeneous problem has $\kappa + 1$ linearly independent solutions; if $\kappa < 0$, then the problem has no non-trivial solutions. If $\kappa \geq -1$, then the non-homogeneous problem is always solvable; otherwise it is solvable if and only if the function $g(t)$ satisfies $(-\kappa - 1)$ additional conditions. In the classic statement of the Riemann problem the index κ is always finite. The problem can be stated in a much more general form, with an index κ that is finite while the above-mentioned conditions concerning solvability and the number of linearly independent solutions are still satisfied. Such problems were treated in the monographs by F.D. Gakhov [2, Ch. VI] and N.I. Mushelishvili [25, Ch. IV].

In the 1960s F.D. Gakhov proposed that I investigate the Riemann boundary problem with infinite index. I managed to obtain some complete results under special assumptions. The results were partially published in papers [3–7,11,13,16,17] and were reflected in F.D. Gakhov's monograph [2]. The present book describes all the results in detail.

We state the problem in this way. Let L be a smooth contour connecting the finite point $z = t_0$ with the point $z = \infty$. Denote by D the domain obtained from the finite plane by cutting along L. One has to find a function $\Phi(z)$ analytic and bounded in D which admits a continuous extension onto L both from the left (denoted as $\Phi^+(t)$) and from the right (denoted as $\Phi^-(t)$) and satisfying the boundary condition

$$\Phi^+(t) = G(t)\Phi^-(t) + g(t)$$

on L. The functions $\log(G(t))$ and $g(t)$ are assumed to satisfy the Hölder condition and, besides, there is the representation

$$\frac{1}{2\pi}\arg G(t) = \varphi(t)|t|^{\rho},$$

where $0 < \rho < \infty$ and $\varphi(t)$ satisfies the Hölder condition and at $t \to \infty$ has a non-zero limit

$$\lambda = \lim_{t\to\infty} \varphi(t).$$

It is obvious that if one defines the index of the problem κ as the increment of $\arg G(t)$ along L divided by 2π, then $\kappa = +\infty$ for $\lambda > 0$ and $\kappa = -\infty$ for $\lambda < 0$.

The results pertaining to the classic case of the Riemann problem make the following assumptions natural: (1) a homogeneous problem with the $+\infty$ index has an infinite set of linearly independent solutions, whereas a non-homogeneous problem is always solvable; (2) a homogeneous problem with the $-\infty$ index has but a trivial solution, whereas a non-homogeneous problem is solvable only when a countable set of conditions about $g(t)$ is satisfied. Both these assumptions prove to be, generally speaking, false: it can happen that a homogeneous problem with the $+\infty$ index has a trivial solution only (see Sect. 28), while a homogeneous problem with the $-\infty$ index has an infinite set of linearly independent solutions (see Sect. 32).

The above-formulated assumptions, however, are valid if some additional conditions are imposed on the contour L and the functions $G(t)$ and $g(t)$. In this case, in order to describe the set of solutions of the problem with the $+\infty$ index, one must use the theory of entire functions of completely regular growth and the theory of analytic functions of completely regular growth within an angle.

The theory of entire functions of completely regular growth was constructed in the 1930s by B.Ya. Levin and A. Pflüger; its exhaustive description is given in B.Ya. Levin's monograph [24]. We shall have to develop a similar theory for functions analytic within an angle. The main difficulty here lies in the necessity to take into consideration, along with the distribution of zeros, the asymptotic behavior of boundary values.

This book consists of seven chapters. The first three are devoted to the theory of analytic functions of the completely regular growth within an angle. This theory is of interest as such, so it will be rendered in all details, though not all of them will be used in the remainder of the book. Chapters 4 to 6 form the core of the book and are devoted to the Riemann boundary problem with an infinite index.

Since the publication of papers [3–17], several works have appeared that are either based on the results rendered in this book or develop them in various directions. We will mention only Refs. [30–33][1], connected with the material of Chapters 1 to 3, and Sect. 49 of book [2] where the material of Chapters 4 to 6 is reviewed[2]

I recollect with profound gratitude my teacher, Academician of Byelorussia Fedor D. Gakhov, who made me interested in the Riemann boundary problem with infinite index, I am indebted to Professors B.Ya. Levin and A.A. Goldberg for their valuable advice and remarks.

N.V. Govorov

[1]See also L.I. Ronkin "Functions of completely regular growth in a half-plane" (Ukrain. Doklady, ser. A, 1985, No.3, pp. 10–13). — Editor's note.

[2]References to later works see in S.V. Rogozin "Boundary problem and special integral equations with an infinite index". In: Proc.Sem. in Boundary Problems in honor of 75 Birthday of F.D. Gakhov, Minsk, BGU Press, 1975, pp. 95–103. — Editor's note.

PART I

GENERAL PROPERTIES OF ANALYTIC AND FINITE ORDER FUNCTIONS IN THE HALF-PLANE

1 Definition of order and indicator of a function holomorphic in an angle. Relations between various definitions of order

Let us introduce the notions of order and indicator of a function. These notions describe growth and decrease of a function at infinity.

Unless it leads to a misunderstanding, we denote the angle ($\alpha \leq \arg z \leq \beta$) and ($\alpha < \arg z < \beta$) by $[\alpha, \beta]$ and (α, β) respectively.

Definition 1.1. Let a nonzero function $f(z)$ be analytic in (α, β), bounded in each sector

$$(\alpha, \beta) \cap \{|z| < r\}$$

and such that the set $\{\mu\}$ of all $\mu > 0$ for which

$$\sup_{\alpha<\theta<\beta} |f(re^{i\theta}| < \exp(r^{\mu}), \quad r > r_{\mu}. \tag{1.1}$$

is non-empty. Denote by $\{\nu\}$ the set of all $\nu > 0$ such that

$$\limsup_{r \to \infty} r^{-\nu} \log |f(re^{i\theta})| \equiv 0, \quad \alpha < \theta < \beta. \tag{1.2}$$

By the order of the function $f(z)$ in the angle (α, β) we mean the quantity

$$\rho = \max\{\inf\{\mu\}, \ \inf\{\nu\}\}.$$

The function

$$h_f(\theta) = \limsup_{r \to \infty} r^{-\rho} \log |f(re^{i\theta})|, \quad \alpha < \theta < \beta, \tag{1.3}$$

is called the indicator of the function $f(z)$. If inequality (1.1) is false for all $\mu > 0$, we say that $\rho = \infty$.

The following theorem makes this definition correct.

Theorem 1.1 *Let $f \not\equiv 0$ and relation (1.1) hold for at least one $\mu > 0$. Then inequality (1.2) is true for all $\nu > \max\{\mu, \pi/(\beta - \alpha)\}$.*

Proof. The function

$$h^{(\nu)}(\theta) = \limsup_{r\to\infty} r^{-\nu} \log|f(re^{i\theta})|, \quad \alpha < \theta < \beta,$$

is non-positive and trigonometrically ν-convex (see [24, p.74]). If $h^{(\nu)}(\theta) \not\equiv 0$ for some ν, $\beta-\alpha > \pi/\nu$, then there exists a θ such that both θ and $\theta+\pi/\nu$ belong to (α,β) and $h^{(\nu)}(\theta) + h^{(\nu)}(\theta+\pi/\nu) < 0$, which is impossible for a trigonometrically ν-convex function (see [24, p.84]).

To study functions that are continuous in a closed angle, we introduce the following definition.

Definition 1.1′. Let a nonzero function $f(z)$ be analytic in (α,β), continuous in $[\alpha,\beta]$, and let for some $\mu > 0$

$$\max_{\alpha\le\theta\le\beta} |f(re^{i\theta})| < \exp(r^{\mu}), \quad r > r_{\mu}.$$

Denote by ρ the most lower bound of all $\nu_* > 0$ such that

$$\limsup_{r\to\infty} r^{-\nu_*} \log|f(re^{i\theta})| \equiv 0, \quad \alpha \le \theta \le \beta.$$

The number ρ is called *the order* of the function $f(z)$, and the function $h(\theta)$ defined by relation (1.3) is called the indicator of f.

As follows from the theorem below definitions 1.1 and 1.1′ are equivalent for a function f which is continuous in the closed angle $[\alpha,\beta]$.

Theorem 1.2 *Let a function $f(z)$ be analytic in an open angle (α,β), continuous in $[\alpha,\beta]$, and let the relations*

$$\max_{\alpha\le\theta\le\beta} |f(re^{i\theta})| < \exp(r^{\mu}), \quad r > r_{\mu},$$

$$\limsup_{r\to\infty} r^{-\nu_*} \log|f(re^{i\theta})| \equiv 0, \quad \alpha \le \theta \le \beta, \tag{1.4}$$

hold for some $\mu > 0$ and $\nu_* > 0$. Then for any ϵ

$$\max_{\alpha\le\theta\le\beta} |f(re^{i\theta})| < \exp(\epsilon r^{\nu_*}), \quad r > r_{\epsilon}.$$

Proof. Choose $\theta_1, \theta_2 \in [\alpha,\beta]$ such that $0 < \theta_2 - \theta_1 < \pi/(2\mu)$. It follows from (1.4) that [3]

$$|f(re^{i\theta_j})| < C_{\epsilon} \exp\left(\frac{\epsilon}{2} r^{\nu_*}\right), \quad 0 < r < \infty, \ j = 1, 2,$$

[3]Here and in what follows the letter C or maybe C with subscripts denotes various positive constants.

for any $\epsilon > 0$. By applying the Phragmen-Lindelöf principle (see [24, p.69]) to the angle $[\theta_1, \theta_2]$ we obtain

$$\max_{\theta_1 < \theta < \theta_2} |f(re^{i\theta})| < C_\epsilon \exp(\epsilon r^{\nu_*}).$$

Now one may prove the theorem by dividing the whole angle $[\alpha, \beta]$ into smaller angles of opening less than $\pi/(2\mu)$.

Corollary 1.1 *Let a function $f(z)$ analytic in $[\alpha, \beta]$ be of order $\rho \geq 0$ in the sense of definition 1.1. Then*

$$\inf\{\nu_*\} \geq \inf\{\mu\} \tag{1.5}$$

or, what is the same

$$\max_{\alpha \leq \theta \leq \beta} |f(re^{i\theta})| < \exp(r^{\rho+\epsilon}), \quad r > r_\epsilon,$$

for any $\epsilon > 0$.

Remark 1. In general the values of both sides of (1.5) may be different. For example, considering the function $f(z) = e^{-z}$ in the angle $[-\pi/2, \pi/2]$, we have $\inf\{\nu_*\} = 1$, though $\inf\{\mu\} = 0$.

Remark 2. If to determine the value ν according to relation (1.2), i.e. such that

$$\limsup_{r\to\infty} r^{-\nu} \log |f(re^{i\theta})| \equiv 0$$

holds in the open angle (α, β) only, it may occur that

$$\inf\{\nu\} < \inf\{\mu\} \tag{1.6}$$

So one cannot replace ν_* by ν in (1.5). For example, consider the function

$$f_0(z) = \exp\left\{\frac{1}{\pi i}\int_0^\infty \frac{u(x)}{x-z}dx\right\}, \quad 0 < \theta < \pi/2,$$

where

$$u(x) = \begin{cases} x^\rho, & x \in \bigcup_{n=1}^\infty [n - 2^{-n}, n + 2^{-n}] \\ 0, & \text{otherwise} \end{cases}$$

and $\rho > 0$ is a positive value. Since $|x - z| \geq x \sin\theta$, $0 < \theta \leq \pi/2$, we have

$$|\log|f_0(re^{i\theta})|| \leq \frac{1}{\pi \sin\theta}\int_0^\infty \frac{u(x)}{x}dx, \quad 0 < \theta \leq \pi/2,$$

and, hence,

$$\lim_{r\to\infty} r^{-\nu}\log|f_0(re^{i\theta})|\equiv 0,\quad 0<\theta\le\pi/2,$$

for any $\nu>0$.

On the other hand, since

$$\log|f_0(re^{i\theta})| = \frac{1}{\pi}\int_0^\infty \frac{u(x)r\sin\theta dx}{x^2+r^2-2rx\cos\theta} \le$$

$$\le \frac{1}{\pi}\int_0^{2r} \frac{(2r)^\rho r\sin\theta dx}{x^2+r^2-2rx\cos\theta} + \frac{1}{\pi}\int_{2r}^\infty \frac{u(x)r\sin\theta dx}{r\sin\theta(x/2)} \le$$

$$\le (2p)^\rho + \frac{2}{\pi}\int_0^\infty \frac{u(x)}{x}dx,$$

we have

$$\sup_{0<\theta\le\pi/2} \log|f_0(re^{i\theta})| < C_\mu r^\mu$$

for any $\mu>\rho$ and, besides, $\log|f_0(n)| = n^\rho$, $n=1,2,\ldots$. So the left-hand side of (1.6) for the function f_0 is zero whereas the right-hand side equals ρ.

The following statement is an addition to Corollary 1.1.

Theorem 1.3 *Let $f(z)$ be a nonzero function analytic in the angle (α,β) and ρ be its order according either to definition 1.1 or to definition 1.1′. If $\rho>\pi/(\beta-\alpha)$, then $\rho=\inf\{\mu\}$, where $\{\mu\}$ is the set of numbers μ such that*

$$\sup_{\alpha<\theta<\beta} |f(re^{i\theta})| < \exp(r^\mu),\quad r>r_\mu.$$

Proof. Indeed, otherwise we have $\rho=\inf\{\nu\}>\inf\{\mu\}$ and Theorem 1.1 implies

$$\inf\{\nu\}\le\max\{\inf\{\mu\},\pi/(\beta-\alpha)\}.$$

Therefore $\rho\le\pi/(\beta-\alpha)$, which contradicts the hypothesis.

Let $f(z)$ be analytic in the open angle (α,β) and bounded in each sector of the form $(\alpha,\beta)\cap\{z:\ |z|<r\}$, $0<r<\infty$. The Fatou theorem (see [2, p.66]) implies that, for almost all t from the sides of the angle (α,β), there exist the finite boundary values

$$\lim_{z\to t,\ z\in(\alpha,\beta)} f(z) = f(t), \tag{1.7}$$

where z approaches t along any non-tangential path, and, in addition, $f(t)\ne 0$ almost everywhere. In what follows every such function $f(z)$

will be extended to the sides of the angle (α, β) almost everywhere, according to relation (1.7). If t is a point of the exceptional set, we define $|f(t)|$ by setting

$$|f(t)| = \limsup_{z \to t,\ z \in (\alpha,\beta)} |f(z)|,$$

where t approaches z along any path.

After this remark one can see that definitions 1.1 and 1.1′ are equivalent for any function $f(z)$ analytic in (α, β) and bounded in each sector $(\alpha, \beta) \cap \{z : |z| < r\}$.

In the case when $f(z)$ is an entire function and $(\alpha, \beta) = (0, 2\pi)$, definition 1.1 coincides with the usual one (see [24, p.11]):

$$\rho = \limsup_{r \to \infty} (\log r)^{-1} \log \log M_f(r), \quad M_f(r) = \max_{|z|=r} |f(z)|.$$

There is another definition of the order of a function analytic inside an angle (see [24, p.69]), which in general is not equivalent to definitions 1.1 and 1.1′.

Definition 1.2. Let a function $f(z)$ be analytic in (α, β) and continuous in $[\alpha, \beta]$. By its *order in the angle* we mean the value

$$\rho = \limsup_{r \to \infty} (\log r)^{-1} \log \log M_f(r, \alpha, \beta), \quad M_f(r, \alpha, \beta) = \sup_{\alpha \le \theta \le \beta} |f(re^{i\theta})|.$$

The main purpose of introducing definitions 1.1 and 1.1′ is to describe both growth and decrease of a function. For example, the function $f(z) = e^{-z}$, being considered in $[-\pi/2, \pi/2]$, has order 1 in the sense of definition 1.1 (by the way, this function has the same order in the whole complex plane) and indicator $h_f(\theta) = -\cos\theta$, which describes its decrease along the rays $\arg z = \theta$, $-\pi/2 < \theta < \pi/2$. When using definition 1.2 we have $\rho = 0$ and $h_f(\theta) = -\infty$, which gives worse description of the function behavior as $z \to \infty$. Nevertheless the following theorem is true.

Theorem 1.4 *Let a function $f(z)$ be analytic in an angle (α, β), and have order $\rho > \pi/(\beta - \alpha)$ with respect to definition 1.1 or definition 1.2. Then it has the same order according to the other definition.*

This is a direct consequence of Theorem 1.3.

Definition 1.3. Let $f(z)$ have order $\rho > 0$ in an angle (α, β) with respect to definition 1.1. If

$$\sup_{\alpha < \theta < \beta} \log |f(re^{i\theta})| < -r^{\rho - \epsilon}, \quad r > r_\epsilon,$$

for any ϵ, then we also call ρ *the order of decrease* of the function $f(z)$ in the angle (α, β).

If, for any ϵ, one may find a sequence $0 < r_1 < r_2 < \cdots < r_n \to \infty$ such that

$$\sup_{\alpha<\theta<\beta} \log |f(r_n e^{i\theta})| > r_n^{\rho-\epsilon},$$

we call ρ *the order of growth* of the function $f(z)$ in the angle (α, β).

One can easily prove that

1. The order of a function is either its order of growth in (α, β) or its order of decrease in each inner angle $[\alpha + \delta, \beta - \delta]$, $\delta > 0$.

2. Definitions 1.1 and 1.2 are equivalent exactly for those functions $f(z)$ whose order is that of growth.

Everywhere in what follows we consider the order of a function only according to definitions 1.1 and 1.1′ .

We shall use the following three theorems that are simple consequences of the well-known results (see [24, p.79]).

Theorem 1.5 *Let $f(z)$ be analytic and of order ρ in (α, β), with the indicator satisfying the inequality*

$$-\infty \leq h_f(\theta) \leq H(\theta), \quad \alpha \leq \theta \leq \beta,$$

where the function $H(\theta)$ is continuous on the segment $\alpha \leq \theta \leq \beta$. Then

$$\log |f(re^{i\theta})| < (H(\theta + \epsilon)r^\rho + C_\epsilon,$$

for all $\epsilon > 0$ and $\alpha \leq \theta \leq \beta$.

Theorem 1.6 *Let $f(z)$ be analytic and of finite order in the angle (α, β) and let for some $\rho > 0$*

$$\sup_{\alpha\leq\theta\leq\beta} \limsup_{\rho\to\infty} r^{-\rho} \log |f(re^{i\theta})| < 0.$$

Then $f(z)$ is bounded and has order ρ which is the order of decrease.

Theorem 1.7 *Let $f(z)$ have order ρ in the angle (α, β) and its indicator $h_f(\theta)$ be negative for all θ, $\alpha < \theta < \beta$. Then the function $f(z)$ is bounded and its order of decrease is ρ.*

E.Titchmarsh introduced another definition of order.

Definition 1.4. The *order* of a function $f(z)$ analytic in the angle (α, β) is the greater lower bound of those $\nu > 0$ for which

$$\limsup_{\rho\to\infty} r^{-\nu} \log |f(re^{i\theta})| \equiv 0,$$

uniformly with respect to θ, $\alpha < \theta < \beta$.

This definition is equivalent to definition 1.1. To prove this we need the following theorem which is also interesting in itself.

Theorem 1.8 *Let a function $f(z)$ be analytic and bounded in the disk $|z| < r$ and $|f(0)| \leq 1$. Then, for any $N > 1$, there exists an at most countable set of the disks $|z - z_n| \leq \rho_n$, for which*

$$\sum_n \rho_n \leq R/N \tag{1.8}$$

and such that

$$\log|f(z)| \geq -AN \log \sup_{|z|<R} |f(z)|, \tag{1.9}$$

for all

$$z \in D_{f,N,R} = \{z: \ |z| \leq R\} \cap \bigcap_n \{z: \ |z - z_n| \geq \rho_n\},$$

where A is an absolute positive constant. Here, for $|z| = R$, by $f(z)$ we mean the non-tangential boundary value.

Proof. We apply the method which was used by Hayman [38] in a somewhat different problem. Without loss of generality one may assume $R = 1$. Denote $M = \log \sup_{|z|<R} |f(z)|$. We need the following results on bounded analytic functions in the disk $|z| < 1$ (see [28, Ch.2]). Let $f(z)$ be such a function. Then

a) The relation

$$\int_0^{2\pi} |\log|f(e^{i\theta})||d\theta + \sum_m (1 - |\zeta_n|) < \infty, \tag{1.10}$$

holds, where $f(e^{i\theta})$ are the non-tangential boundary values of $f(z)$ on $|z| = 1$ and ζ_n are zeros of $f(z)$ into $|z| < 1$, that are counted according to their multiplicities.

b) The function $\log|f(z)|$ admits the representation

$$\log|f(re^{i\varphi})| = \frac{1}{2\pi}\int_0^{2\pi} \frac{1-r^2}{1+r^2-2r\cos(\varphi-\theta)} \log|f(e^{i\theta})|d\theta -$$

$$-\frac{1}{2\pi}\int_0^{2\pi} \frac{1-r^2}{1+r^2-2r\cos(\varphi-\theta)} d\sigma(\theta) +$$

$$+\sum_n \log\left|\frac{re^{i\varphi}-\zeta_n}{1-\bar{\zeta}_n re^{i\varphi}}\right|. \tag{1.11}$$

Here $0 \leq r < 1, 0 \leq \varphi \leq 2\pi$, and $\sigma(\theta)$ is a bounded non-decreasing function such that $\sigma'(\theta) = 0$ almost everywhere.

Introduce a non-negative Borelian measure μ in the complex plane by setting

$$\mu(E) = \frac{1}{2\pi} \int_{E\cap\{z:\ |z|=1\}} \log^+ |1/f(e^{i\theta})| d\theta +$$

$$+\frac{1}{2\pi} \int_{E\cap\{z:\ |z|=1\}} d\sigma(\theta) + \sum_{\zeta_n \in E} \log |1/\zeta_n|.$$

Substituting $r = 0$ in (1.11), we obtain

$$\log |f(0)| = \frac{1}{2\pi} \int_0^{2\pi} \log^+ |f(e^{i\theta})| d\theta) - \int_{|\zeta|\leq 1} d\mu(\zeta).$$

Since, by the hypothesis, $|f(0)| \geq 1$ we have

$$\mu(\{\zeta :\ |\zeta| \leq 1\}) = \int_{|\zeta|\leq 1} d\mu(\zeta) \leq M. \tag{1.12}$$

Let

$$K(\zeta, z) = \begin{cases} \dfrac{1-|z|^2}{|\zeta - z|^2} & \text{for} \quad |\zeta| = 1,\ |z| < 1, \\ \left(\log\left|\dfrac{1-\bar{\zeta}z}{\zeta - z}\right|\right) \Big/ \log\dfrac{1}{|\zeta|} & \text{for} \quad 0 < |\zeta| < 1,\ |z| < 1, \\ 0, & \text{otherwise.} \end{cases}$$

One may transform representation (1.11) to the form

$$\log |f(re^{i\varphi})| = \frac{1}{2\pi} \int_0^{2\pi} \frac{(1-r^2) \log^+ |f(e^{i\theta})|}{1 + r^2 - 2r\cos(\varphi - \theta)} d\theta -$$

$$- \int_{|\zeta|\leq 1} K(\zeta, re^{i\varphi}) d\mu(\zeta),$$

which implies

$$\log |f(z)| \geq - \int_{|\zeta|\leq 1} K(\zeta, z) d\mu(\zeta), \quad |z| < 1. \tag{1.13}$$

We say that a point $z_0 \in \{z :\ |z| \leq 1\}$ is *light* if

$$\mu(\{\zeta :\ |\zeta - z_0| \leq \rho\}) \leq 24NM\rho,$$

for any $\rho > 0$, otherwise we say that z_0 is *heavy*.

Let us prove that the set of all heavy points is contained in a union of disks $|z - z_n| \leq \rho_n$ such that

$$\sum_n \rho_n \leq 1/(4N).$$

Indeed, let z be a heavy point. Then there is a number $\rho_z > 0$ such that

$$\mu(\{\zeta : |\zeta - z| \leq \rho_z\}) > 24NM\rho_z,$$

and (1.12) yields $\rho_z < 1/(24N)$. The union of disks $\{\zeta : |\zeta - z| \leq \rho_z\}$ is a covering of the set of all heavy points by disks of bounded radii. It is known (see, for example [23, p.246], that every such covering contains at most a countable subcovering such that each heavy point belongs to at most six disks. Denoting the disks from this subcovering by

$$\{\zeta : |\zeta - z_n| \leq \rho_n\}$$

we have

$$\sum_n (24NM\rho_n) < \sum_n \mu(\{z : |z - z_n| \leq \rho_n\}) \leq$$

$$\leq 6\mu(\{\zeta : |\zeta| \leq 1\}) \leq 6M,$$

which implies (1.8).

To estimate $\log|f(z)|$ from below at an arbitrary light point z, $|z| \leq 1$, we shall use inequality (1.13). Set

$$\int_{|\zeta|\leq 1} K(\zeta, z)d\mu(\zeta) =$$

$$= \int_{|\zeta - z| < \frac{1}{2}(1-|z|)} + \int_{|\zeta - z| \geq \frac{1}{2}(1-|z|)} = I_1 + I_2,$$

and

$$D_k = \{\zeta : 2^{k-1}(1 - |z|) \leq |\zeta - z| < 2^k(1 - |z|)\}, \quad k = 0, \pm 1, \ldots.$$

Since the point z is light, we have

$$\mu(D_k) \leq 24MN2^k(1 - |z|). \tag{1.14}$$

Let us estimate the integral I_1. If $|\zeta - z| < \frac{1}{2}(1 - |z|)$, then

$$|1 - \bar{\zeta}| \leq \frac{5}{2}(1 - |z|) \text{ and } \log(1/|\zeta|) \geq 1 - |\zeta| \geq \frac{1}{2}(1 - |z|).$$

Therefore $K(\zeta, z) \leq 2\log(5 \cdot 2^{-k})/(1-|z|)$, for $\zeta \in D_k$, $k < 0$. Now with accounting (1.14) we obtain

$$I_1 = \sum_{k=-\infty}^{-1} \int_{D_k} K(\zeta, z) d\mu(\zeta) \leq \sum_{k=-\infty}^{-1} \frac{2\log(5 \cdot 2^{-k})}{1-|z|} \mu(D_k) \leq$$

$$\leq 48NM \sum_{k=-\infty}^{-1} 2^k \log(5 \cdot 2^{-k}) = A_1 NM,$$

for some constant A_1.

Now let us estimate the integral I_2. Since for $|\zeta| < 1$, $|z| < 1$ we have

$$\log\left|\frac{1-\bar{\zeta}z}{\zeta - z}\right| = \frac{1}{2}\log\left\{1 + \frac{(1-|\zeta|^2)(1-|z|^2)}{|\zeta - z|^2}\right\} \leq$$

$$\leq \frac{(1-|\zeta|^2)(1-|z|^2)}{2|\zeta - z|^2}$$

and $\log(1/|\zeta|) \geq 1 - |\zeta|$, then

$$K(\zeta, z) \leq \frac{1-|z|^2}{|\zeta - z|^2} \leq \frac{2}{2^{2k-2}(1-|z|)}, \quad \text{for } \zeta \in D_k, \ k \geq 0.$$

Accounting (1.14) we obtain

$$I_2 = \sum_{k=1}^{\infty} \int_{D_k} K(\zeta, z) d\mu(\zeta) \leq$$

$$\leq \sum_{k=0}^{\infty} \frac{2}{2^{2k-2}(1-|z|)} \mu(D_k) \leq$$

$$\leq 48NM \sum_{k=0}^{\infty} \frac{1}{2^{k-2}} = 384NM.$$

So we have proved the desired estimate (1.9) for all light points z, $|z| < 1$. Clearly, this estimate is still valid for all boundary points $z = e^{i\theta}$, such that for some r_θ the whole interval $(r_\theta e^{i\theta}, e^{i\theta})$ consists of light points. Consider the disks $\{\zeta : \ |\zeta - z_n| \leq \rho_n$ that form the subcovering and choose those z_n for which $|z_n| \geq 1/2$. For every such z_n consider the disk $\{\zeta : \ |\zeta - e^{i \arg z_n}| \leq 2\rho_n\}$ and add all these disks to the subcovering. We obtain a new set of disks and the sum of all the radii is less than $1/(4N) + 2/(4N) \leq 1/N$. If the point $z = e^{i\theta}$ does not belong to the union of these disks, then the whole interval $(\frac{3}{4}e^{i\theta}, e^{i\theta})$ consists of light points and inequality (1.9) holds at the point z.

Remark. Theorem 1.8 was substantially improved by the author (see [12]) (both with respect to the estimate of the constant A and the location of the exceptional disks $|z - z_n| < \rho_n$) as well as being extended to subharmonic functions in the disk. For other generalizations of this theorem see A. F. Grishin [21] and L. S. Kudina [22] [4].

Lemma 1.1 *Let $\varphi(z)$ be analytic and bounded in the domain*

$$B = \left\{z : \frac{1}{2}R < |z| < \frac{3}{2}R\right\} \cap \{z : \Im z > 0\},$$

and also $|\varphi(Ri)| \geq 1$. Then there is at most a countable set of disks $|z - z_n| \leq \rho_n$ with $\sum_n \rho_n < R/4$ and such that

$$\log|\varphi(z)| > -D \log \sup_{z\in B} |\varphi(z)|$$

outside the union of these disks. Here D is an absolut constant.

Proof. Denote

$$B_* = B \cap \left\{z : \left|z \pm \frac{1}{2}R\right| > \frac{1}{32}R\right\} \cap \left\{z : \left|z \pm \frac{3}{2}R\right| > \frac{1}{32}R\right\}.$$

Let $\zeta = \omega(z)$ be a conformal mapping of the domain B onto $|\zeta| < R$. Since $\omega(z)$ is analytic in the closed domain $\bar{B}_*$, we have

$$\min_{z\in \bar{B}_*} |\omega'(z)| > C > 0, \tag{1.15}$$

Where C is a universal constant. By Theorem 1.8 there exists an at most countable set of disks

$$Q = \bigcup_n \{\zeta : |\zeta - \zeta_n| \leq t_n\}, \quad \sum_n t_n < CR/8,$$

such that

$$\log|\varphi(\omega^{-1}(\zeta))| > -\frac{8A}{C} \log \sup_{z\in B} |\varphi(z)|$$

for $|\zeta| \leq R$ and $\zeta \notin Q$.

[4]see also A.N. Fridman "Estimatis from below of subharmonic functions" (Ukr.Mat.Zhurn. 1980, Vol.32, No.5, pp.701–706) and V.Ya. Eiderman "On radii of exclusive disks in estimating from below the modulas of a function of bounded type" (Uspekhie Mat.Nauk, 1981, Vol.36, No.6, pp.233–234). — Editor's Note.

Let $P \subset B$ be the image of the set Q under the mapping $z = \omega^{-1}(\zeta)$. It remains to prove that one may cover P by the set of disks with the sum of radii less than $R/4$. According to estimate (1.15) one may cover the set $P_* = P \cap B_*$ by disks with the sum of radii less than $(CR/8)(1/C) = R/8$. As to the set $P_{**} = P \backslash P_*$, it lies in the union of the disks

$$\left|z \pm \frac{1}{2}R\right| \leq \frac{1}{32}R, \quad \left|z \pm \frac{3}{2}R\right| \leq \frac{1}{32}R$$

whose sum of radii is $R/8$.

Corollary 1.2 *Let a function $\varphi(z)$ satisfy the conditions of Lemma 1.1. Then there exists a number R_φ, $\frac{1}{2}R < R_\varphi < \frac{3}{2}R$, such that*

$$\inf\{\log|\varphi(z)| : \ |z| = R_\varphi, \ \Im z \geq 0\} > -D \sup_{z \in B} \log|\varphi(z)|.$$

Here $D > 0$ is a universal constant.

Theorem 1.9 *Let a function $f(z)$ be analytic in the angle (α, β) and there exist $\mu > 0$ such that*

$$\sup_{\alpha < \theta < \beta} \log|f(re^{i\theta})| < r^\mu, \quad r > r_\mu.$$

Then for each ν we have either

$$\limsup_{r \to \infty} r^{-\nu} \log|f(re^{i\theta})| \neq 0 \tag{1.16}$$

for at least one θ, $\alpha < \theta < \beta$, or

$$\limsup_{r \to \infty} r^{-\nu} \log|f(re^{i\theta})| \equiv 0, \quad \alpha \leq \theta \leq \beta, \tag{1.17}$$

where the upper limit is uniform with respect to θ.

Proof. Without loss of generality we may assume $\alpha = 0$, $\beta = \pi$. Let relation (1.17) hold. Then the only fact we have to prove is the uniformity of the upper limit in (1.17). Let D be the constant from Lemma 1.1. For any $\epsilon > 0$ Theorem 1.2 yields

$$\sup_{0 \leq \theta \leq \pi} \log|f(re^{i\theta})| < \frac{\epsilon}{2D}\left(\frac{2}{3}\right)^{2\nu} r^{\nu'}, \quad r > r_\epsilon, \tag{1.18}$$

On the other hand relation (1.17) with $\theta = \pi/2$ gives

$$\log|f(R_n i)| > -\frac{\epsilon}{2D}\left(\frac{2}{3}\right)^\nu R_n^\nu \tag{1.19}$$

for some sequence $R_n \to \infty$.

Consider the functions

$$f_n(z) = \exp\Big(\frac{\epsilon}{2D}\Big(\frac{2}{3}\Big)^\nu R_n^\nu\Big) f(z)$$

in the semi-annuli

$$Q_n = \Big\{z : \frac{1}{2}R_n < |z| < \frac{3}{2}R_n\Big\} \cap \Big\{z : \Im z > 0\Big\}.$$

According to (1.18) we have

$$\sup_{z \in Q_n} \log |f_n(z)| \le$$

$$\le \frac{\epsilon}{2D}\Big(\frac{2}{3}\Big)^\nu \Big(\Big(\frac{2}{3}\Big)^\nu R^\nu + R_n^\nu\Big) < \frac{\epsilon}{D}\Big(\frac{2}{3}\Big)^\nu R_n^\nu, \quad n > n_0,$$

and according to (1.19) we have $\log |f_n(R_n i)| > 0$. Then by virtue of Lemma 1.1 we obtain

$$\inf_{0 \le \theta \le \pi} \log |f_n(r_n e^{i\theta})| > -\epsilon(2/3)^\nu R_n^\nu$$

for some r_n, $R_n/2 < r_n < 3R_n/2$, or

$$\inf_{0 \le \theta \le \pi} \log |f_n(R_n e^{i\theta})| > -\epsilon(1 + \frac{1}{2D})(\frac{1}{2})^\nu R_n^\nu > -2\epsilon(\frac{1}{3})^\nu r_n^\nu, \tag{1.20}$$

since we may always assume $D > 1$. Relations (1.18) and (1.20) mean that approaching the upper limit in (1.17) is uniform with respect to θ.

Theorem 1.10 *Definitions (1.1) and (1.4) of the order of a function analytic in an angle (α, β) are equivalent.*

Proof. Let ρ be the order of the function $f(z)$ in the angle (α, β) according to definition 1.4. Then, for any $\epsilon > 0$,

$$\limsup_{r \to \infty} r^{-\rho-\epsilon} \log |f(re^{i\theta})| \equiv 0, \quad \alpha < \theta < \beta, \tag{1.21}$$

where the limit is uniform with respect to θ. Or, which is equivalent,

$$\limsup_{r \to \infty} r^{-\rho-\epsilon} \log |f(re^{i\theta})| \equiv 0, \quad \alpha \le \theta \le \beta \tag{1.22}$$

uniformly with respect to θ. By Theorem 1.9 there is at least one θ_0, $\alpha \le \theta_0 \le \beta$ such that

$$\limsup_{r \to \infty} r^{-\rho+\epsilon} \log |f(re^{i\theta_0})| = a \ne 0. \tag{1.23}$$

Since the approaching to the upper limit is uniform, one obtains

$$\sup_{\alpha\leq\theta\leq\beta} \log|f(re^{i\theta})| < r^{\rho+\epsilon}, \quad r > r_\epsilon. \tag{1.24}$$

It follows from (1.22)–(1.24) that ρ is the order of $f(z)$ according to definition 1.1 and, hence, according to definition 1.1$'$.

Conversely, let $f(z)$ have order ρ according to definition 1.1$'$ (which is equivalent to definition 1.1). Then relations (1.22) and (1.23) hold for any $\epsilon > 0$ (the latter relation holds for at least one value θ_0). Whence, by using Theorem 1.9, we obtain that ρ is the order of $f(z)$ according to definition 1.4.

2 Generalized Nevanlinna and Carleman formulas

Nevanlinna [37] and Carleman [34] have obtained relations between the logarithm of modulus of a function analytic in the closed half-disk:

$$Q_R = \{z : \; |z| < R, \; \Im z > 0\}$$

and the location of its zeros.These relations have various applications to the theory of entire and meromorphic functions (see [24, p.19]). Every function $f(z)$ analytic in the closure of Q_R is analytic and bounded inside $Q_{R'}$ for some $R' > R$. We extend the Nevanlinna and Carleman formulas to a wider class of functions analytic and bounded in $Q_{R'}$ for some $R' > R$.

Let

$$g_R(\zeta, z) = \log\left|\frac{R^2 - \bar{\zeta}z}{R(\zeta - z)} \cdot \frac{R(\bar{\zeta} - z)}{R^2 - \zeta z}\right| \tag{2.1}$$

be the Green function of the half-disk Q_R. Its normal derivative on the boundary of Q_R admits the representation

$$\frac{\partial g_R}{\partial n_\zeta}|d\zeta| = \left\{\frac{R^2 - |z|^2}{|\zeta - z|^2} - \frac{R^2 - |z|^2}{|\bar{\zeta} - z|^2}\right\}d\theta,$$

$$\zeta = Re^{i\theta}, \quad 0 < \theta < \pi; \tag{2.2}$$

$$\frac{\partial g_R}{\partial n_\zeta}|d\zeta| = 2\left\{\frac{r\sin\psi}{|z - t|^2} \quad \frac{R^2 r\sin\psi}{|R^2 - zt|^2}\right\}dt,$$

$$\zeta = t \in (-R, R), \quad z = re^{i\psi}, \tag{2.3}$$

(see [19, p.12])

The generalization of the Nevanlinna formula is as follows:

Theorem 2.1 *Let a function $f(z)$ be analytic and bounded in the half-disk $Q_{R'}$ for some $R' > R$. Then, for $z = re^{i\psi} \in Q_R$, we have*

$$\log|f(z)| = \sum_{z_k \in Q_R} \log\left|\frac{R(z - z_k)}{R^2 - \bar{z}_k z} \cdot \frac{R^2 - z_k z}{R(z - \bar{z}_k)}\right| +$$

$$+\frac{1}{2\pi}\int_0^\pi \left(\frac{R^2 - |z|^2}{|Re^{i\theta} - z|^2} - \frac{R^2 - |z|^2}{|Re^{-i\theta} - z|^2}\right)\log|f(Re^{i\theta})|d\theta +$$

$$+\frac{1}{\pi}\int_{-R}^{R}\left(\frac{r\sin\psi}{|t - z|^2} - \frac{R^2 r\sin\psi}{|R^2 - zt|^2}\right)\log|f(t)|dt +$$

$$+\frac{1}{\pi}\int_{-R}^{R}\left(\frac{r\sin\psi}{|t - z|^2} - \frac{R^2 r\sin\psi}{|R^2 - zt|^2}\right)d\varphi(t) \tag{2.4}$$

Here z_k are zeros of the function $f(z)$, $\varphi(t)$ is some non-increasing function on $[-R,R]$ such that $\varphi'(t)=0$ almost everywhere and $f(t)$ are non-tangential boundary values of $f(t)$ on $[-R,R]$. All series and integrals in this relation are absolutely convergent. The function $\varphi(t)$ is defined up to a constant and

$$\varphi(t_2)-\varphi(t_1)=$$

$$\lim_{y\to+0}\int_{t_1}^{t_2}\log|f(x+iy)|dx-\lim_{y\to+0}\int_{t_1}^{t_2}\log|f(x)|dx \tag{2.5}$$

for all continuity points t_1, t_2, $(t_1<t_2)$ of the function $\varphi(t)$.

In what follows the function $\varphi(t)$ is called a *singular boundary function.* The classical Nevanlinna formula differs from formula (2.4) by the absence of the function $\varphi(t)$ and by the fact that the number of the zeros $z_k\in Q$ is finite.

Proof of formula (2.4) is principally the same as of the similar formula for functions bounded in a disk (see [28, p.111]). Therefore we only outline it and omit the details.

Apply the classical Nevanlinna formula (see, for example, [19, p.16]) to the function $f(z+i\epsilon)$ in the half-disk $Q_{R'}$, assuming $\epsilon>0$ and $R'-R>0$ to be small enough. We obtain

$$\log|f(z+i\epsilon)|=\frac{1}{2\pi}\int_{\partial Q_{R'}}\log|f(\zeta+i\epsilon)|\frac{\partial g_{R'}(\zeta,z)}{\partial n_\zeta}|d\zeta|-$$

$$-\sum_{z_k-i\epsilon\in Q_{R'}}g_{R'}(z_k-i\epsilon,z) \tag{2.6}$$

One may rewrite this relation as

$$\frac{1}{2\pi}\int_{\partial Q_{R'}}\log^+\frac{1}{|f(\zeta+i\epsilon)|}\frac{\partial g_{R'}(\zeta,z)}{\partial n_\zeta}|d\zeta|+\sum_{z_k-i\epsilon\in Q_{R'}}g_{R'}(z_k-i\epsilon,z)=$$

$$=\frac{1}{2\pi}\int_{\partial Q_{R'}}\log^+|f(\zeta+i\epsilon)|\frac{\partial g_{R'}(\zeta,z)}{\partial n_\zeta}|d\zeta|-\log|f(z+i\epsilon)| \tag{2.7}$$

Fix $z\in Q_R$ such that $f(z)\neq 0$. Since the function f is bounded in Q'_R the right-hand side of (2.7) has a bounded upper limit as $\epsilon\to 0$. Since the both summands in the left-hand side of (2.7) are non-negative, each is bounded by a quantity independent of ϵ. As can be easily seen, the second summand in the left-hand side will decrease if one sums over $z_k\in Q_R\cap\{z:\ \Im z>0\}$ instead of $z_k-i\epsilon\in Q'_R$ and changes g'_R by g_R. This yields the convergence of the series $\sum_{z_k \in Q_R} g_R(z_k, z)$, which up to the sign is the same as the first term in the right-hand side of (2.4). Note, that

$$\sum_{z_k \in Q_R} \Im z_k < \infty. \tag{2.8}$$

Indeed, if there are infinitely many points $z_k \in Q_R$, then $\Im z_k \to 0$. A simple calculation shows that the ratio of $g_R(z_k, z)$ and $\Im z_k$ approaches a finite limit as $\Im z_k \to 0$, but $z_k \not\to \pm R$.

Denote the first summand in the left-hand side of (2.7) by $I_{R',\epsilon}$. It may decrease only if one changes the integration over $\partial Q_{R'}$, by the integration over $[-R, R]$. Relation (2.3) yields the derivative $\partial g_{R'}(\zeta, z)/\partial n_\zeta$ to be bounded from below on $[-R, R]$ by a positive constant $C_{R,R'}$, independent of ζ. Therefore

$$I_{R',\epsilon} \geq \frac{C_{R,R'}}{2\pi} \int_{-R}^{R} \log^+ \frac{1}{|f(\zeta + i\epsilon)|} |d\zeta|$$

Since $I_{R',\epsilon}$ is bounded as $\epsilon \to 0$, the family of the measures

$$m_\epsilon(E) = \int_E \log^+ \frac{1}{|f(\zeta + i\epsilon)|} |d\zeta|, \quad E \subset [-R, R]$$

is also uniformly bounded. Applying the Helly theorem we can choose a sequence $m_{\epsilon_n}(E)$ convergent to a measure $m(E)$, and by the Fatou theorem

$$m(E) \geq \int_E \log^+ \frac{1}{|f(\zeta)|} |d\zeta|$$

Define, up to a constant, a non-decreasing function $\varphi(t)$ by setting

$$\varphi(t_2) - \varphi(t_1) = \int_{t_1}^{t_2} \log^+ \frac{1}{|f(\zeta)|} |d\zeta| - m([t_1, t_2]), \quad t_1 < t_2$$

Substituting R' with R in (2.6) and passing to the limit as $\epsilon = \epsilon_n \to 0$, we obtain

$$\log|f(z)| = \frac{1}{2\pi} \int_{\partial Q_R} \log|f(\zeta)| \frac{\partial g_R(\zeta, z)}{\partial n_\zeta} |d\zeta| +$$

$$+ \frac{1}{2\pi} \int_{-R}^{R} \frac{\partial g_R(t, z)}{\partial n_t} d\varphi(t) - \sum_{z \in Q_R} g_R(z_k, z)$$

Taking account of equalities (2.1)–(2.3), this relation coincides with (2.4).

To prove that $\varphi'(t) = 0$ almost everywhere let us compare the boundary values as $t \to z_0 \in (-R, R)$ of the both sides of (2.4). Denote the four summands in the right-hand side of (2.4) by $A_1(z)$, $A_2(z)$, $A_3(z)$, $A_4(z)$ respectively. One can easily see, that, as $z \to t_0$,

$$A_1(z) - \sum_{z_k \in Q_R} \log\left|\frac{z - z_k}{z - \bar{z}_k}\right| \to 0, \tag{2.9}$$

$$A_2(z) \to 0, \tag{2.10}$$

$$A_3(z) - \frac{1}{\pi}\int_{-R}^{R} \frac{r \sin\psi}{|z - t|^2} \log|f(t)|dt \to 0, \tag{2.11}$$

$$A_4(z) - \frac{1}{\pi}\int_{-R}^{R} \frac{r \sin\psi}{|z - t|^2} d\varphi(t) \to 0. \tag{2.12}$$

The sum in (2.9) is the logarithm of a Blaschke product for the half-plane (which converges by virtue of (2.8)) and, hence, its non-tangential boundary values for almost all $t \in (-R, R)$ are equal to zero. Integrals from the left-hand side of (2.11) and (2.12) are respectively Poisson-Lebesgue and Poisson-Stiltjes integrals for the half-plane. Therefore their non-tangential boundary values are $\log|f(t_0)|$ and $\varphi'(t_0)$ respectively for almost all $t_0 \in (-R, R)$. Thus the non-tangential boundary values of the right-hand side of (2.4) equal $\log|f(t_0)| + \varphi'(t_0)$ for almost all $t_0 \in (-R, R)$. This implies $\varphi'(t_0) = 0$ almost everywhere.

To obtain relation (2.5) put $z = x + iy$ in (2.4) and integrate by x from t_1 to t_2. Then passing to the limit as $y \to +0$ and using relations (2.9)–(2.12) and also well-known properties of the logarithm of Blaschke products and Poisson-Lebesgue and Poisson-Stiltjes integrals (see,for example, [20, 22] we obtain

$$\int_{t_1}^{t_2} A_1(x + iy)dx \to 0, \quad \int_{t_1}^{t_2} A_3(x + iy)dx \to \int_{t_1}^{t_2} \log|f(t)|dt \to 0,$$

$$\int_{t_1}^{t_2} A_4(x + iy)dx \to \int_{t_1}^{t_2} d\varphi(t).$$

Now let us give a generalization of the Carleman formula.

Theorem 2.2 *Let a function $f(z)$ be analytic and bounded in the half-disk*

$Q_{R'}$ for some $R' > R$. For integer $q \geq 1$ and $\lambda \in (0, R)$ we have [5]

$$\frac{1}{q} \sum_{\lambda < r_n \leq R} \left(\frac{1}{r_n^q} - \frac{r_n^q}{R^{2q}} \right) \sin q\theta_n =$$

$$= \frac{1}{\pi R^q} \int_0^{\pi} \log |f(Re^{i\theta})| \sin q\theta d\theta +$$

$$+ \frac{1}{2\pi} \int_{\lambda \leq |x| \leq R} \left(\frac{1}{x^{q+1}} - \frac{x^{q-1}}{R^{2q}} \right) \log |f(x)| dx +$$

$$+ \frac{1}{2\pi} \int_{\lambda \leq |x| \leq R} \left(\frac{1}{x^{q+1}} - \frac{x^{q-1}}{R^{2q}} \right) d\varphi(x) + A_\lambda^{(q)}(f, R), \tag{2.13}$$

here $r_n e^{i\theta_n}$ are the zeros of $f(z)$, $\varphi(x)$ is the singular function, which is defined by relation (2.5) and $A_\lambda^{(q)}(f, R)$ has the form

$$A_\lambda^{(q)}(f, R) = c(\lambda, f, q) + R^{-2q} d(\lambda, f, q), \tag{2.14}$$

where c and d are independent of R. All series and integrals are absolutely convergent.

Remark. The remainder term $A_\lambda^{(q)}(f, R)$ admits the representation

$$A_\lambda^{(q)}(f, R) = \frac{1}{2\pi} \int_0^{\pi} \Im \Big(\log f(\lambda e^{i\theta}) \Big(\frac{e^{-iq\theta}}{\lambda^q} - \frac{\lambda^q e^{iq\theta}}{R^{2q}} \Big) \Big) d\theta.$$

Relation (2.13) was obtained by Nevanlinna [37] (and by Carleman [34], when $q = 1$) for functions analytic in the closure of Q_R, in this case the integral which includes the boundary singular function does not appear. By the generalized Carleman formula we mean relation (2.13) with $q = 1$. Let us write it down separately

$$\sum_{\lambda < r_k \leq R} \left(\frac{1}{r_n} - \frac{r_n}{R^2} \right) \sin \theta_n = \frac{1}{\pi R} \int_0^{\pi} \log |f(Re^{i\theta})| \sin \theta d\theta +$$

$$+ \frac{1}{2\pi} \int_\lambda^R \left(\frac{1}{x^2} - \frac{1}{R^2} \right) \log |f(x) f(-x)| dx +$$

[5] In what follows we write for simplicity ($0 \leq a < b < \infty$)

$$\int_{a \leq |x| \leq b} g(x) d\beta(x) = \int_a^b g(x) d\beta(x) + \int_{-b}^{-a} g(x) d\beta(x),$$

and assume $\pm\lambda$ to be points of continuity of the function $\varphi(x)$.

$$+\frac{1}{2\pi}\int_{\lambda}^{R}\left(\frac{1}{x^2}-\frac{1}{R^2}\right)d(\varphi(x)-\varphi(-x))+A_{\lambda}(f,R), \tag{2.15}$$

where

$$A_{\lambda}(f,R)=c(\lambda,f)+R^{-2}d(\lambda,f) \tag{2.16}$$

and c and d are independent of R.

Proof of formula (2.15) follows the scheme which was used by Nevanlinna in [37]. Consider the question

$$U_0(z)=\sum_{z_k\in Q_{\lambda}}\log\left|\frac{z-z_k}{z-\bar{z}_k}\right|+\frac{1}{\pi}\int_{-\lambda}^{\lambda}\frac{r\sin\psi}{|z-t|^2}\log|f(t)|dt+$$

$$+\frac{1}{\pi}\int_{-\lambda}^{\lambda}\frac{r\sin\psi}{|z-t|^2}d\varphi(t),\quad z=re^{i\psi}.$$

The convergence of the series and the integrals in this definition follows from relation (2.4) for the function $f(z)$. The function $U_0(z)$ is the logarithm of the modulus of a function that is bounded and analytic in the half-plane $\Im z>0$. Applying Theorem 2.1 to this function and using (2.1)-(2.3) we obtain

$$U_0(z)=-\sum_{z_k\in Q_{\lambda}}g_R(z_k,z)+\frac{1}{2\pi}\int_0^{\pi}\frac{\partial g_R(Re^{i\theta},z)}{\partial n}U_0(Re^{i\theta})Rd\theta+$$

$$\frac{1}{2\pi}\int_{-\lambda}^{\lambda}\frac{\partial g_R(t,z)}{\partial n}\log|f(t)|dt+\frac{1}{2\pi}\int_{-\lambda}^{\lambda}\frac{\partial g_R(t,z)}{\partial n}d\varphi(t).$$

Subtraction of $U_0(z)$ from (2.4) yields

$$\log|f(z)|-U_0(z)=-\sum_{z_k\in Q_R\setminus Q_{\lambda}}g_R(z_k,z)+$$

$$+\frac{1}{2\pi}\int_0^{\pi}\frac{\partial g_R(Re^{i\theta},z)}{\partial n}+(\log|f(Re^{i\theta})|-U_0(Re^{i\theta}))Rd\theta+$$

$$+\frac{1}{2\pi}\int_{\lambda<|t|<R}\frac{\partial g_R(t,z)}{\partial n}\log|f(t)|dt+\frac{1}{2\pi}\int_{\lambda<|t|<R}\frac{\partial g_R(t,z)}{\partial n}d\varphi(t) \tag{2.17}$$

The function $\log|f(z)|-U_0(z)$ is harmonic in Q_{λ} and continuous in $Q_{\lambda}\cup(-\lambda,\lambda)$. It vanishes on $(-\lambda,\lambda)$. By the symmetry principle this function admits the extension into the disk Q_{λ}, and this extension is odd with respect to $\arg z$. Therefore we have the absolutely convergent Fourier expansion.

$$\log|f(re^{i\psi})|-U_0(re^{i\psi})=$$

$$= \sum_{q=1}^{\infty} c_q r^q \sin q\psi, \quad 0 \le r < \lambda, \quad 0 \le \psi \le \pi \tag{2.18}$$

Direct calculation gives

$$g_R(z_k, re^{i\psi}) =$$

$$= 2\sum_{q=1}^{\infty}\Big(\frac{1}{r_k^q} - \frac{r_k^q}{R^{2q}}\Big)\frac{\sin q\theta_k}{q} r^q \sin q\psi,$$

$$0 \le r < \lambda < r_k < R, \quad 0 < \theta_k, \quad \psi < \pi; \tag{2.19}$$

$$\frac{\partial g_R}{\partial n}(Re^{i\theta}, re^{i\psi}) = \frac{4}{R}\sum_{q=1}^{\infty}\frac{\sin q\theta}{R^q} r^q \sin q\psi,$$

$$0 \le r < R, \quad 0 < \theta, \psi < \pi; \tag{2.20}$$

$$\frac{\partial g_R}{\partial n}(t, re^{i\psi}) = \sum_{q=1}^{\infty}\Big(\frac{1}{t^{q+1}} - \frac{t^{q-1}}{R^{2q}}\Big) r^2 \sin q\psi,$$

$$0 \le r < \lambda < |t| < R, \quad 0 < \theta, \psi < \pi. \tag{2.21}$$

Having relations (2.18)–(2.21), one may compare the Fourier coefficients with respect to ψ (here $z = re^{i\psi}$, $0 < r < \lambda$) in the both sides of (2.17) (it should be noted that there is no need to substitute series (2.19)–(2.21) into (2.17) and integrate them termwise — it suffices to multiply (2.17) by $\sin q\psi$ and then integrate with respect to ψ from 0 to π, due to the Fubini theorem one may change the order of integrals). For $q = 1, 2, \ldots.$ we have

$$c_q = -2 \sum_{z_k \in Q_R \setminus Q_\lambda} \Big(\frac{1}{r_k^q} - \frac{r_k^q}{R^{2q}}\Big)\frac{\sin q\theta_k}{q} +$$

$$+ \frac{2}{\pi R^q}\int_1^{\pi} \sin q\theta \Big(\log|f(Re^{i\theta})| - U_0(Re^{i\theta})\Big) d\theta +$$

$$+ \frac{1}{\pi}\int_{\lambda<|t|<R}\Big(\frac{1}{t^{q+1}} - \frac{t^{q-1}}{R^{2q}}\Big)\log|f(t)|dt +$$

$$+ \frac{1}{\pi}\int_{\lambda<|t|<R}\Big(\frac{1}{t^{q+1}} - \frac{t^{q-1}}{R^{2q}}\Big)d\varphi(t).$$

After setting

$$A_\lambda^{(q)}(f, R) = -\frac{1}{2}c_q - \frac{1}{\pi R^q}\int_0^{\pi} \sin q\theta \cdot U_0(Re^{i\theta}) d\theta,$$

this relation turns into desired formula (2.13). The function $U_0(z)$ is harmonic in the domain $|z| > \lambda$, tends to zero as $z \to \infty$ and vanishes for real z, $|z| > \lambda$. Therefore the Fourier expansion of the function $U_0(Re^{i\theta})$ with respect to θ has the form

$$U_0(Re^{i\theta}) = \sum_{q=1}^{\infty} d_q R^{-q} \sin q\theta.$$

Hence

$$A_\lambda^{(q)}(f, R) = -\frac{1}{2}c_q - \frac{1}{2R^{2q}}d_q.$$

We have obtained equality (2.13).

3 Canonical representation of a function of finite order in the half-plane

When studying the theory of entire functions of finite order an important role is played by Hadamard factorization theorem. According to this theorem any entire function $f(z)$ of finite order ρ admits a representation of the form

$$f(z) = z^m e^{P(z)} \prod_n E\Big(\frac{z}{z_n}, [\rho]\Big) \tag{3.1}$$

where m is non-negative integer, $P(z)$ is a polynomial of degree $\leq [\rho]$ and $E(u, q)$ is the canonical Weierstrass factor

$$E(u,q) = \begin{cases} 1-u, & q = 0 \\ (1-u)\exp\Big(u + \dfrac{u^2}{2} + \cdots + \dfrac{u^q}{q}\Big), & q = 1, 2, \ldots \end{cases} \tag{3.2}$$

Representation (3.1) is called the canonical representation of the entire function of finite order.

In this section we obtain another representation which plays a similar role when studying functions that are bounded and of finite order in the half-plane $\Im z > 0$. First let us consider separately the relatively simple special case when the function $f(z)$ is either bounded in $\Im z > 0$ or is of order < 1.

Let a function $f(z)$ be analytic and bounded in $\Im z > 0$. As it is known (see, for example, [20, pp.189–190]), every such function may be represented in the form

$$f(z) = e^{i(a_0 + a_1 z)} \Big(\frac{z-i}{z+i}\Big)^m \prod_n \Big(\frac{z - z_n}{z - \bar{z}_n} \cdot \frac{|z_n^2 + 1|}{z_n^2 + 1}\Big) \times$$

$$\times \exp\Big\{\frac{1}{\pi i}\int_{-\infty}^{\infty} \frac{(tz+1)\log|f(t)|}{(t^2+1)(t-z)} dt + \frac{1}{\pi i}\int_{-\infty}^{\infty} \frac{(tz+1)d\varphi(t)}{(t^2+1)(t-z)}\Big\}, \tag{3.3}$$

where a_0, a_1 are real, $a_1 \geq 0$, m is a non-negative integer, $z_n \neq i$ are zeros of the function $f(z)$, $|f(t)|$ is bounded and $\varphi(t)$ is a singular boundary function. In addition all integrals and infinite products in (3.3) are absolutely convergent. The latter is equivalent to fulfillment of the following three conditions:

$$\sum_n \frac{\Im z_n}{|z_n|^2 + 1} < \infty,$$

$$\int_{-\infty}^{\infty} \frac{|\log|f(t)||}{1+t^2} < \infty, \qquad \int_{-\infty}^{\infty} \frac{|d\varphi(t)|}{1+t^2} < \infty, \tag{3.4}$$

This yields the following statement.

Theorem 3.1 *Let a function $f(z)$ be analytic in the half-plane $\Im z > 0$ and either be bounded or have order $\rho < 1$. Then $f(z)$ admits the representation*

$$f(z) = e^{i(a_0+a_1 z)} \prod_{|z_n|\le 1} \frac{z - z_n}{z - \bar{z}_n} \prod_{|z_n|>1} \frac{1 - z/z_n}{1 - z/\bar{z}_n} \times$$

$$\times \exp\Big\{\frac{1}{\pi i}\int_{-\infty}^{\infty} \frac{(tz+1)\log|f(t)|}{(t^2+1)(t-z)}dt + \frac{1}{\pi i}\int_{-\infty}^{\infty}\frac{(tz+1)d\varphi(t)}{(t^2+1)(t-z)}\Big\}, \tag{3.5}$$

where a_0, a_1 are real, $a_1 \ge 0$, z_n are zeros of the function $f(z)$, $\varphi(t)$ is a singular boundary function. Besides all integral and infinite products in (3.5) are absolutely convergent and relations (3.4) hold. In addition, if $f(z)$ has order $\rho < 1$, then $a_1 = 0$.

Proof. First assume $f(z)$ to be bounded in $\Im z > 0$. The first of conditions (3.4) is equivalent to

$$\sum_{|z_n|\le 1} \Im z_n < \infty, \quad \sum_{|z_n|>1} |z_n|^{-2}\Im z_n < \infty. \tag{3.6}$$

The first of these inequalities yields the absolute convergence of the products

$$\prod_{|z_n|\le 1} \frac{z - z_n}{z - \bar{z}_n}, \quad \prod_{|z_n|\le 1,\ z_n\ne i} \frac{|z_n^2+1|}{z_n^2+1}, \tag{3.7}$$

whereas the second inequality yields the absolute convergence of the products

$$\prod_{|z_n|>1} \frac{1 - z/z_n}{1 - z/\bar{z}_n}, \quad \prod_{|z_n|>1} \frac{|1+z_n^{-2}|}{1+z_n^{-2}}. \tag{3.8}$$

The products in (3.7) and (3.8) that do not contain z are unimodular. So one may pass from representation (3.3)to representation (3.5) changing, if it is needed, the value of the coefficient a_0.

Now let $f(z)$ be bounded and in addition have order $\rho < 1$. Representation (3.3) implies $f(z) = e^{i(a_0+a_1 z)} f_1(z)$ where $f_1(z)$ is a bounded function. If $a_1 > 0$, then

$$\limsup_{r\to\infty} r^{-\rho-\epsilon}\log|f(re^{i\psi})| \equiv -\infty,$$

for $0 < \epsilon < 1 - \rho$. This contradicts the definition of order.

At last, let the function $f(z)$ have order $\rho < 1$, and may be unbounded. Fix ϵ, $0 < \epsilon < 1 - \rho$ and consider the function

$$f_1(z) = f(z)\exp(-(-iz)^{\rho+\epsilon}).$$

This function is bounded and of order $\rho+\epsilon<1$ in the half-plane $\Im z>0$,so it admits representation (3.5) with $a_1=0$. We use this representation and note that the singular boundary function for $f_1(z)$ coincides with that for $f(z)$ and also

$$\log|f_1(t)|=\log|f(t)|-|t|^{\rho+\epsilon}\cos\frac{\pi}{2}(\rho+\epsilon),$$

and

$$\exp(-(-iz)^{\rho+\epsilon})=\exp\Big(-\frac{\cos(\pi(\rho+\epsilon)/2)}{\pi i}\int_{-\infty}^{\infty}\frac{(tz+1)|t|^{\rho+\epsilon}}{(t^2+1)(t-z)}dt\Big).$$

We have obtained the desired representation (3.5) with $a_1=0$ for the function $f(z)$.

To formulate the main theorem on the canonical representation of a function of an arbitrary finite order [6], let us (following [37]) define the *canonical Nevanlinna factor:*

$$D_q(u,v)=\frac{E(u/v,q)}{E(u/\bar{v},q)},$$

where $E(u,q)$ is the canonical Weierstrass factor, defined by (3.2). The theorem is formulated as follows:

Theorem 3.2 *Every function $f(z)$ analytic and of a finite order ρ in the half-plane $\Im z>0$ admits the representations*

$$f(z)=e^{i(a_0+a_1z+\cdots+a_qz^q)}\prod_{|z_n|\le 1}\frac{z-z_n}{z-\bar{z}_n}\prod_{|z_n|>1}D_q(z,z_n)\times$$
$$\times\exp\Big\{\frac{1}{\pi i}\int_{-\infty}^{\infty}\frac{(tz+1)^{q+1}\log|f(t)|}{(t^2+1)^{q+1}(t-z)}dt+$$
$$+\frac{1}{\pi i}\int_{-\infty}^{\infty}\frac{(tz+1)^{q+1}d\varphi(t)}{(t^2+1)^{q+1}(t-z)}\Big\},\tag{3.9}$$

where $q=[\rho]$, $a_0,\dots,a_q$ are real, $z_n=r_ne^{i\theta_n}$, $0<\theta_n<\pi$, are zeros of the function $f(z)$, $\varphi(t)$ is a singular boundary function. All integrals and infinite products in (3.9)are absolutely convergent. The following relations hold

$$\sum_{r_n\le 1}r_n\sin\theta_n<\infty,\quad \sum_{r_n>1}r_n^{-\mu-\epsilon}\sin\theta_n<\infty,\tag{3.10}$$

[6] Generalizations to the case of infinite order see in I.O. Khachatrian "Representation of meromorphic functions of infinite order in the half-plane" (Izv.AN Arm.SSR, 1965, Vol.18, No.2, pp.15–25) and in A.I. Heifiz "Representation of infinite order functions analitic in an open half-plane" (Ibid., 1971, Vol.6, No.6, pp.472–476). — Editor's Note.

$$\int_{-\infty}^{\infty} \frac{|\log|f(t)||}{1+|t|^{1+\mu+\epsilon}} dt < \infty, \quad \int_{-\infty}^{\infty} \frac{|d\varphi(t)|}{1+|t|^{1+\mu+\epsilon}} < \infty, \tag{3.11}$$

where $\mu = \max(\rho, 1)$ *and* ϵ *is an arbitrary positive number.*

Remark 1. If either $\rho < 1$ or $f(z)$ is bounded, relation (3.9) is a consequence of Theorem 3.1 and relations (3.10), (3.11) hold for $\epsilon = 0$ as well. (see(3.4)).

Remark 2. For functions analytic in $\Im z > 0$ (and more generally—for meromorphic functions), but for some other definition of order, representation (3.9) with $\varphi(t) \equiv 0$ was obtained by Nevanlinna (see [37]). To prove Theorem 3.2 we need some auxiliary lemmas.

Lemma 3.1 *If* $f(z)$ *is analytic and of finite order in* $\Im z > 0$*, then relations (3.10), (3.11) hold.*

Proof. The first of relations (3.10) is a consequence of Theorem 2.1. Indeed, since $f(z)$ is boundeed in the half-disk $Q_R = \{z : |z| < R, \Im z > 0\}$, $R > 1$, the series in the right-hand side of (2.4) is absolutely convergent. As was mentioned when proving Theorem 2.1, this yields (2.8).

To complete the proof we apply the generalized Carleman formula (2.15) with $\lambda = 1$ and $R = r$ to the function $f(z)$. Since $\log|f| = \log^+|f| - \log^+|1/f|$ and the function $\varphi(t)$ is non-increasing, we have

$$\sum_{1<r_n\leq r} \left(\frac{1}{r_n} - \frac{r_n}{r^2}\right) \sin\theta_n +$$

$$\frac{1}{2\pi}\int_1^r \left(\frac{1}{t^2} - \frac{1}{r^2}\right)\left(\log^+\frac{1}{|f(t)|} + \log^+\frac{1}{|f(-t)|}\right)dt +$$

$$+\int_1^r \left(\frac{1}{t_n} - \frac{1}{r^2}\right)(|d\varphi(t)| + d\varphi(-t)) =$$

$$= \frac{1}{\pi r}\int_0^\pi \log|f(re^{i\theta})| \sin\theta d\theta +$$

$$+\frac{1}{2\pi}\int_1^r \left(\frac{1}{t_n} - \frac{1}{r^2}\right)(\log^+|f(t)| + \log^+|f(-t)|)dt + A_1(f,r). \tag{3.12}$$

By the definition of the order

$$\sup_{0<\theta<\pi} \log^+|f(re^{i\theta})| < C_\epsilon r^{\rho+\epsilon/2}, \quad r \geq 1. \tag{3.13}$$

Therefore the right-hand side of (3.12) is less then $C_{\epsilon 1} r^{\rho-1+\epsilon/2}$. Hence we have

$$\sum_{1<r_n\le r/2} \frac{1}{r_n}\sin\theta_n \le$$

$$\le \frac{4}{3}\sum_{1<r_n\le r/2}\Big(\frac{1}{r_n}-\frac{r_n}{r^2}\Big)\sin\theta_n < C_{\epsilon 2} r^{\rho-1+\epsilon/2}, \tag{3.14}$$

$$\int_1^{r/2}\frac{1}{t^2}\Big(\log^+\frac{1}{|f(t)|}+\log^+\frac{1}{|f(-t)|}\Big)dt \le$$

$$\le \frac{4}{3}\int_1^r \frac{1}{t^2}\Big(\frac{1}{t^2}-\frac{1}{r^2}\Big)\Big(\log^+\frac{1}{|f(t)|}+\log^+\frac{1}{|f(-t)|}\Big)dt < C_{\epsilon 3} r^{\rho-1+\epsilon/2},$$

$$\int_1^{r/2}\frac{1}{t^2}(|d\varphi(t)|+d\varphi(-t)) \le$$

$$\le \frac{4}{3}\int_1^r\Big(\frac{1}{t^2}-\frac{1}{r^2}\Big)(|d\varphi(t)|+d\varphi(-t)) < C_{\epsilon 3} r^{\rho-1+\epsilon/2}, \tag{3.15}$$

where $C_{\epsilon j}$ are some constants. Setting

$$s(r) = \sum_{1<r_n\le r} r_n^{-1}\sin\theta_n,$$

we obtain

$$\sum_{1<r_n\le R}\frac{\sin\theta_n}{r_n^{\rho+\epsilon}} =$$

$$= \int_1^R \frac{ds(r)}{r^{\rho+\epsilon-1}} = \frac{s(R)}{R^{\rho+\epsilon-1}} + (\rho+\epsilon-1)\int_1^R \frac{s(r)}{r^{\rho+\epsilon}}dr.$$

Now using (3.14) we obtain the second relation (3.10). Setting

$$s_1(r) = \int_1^r t^2\Big(\log^+\frac{1}{|f(t)|}+\log^+\frac{1}{|f(-t)|}\Big)dt,$$

we obtain

$$\int_1^R t^{-\rho+1-\epsilon}\Big(\log^+\frac{1}{|f(t)|}+\log^+\frac{1}{|f(-t)|}\Big)dt = \int_1^R t^{-\rho-1-\epsilon}ds_1(t)$$

$$= R^{-\rho-\epsilon+1}s_1(R) + (\rho+\epsilon-1)\int_1^R r^{-\rho-\epsilon}s_1(r)dr$$

and using (3.15) we see that

$$\int_{-\infty}^{\infty} \frac{\log^+ |1/f(t)|}{1+|t|^{\rho+\epsilon+1}} dt < \infty. \tag{3.16}$$

Since $|\log|f|| = \log^+|f| + \log^+|1/f|$, then, accounting for (3.13), we obtain the first of relations (3.11). The second one may be obtained in the same way as (3.16)

Lemma 3.2 *Let a function $f(z)$ be analytic, bounded and of a finite order in the upper half-plane $\Im z > 0$. Then all integrals and infinite products in the right-hand side of (3.9) are absolutely convergent.*

Proof. The absolute convergence of the both integrals from (3.9) is a direct consequence of (3.11). The relation

$$\log \frac{z-z_k}{z-\bar{z}_k} \sim -2i\frac{\Im z_k}{z-z_k}, \quad \Im z_k \to 0,$$

together with (3.10) provide the infinite product over $|z_n| \le 1$ To prove that the infinite product over $|z_n| > 1$ is also absolutely convergent we denote $z = re^{i\theta}$, $z_n = r_n e^{i\theta_n}$ and split this product into two parts

$$\prod_{|z_n|>1} D_q(z,z_n) = \prod_{1<|z_n|<2|z|} D_q(z,z_n) \times \prod_{|z_n|>2|z|} D_q(z,z_n) = \Pi_1(z)\Pi_2(z) \tag{3.17}$$

To prove the absolut convergence of the product $\Pi_2(z)$ we take into consideretion that for $|z_n| > |2z|$, we have

$$\log D_q(z,z_n) = \log E(z/z_n, q) - \log E(z/\bar{z}_n, q) =$$

$$= -\sum_{s=q+1} \frac{1}{s}\left(\left(\frac{z}{z_n}\right)^s - \left(\frac{z}{\bar{z}_n}\right)^s\right) = 2i \sum_{s=q+1} \frac{z^s}{s} \cdot \frac{\sin s\theta_n}{r_n^s}$$

Using the inequality $|\sin s\theta_n| < s|\sin\theta_n|$, we obtain

$$|\log D_q(z,z_n)| \le 4|z|^{q+1} r_n^{-q-1} \sin\theta_n, \quad |z| \le 1/2r_n. \tag{3.18}$$

Now the second of conditions (3.10) yields the desired result.

The product $\Pi_1(z)$ can also contain infinitely many factors because the set $\{z_n\}$ may have finite condensation points on the real axis. Therefore it suffices to prove the absolut convergence of the product

$$\tilde{\Pi}_1(z) = \prod_{1<|z_n|<2|z|,\, z_n\in\Delta(z)} D_q(z,z_n),$$

where $\Delta(z) = \{\zeta : 0 < \min(\arg\zeta, \pi - \arg\zeta) < \frac{1}{4}\min(\theta, \pi-\theta)\}$ To study this product we need the following lemma.

Lemma 3.3 *Let $z = re^{i\theta} \neq 0$, $z_0 = r_0 e^{i\theta_0} \neq 0$ and the relation*

$$0 < \eta \leq \theta \leq \pi - \eta, \quad \theta_0 \in (0, \eta/2) \cup (\pi - \eta/2, \pi). \tag{3.19}$$

holds. Then

$$\log D_q(z, z_0) =$$

$$= \begin{cases} \dfrac{2iz^{q+1}}{r_0^q(r_0 - z)} \sin\theta_0(1 + \beta_q(z, z_0)), & \theta_0 \in (0, \eta/2), \\ (-1)^q \dfrac{2iz^{q+1}}{r_0^q(r_0 + z)} \sin\theta_0(1 + \beta_q(z, z_0)), & \theta_0 \in (\pi - \eta/2, \pi), \end{cases} \tag{3.20}$$

where

$$|\beta_q(z, z_0)| \leq 2(q+1)\sin\theta_0 / \sin(\eta/2). \tag{3.21}$$

Proof. We shall prove only the first relation in (3.20) and estimate (3.21). We have

$$\log D_q(z, z_0) =$$

$$= \int_{-\theta_0}^{\theta_0} \frac{\partial}{\partial\alpha}\Big(\log\big(1 - \frac{z}{r_0}e^{-i\alpha}\big) + \frac{z}{r_0}e^{-i\alpha} + \cdots + \frac{z^q}{qr_0^q}e^{-iq\alpha}\Big)d\alpha =$$

$$= \frac{iz^{q+1}}{r_0^q} \int_{-\theta_0}^{\theta_0} \frac{e^{-i(q+1)\alpha}}{r_0 - ze^{-i\alpha}} d\alpha =$$

$$= \frac{2iz^{q+1}\sin\theta_0}{r_0^q(r_0 - z)} \cdot \frac{r_0 - z}{2\sin\theta_0} \int_{-\theta_0}^{\theta_0} \frac{e^{-i(q+1)\alpha}}{r_0 - ze^{-i\alpha}} d\alpha.$$

Hence

$$\beta_q(z, z_0) = \frac{r_0 - z}{2\sin\theta_0} \int_{-\theta_0}^{\theta_0} \frac{e^{-i(q+1)}}{r_0 - ze^{-i\alpha}} d\alpha - 1 =$$

$$= \frac{1}{2\sin\theta_0} \int_{-\theta_0}^{\theta_0} \frac{(r_0 - z)e^{-i(q+1)\alpha} - r_0 e^{i\alpha} + z}{r_0 - ze^{-i\alpha}} d\alpha.$$

Since

$$|(r_0 - z)e^{-i(q+1)\alpha} - r_0 e^{i\alpha} + z| =$$

$$= \Big|\int_0^\alpha (-i(q+1)(r_0 - z)e^{-i(q+1)\tau} - ir_0 e^{i\tau})d\tau\Big| \leq$$

$$\leq |\alpha|(q+1)\{|r_0 - z| + r_0\} \leq$$

$$\leq |\alpha|(q+1)\{|r_0 - ze^{-i\alpha}| + 2r_0\},$$

we get

$$|\beta_q(z,z_0)| \le \frac{q+1}{2\sin\theta_0}\int_{-\theta_0}^{\theta_0}|\alpha|\Big(1+\frac{2r_0}{|r_0-ze^{-i\alpha}|}\Big)d\alpha.$$

By (3.19)

$$|r_0-ze^{-i\alpha}| = |r_0e^{i(\alpha-\theta)}-r| \ge r_0\sin(\theta-\alpha) \ge$$

$$\ge r_0\sin(\eta-\theta_0) \ge r_0\sin(\eta/2),$$

and we have

$$|\beta_q(z,z_0)| \le \frac{q+1}{2\sin\theta_0}\int_{-\theta_0}^{\theta_0}|\alpha|\Big\{1+\frac{2}{\sin(\eta/2)}\Big\}d\alpha \le$$

$$\le \frac{(q+1)\theta_0^2}{2\sin\theta_0}\cdot\frac{3}{\sin(\eta/2)} = (q+1)\sin\theta_0\Big(\frac{\theta_0}{\sin\theta_0}|Bigr)^2\frac{3}{2\sin(\eta/2)} \le$$

$$\le (q+1)\sin\theta_0\cdot\Big(\frac{\pi\sqrt{2}}{4}\Big)^2\frac{3}{2\sin(\eta/2)},$$

which implies (3.21) It follows now from (3.19) that there exists a constant C such that

$$|\log D_q(z,z_n)| \le C\frac{|z|^{q+2}}{|r_n-z|}\frac{\sin\theta_n}{r_n^{q+1}}$$

for all $z_n \in \Delta(z), |z_n| < 2|z|$. Now to complete the proof of lemma 3.2 it suffices to aply inequality (3.10).

Lemma 3.4 *Let a function $f(z)$ be analytic, bounded and of a finite order ρ in the half-plane $\Im z > 0$. Set*

$$g(z) = \prod_{|z_n|\le 1}\frac{z-z_n}{z-\bar{z}_n}\cdot\prod_{|z_n|>1}D_q(z,z_n)\times$$

$$\times\exp\Big\{\frac{1}{\pi i}\int_{-\infty}^{\infty}\frac{(tz+1)^{q+1}\log|f(t)|}{(t^2+1)^{q+1}(t-z)}dt+$$

$$+\frac{1}{\pi i}\int_{-\infty}^{\infty}\frac{(tz+1)^{q+1}d\varphi(t)}{(t^2+1)^{q+1}(t-z)}\Big\}, \tag{3.22}$$

where $\{z_n\}$ are zeros of the function $f(z)$ in the halfplane $\Im z > 0$ and $q = [\rho]$. Then

$$\sup_{0<\theta<\pi}\log|g(re^{i\theta})| < r^{\rho+\epsilon}, \quad r > r_\epsilon. \tag{3.23}$$

for any $\epsilon > 0$. (By virtue of Lemma 3.2 all integrals and infinite products are absolutely convergent).

Proof. If $\rho < 1$, the lemma is a direct consequence of Theorem 3.1. So we assume $\rho \geq 1$. The absolute value of the first product in (3.22) does not exeed to one. Denote the second product by $S(z)$. To control it we need the following estimate of the canonical Nevanlinna factor:

$$\log|D_q(z, z_n)| \leq 2^{q+1}|z|^q r_n^{-q} \sin\theta_n, \quad |z| \geq \frac{1}{2} r_n. \tag{3.24}$$

This inequality may be considered as a complement to inequality (3.18). Relation (3.24) is obtained from the following sequence of inequalities:

$$\log|D_q(z, z_n)| = \log\left|\frac{1 - z/z_n}{1 - z/\bar{z}_n}\right| + 2\sum_{k=1}^{q} \frac{\sin k\theta_n}{k r_n^k} \Im(z^k) <$$

$$< 2\sum_{k=1}^{q} \frac{\sin k\theta_n}{k r_n^k} \Im(z^k) \leq$$

$$\leq 2\sin\theta_n \sum_{k=1}^{q} \left(\frac{|z|}{r_n}\right)^k < 2^{q+1}\left(\frac{|z|}{r_n}\right)^q.$$

Estimates (3.18), (3.24) yield (with $q = [\rho]$)

$$\log|S(z)| = \sum_{|z_n|<2|z|} \log|D_q(z, z_n)| +$$

$$+ \sum_{|z_n|\geq 2|z|} \log|D_q(z, z_n)| \leq$$

$$\leq 2^{q+1}|z|^q \sum_{|z_n|<2|z|} r_n^{-q} \sin\theta_n +$$

$$+ 4|z|^{q+1} \sum_{|z_n|\geq 2|z|} r_n^{-q-1} \sin\theta_n =$$

$$= 2^{q+1}|z|^q \sum_{|z_n|<2|z|} r_n^{\rho+\epsilon-q} \frac{\sin\theta_n}{r_n^{\rho+\epsilon}} +$$

$$+ 4|z|^{q+1} \sum_{|z_n|\geq 2|z|} r_n^{\rho+\epsilon-q-1} \frac{\sin\theta_n}{r_n^{\rho+\epsilon}} \leq$$

$$\leq 2^{\rho+\epsilon+1}|z|^{\rho+\epsilon} \sum_{|z_n|<2|z|} \frac{\sin\theta_n}{r_n^{\rho+\epsilon}} +$$

$$+2^{\rho+\epsilon-q+1}|z|^{\rho+\epsilon}\sum_{|z_n|\geq 2|z|}\frac{\sin\theta_n}{r_n^{\rho+\epsilon}}.$$

According to the second relation in (3.10) we obtain

$$\log|S(z)|\leq C_{q,\epsilon}|z|^{\rho+\epsilon},\quad \Im z>0,\quad |z|>1.$$

Now estimate the function

$$u(z)=\Re\Big\{\frac{1}{\pi i}\int_{-\infty}^{\infty}\frac{(zt+1)^{q+1}\log|f(t)|}{(t^2+1)^{q+1}(t-z)}dt\Big\}.$$

We have

$$\frac{1}{\pi i}\int_{-\infty}^{\infty}=\frac{1}{\pi i}\int_{|t|<\frac{1}{2}|z|}+\frac{1}{\pi i}\int_{\frac{1}{2}|z|\leq|t|\leq 2|z|}+$$

$$+\frac{1}{\pi i}\int_{|t|>2|z|}=I_1+I_2+I_3.$$

Using the first condition in (3.11) we have

$$|I_1|\leq\frac{1}{\pi}\int_{|t|<\frac{1}{2}|z|}\frac{(|t||z|+1)^{q+1}|\log|f(t)||dt}{(t^2+1)^{q+1}\Big(\frac{1}{2}|z|\Big)}\leq$$

$$\leq\frac{1}{\pi}\int_{|t|<\frac{1}{2}|z|}\frac{|\log|f(t)||dt}{(|t|+1)^{\rho+\epsilon+1}}\times$$

$$\times\max_{|t|<\frac{1}{2}|z|}\frac{(1+|t|)^{\rho+\epsilon+1}(|t||z|+1)^{q+1}}{(t^2+1)^{q+1}\Big(\frac{1}{2}|z|\Big)}\leq$$

$$\leq C_{q,\epsilon}|z|^{\rho+\epsilon}\int_{-\infty}^{\infty}(1+|t|)^{-\rho-\epsilon-1}|\log|f(t)||dt;$$

$$|I_3|\leq\frac{1}{\pi}\int_{|t|>2|z|}\frac{|\log|f(t)||dt}{(1+|t|)^{\rho+\epsilon+1}}\times$$

$$\times\max_{|t|>2|z|}\frac{(1+|t|)^{\rho+\epsilon+1}(|t||z|+1)^{q+1}}{(t^2+1)^{q+1}(|t|/2)}\leq$$

$$\leq C_{q,\epsilon}|z|^{\rho+\epsilon}\int_{-\infty}^{\infty}(1+|t|)^{-\rho-\epsilon-1}|\log|f(t)||dt,$$

and further

$$I_2=\frac{1}{\pi i}\int_{\frac{1}{2}|z|\leq|t|\leq 2|z|}\frac{(zt+1)^{q+1}-(t^2+1)^{q+1}}{(t^2+1)^{q+1}(t-z)}\log|f(t)|dt+$$

$$+\frac{1}{\pi i}\int_{\frac{1}{2}|z|\leq|t|\leq 2|z|}\frac{\log|f(t)|}{t-z}dt = I_1^{(1)} + I_2^{(2)}.$$

As can be easily seen,

$$\left|\frac{(zt+1)^{q+1}-(t^2+1)^{q+1}}{t-z}\right| \leq C_q|z|^{2q+1}$$

for $\frac{1}{2}|z| < |t| < 2|z|$ (since the expression in the left-hand side is a polynomial in both z and t). Therefore

$$|I_2^{(1)}| \leq \frac{1}{\pi i}\int_{\frac{1}{2}|z|\leq|t|\leq 2|z|}\frac{C_q|z|^{2q+1}}{(t^2+1)^{q+1}}|\log|f(t)||dt \leq$$

$$\leq C_{q,\epsilon}^{(1)}|z|^{\rho+\epsilon}\int_{\infty}^{\infty}\frac{|\log|f(t)||}{(1+|t|)^{\rho+\epsilon+1}}dt.$$

At last, combining the definition of order and the Poisson formula for the half-plane, we obtain

$$\Re I_2^{(2)} = \frac{1}{\pi}\int_{\frac{1}{2}|z|\leq|t|\leq 2|z|}(\log|f(t)|)\Im\frac{1}{t-z}dt \leq$$

$$\leq \sup_{\frac{1}{2}|z|\leq|t|\leq 2|z|}\log|f(t)| < C_\epsilon r^{\rho+\epsilon}. \tag{3.25}$$

Thus, $|u(z)| < C_{q,\epsilon}r^{\rho+\epsilon}$. The function

$$V(z) = \Re\left\{\frac{1}{\pi i}\int_{-\infty}^{\infty}\frac{(zt+1)^{q+1}d\varphi(t)}{(t^2+1)^{q+1}(t-z)}\right\}$$

may be estimated similarly. The only difference is , since the function $\varphi(t)$ is non-increasing, that the real part of the integral which corresponds to $I_2^{(2)}$ is non-negative. This makes the estimate, which corresponds to estimate (3.25), trivial.

Now we proceed directly to the **proof of Theorem 3.2.**

Let the function $f(z)$ be analytic and of order $\rho \geq 1$ in the half-plane $\Im z > 0$. Define the function $g(z)$ by equality (3.22) and consider the function

$$h(z) = f(z)/g(z).$$

Let us prove that $h(z)$ is a non-vanishing entire function and $|h(t)| = 1$ for real t.

For every $R > 0$ compare representation of $\log|f(z)|$ in the half-disk Q_R, given by the generalized Nevanlinna formula (2.4), and the representation of $\log|g(z)|$, that follows from (3.22). After subtracting we have

$$\log|h(z)| = \sum_{z_n \in Q_1} \log\left|\frac{R^2 - z_n z}{R^2 - \bar{z}_n z}\right| +$$

$$+ \sum_{z_n \in Q_R \setminus Q_1} \left\{\log\left|\frac{R^2 - z_n z}{R^2 - \bar{z}_n z}\right| + 2\sum_{k=1}^{q} \frac{\sin k\theta_n}{k r_n^k} \Im(z^k)\right\} -$$

$$-\frac{1}{\pi}\int_{-R}^{R}\left\{ \Im\left(\frac{(tz+1)^{q+1}}{(t^2+1)^{q+1}(t-z)} - \frac{1}{t-z}\right) + \frac{R^2 \Im z}{|R^2 - zt|^2}\right\}\times$$

$$\times(\log|f(t)|dt + d\varphi(t)) - \sum_{z_n \in \{\Im z > 0\} \setminus Q_R} \log|D_q(z, z_n)| +$$

$$+\frac{1}{2\pi}\int_0^{\pi}\left(\frac{R^2 - |z|^2}{|Re^{i\theta} - z|^2} - \frac{R^2 - |z|^2}{|Re^{-i\theta} - z|^2}\right)\log|f(Re^{i\theta})|d\theta +$$

$$+\frac{1}{\pi}\int_{|t|>R} \Im\frac{(tz+1)^{q+1}}{(t^2+1)^{q+1}(t-z)}(\log|f(t)|dt + d\varphi(t)).$$

As can be easily seen, each term in the right-hand side of this equality is harmonic inside the half-disk Q_R, continuous on $Q_R \cup (-R, R)$ and vanishes on $(-R, R)$. (One should note, that the expression in the square brackets under the integral sign $\int_{-R}^{R}$ is a polynomial with respect to z with real coefficients). Thus the function $\log|h(z)|$ is harmonic in Q_R, continuous in $Q_R \cup (-R, R)$ and vanishing on $(-R, R)$. By the symmetry principle one may extend the function $\log|h(z)|$ to a harmonic function in the disk $|z| < R$, and since R is arbitrary, into the whole plane. Therefore $h(z)$ is a non-vanishing entire function and $|h(t)| = 1$ for real t.

Let us estimate the growth of the function $h(z)$. Since the order of the initial function $f(z)$ is $\rho \geq 1$,then, for every $\epsilon > 0$, there exists a sequence $R_n \to \infty$ such that

$$\inf_{0<\theta<\pi} \log|f(R_n e^{i\theta})| > -R_n^{\rho+\epsilon}.$$

(we use definition 1.4). On the other hand, Lemma 3.3 yields estimate (3.23) of the function $\log|g(re^{i\theta})|$. Therefore

$$\inf_{0<\theta<\pi} \log|h(R_n e^{i\theta})| > -2R_n^{\rho+\epsilon}. \tag{3.26}$$

We apply the Carleman formula (2.15) with $R = R_n$, $\lambda = 1$ to the function $h(z)$. (In this case one may use the classical Carleman formula, due to the absence of the boundary singular function). Since $h(z)$ does not vanish and $|h(t)| = 1$, we obtain

$$\frac{1}{\pi R_n}\int_0^\pi \log^+ |h(R_n e^{i\theta})| \sin\theta d\theta =$$

$$= \frac{1}{\pi R_n}\int_0^\pi \log^+\left|\frac{1}{h(R_n e^{i\theta})}\right| \sin\theta d\theta - A_1(h, R_n),$$

whence by (3.26)

$$\int_0^\pi \log^+ |h(R_n e^{i\theta})| \sin\theta d\theta < C_\epsilon R_n^{\rho+\epsilon} \tag{3.27}$$

Now apply the Nevanlinna formula (2.4) with $R = R_n$ to $h(z)$. The properties of the function $h(z)$ enable this formula to take the form

$$\log|h(z)| =$$

$$= \frac{1}{2\pi}\int_0^\pi \left(\frac{R_n^2 - |z|^2}{|R_n e^{i\theta} - z|^2} - \frac{R_n^2 - |z|^2}{|R_n e^{-i\theta} - z|^2}\right) \log|h(R_n e^{i\theta})| d\theta.$$

After a transformation of the integral kernel we have

$$\log|h(re^{i\psi})| =$$

$$= \frac{1}{2\pi}\int_0^\pi \frac{(R_n^2 - r^2) 4 R_n r \sin\psi \cdot \sin\theta}{|R_n e^{i\theta} - re^{i\psi}|^2 |R_n e^{-i\theta} - re^{i\psi}|^2} \log|h(R_n e^{i\theta})| d\theta.$$

This implies the estimate

$$|\log|h(z)|| \leq \frac{12}{\pi}\int_0^\pi \sin\theta |\log h(R_n e^{i\theta})|| d\theta$$

for $|z| < R_n/z$, $\Im z > 0$. Taking into account relations (3.26) and (3.27), we obtain the estimate $|\log|h(z)|| \leq C_\epsilon R_n^{\rho+\epsilon}$ for $z \in Q_{R_n/2}$. Since $|h(t)| = 1$ for real t, the latter estimate is still valid in the whole disk $|z| \leq R_n/2$, thus we have obtained a sequence $t_n \to \infty$ $(t_n = R_n/2)$ such that

$$\max_{0 \leq \theta \leq 2\pi} |\log|h(t_n e^{i\theta})|| < C_\epsilon t_n^{\rho+\epsilon}. \tag{3.28}$$

Now let us prove that $\log|h(z)|$ is a polynomial of degree not exceeding $[\rho]$. Indeed, the Swartz formula yields

$$\log h(z) =$$

$$= \frac{1}{2\pi}\int_0^{2\pi} \frac{Re^{i\theta}+z}{Re^{i\theta}-z}\log|h(Re^{i\theta})|d\theta + \text{const}, \quad |z|<R.$$

By differentiating $[\rho]+1$ times we obtain

$$(\log h(z))^{([\rho]+1)} = \frac{([\rho]+1)!}{2\pi}\int_0^{2\pi} \frac{2Re^{i\theta}}{(Re^{i\theta}-|z|)^{[\rho]+2}}\log|h(Re^{i\theta})|d\theta,$$

whence

$$|(\log h(z))^{([\rho]+1)}| \le \frac{([\rho]+1)!2R}{(R-|z|)^{[\rho]+2}}\max_{0\le\theta\le 2\pi}|\log|h(Re^{i\theta})||.$$

Setting $R=t_n$ in the right-hand side of the inequality, passing to the limit as $n\to\infty$ and using (3.28), we obtain $(\log h(z))^{[\rho]+1}\equiv 0$, so $\log h(z)$ is a polynomial of degree non greater then $[\rho]$. Since $\log|h(t)|=0$ for real t, we get $\log h(z) = i(a_0 + a_1 z, +\cdots, +a_{[\rho]}z^{\rho})$, where $a_0,\dots,a_{[\rho]}$ are real coefficients. This completes the proof of the theorem.

In some cases the following representation is more convenient than (3.9).

Theorem 3.3 *Each function $f(z)$ analytic and of finite order ρ in the half-plane admits the representation*

$$f(z) = \exp\Big\{i\sum_{k=0}^{q} b_k z^k +$$

$$+\frac{1}{\pi i}\Big(\int_{-1}^{1}\frac{\log|f(t)|}{t-z}dt + \int_{-1}^{1}\frac{d\varphi(t)}{t-z}\Big)\Big\}\prod_{|z_n|<1}\frac{z-z_n}{z-\bar z_n}\times$$

$$\times\exp\Big\{\frac{z^{q+1}}{\pi i}\Big(\int_{|t|\ge 1}\frac{\log|f(t)|}{t^{q+1}(t-z)}dt + \int_{|t|\ge 1}\frac{d\varphi(t)}{t^{q+1}(t-z)}\Big)\Big\}\times$$

$$\times\prod_{|z_n|\ge 1} D_q(z,z_n), \tag{3.29}$$

Here $q=[\rho]$, $b_0,\dots,b_q$ are some positive constants, z_n are the zeros of $f(z)$ in $\Im z>0$ and $\varphi(t)$ is a singular boundary function. All integrals and infinite products in (3.29) are absolutely convergent and also relations (3.10), (3.11) hold.

Proof. Denote by $g(z)$ the ratio of the right-hand sides of (3.9) and (3.29). Setting

$$P(t,z) = \frac{1}{t-z}\Big\{\Big(\frac{z}{t}\Big)^{q+1} - \Big(\frac{tz+1}{t^2+1}\Big)^{q+1}\Big\},$$

$$Q(t,z)=\frac{1}{t-z}\Big\{1-\Big(\frac{tz+1}{t^2+1}\Big)^{q+1}\Big\},$$

we obtain

$$g(z)=\exp\Big\{i\sum_{k=0}^{q}(a_k-b_k)z^k-$$

$$-\frac{1}{\pi i}\Big(\int_{|t|\geq 1}P(t,z)\log|f(t)|dt+\int_{|t|\geq 1}P(t,z)d\varphi(t)+$$

$$+\int_{|t|\leq 1}Q(t,z)\log|f(t)|dt+\int_{|t|\leq 1}Q(t,z)d\varphi(t)\Big)\Big\}. \tag{3.26}$$

Direct calculation shows that both $P(t,z)$ and $Q(t,z)$ are polynomials with respect to z of degree not greater than q and with real coefficients. Therefore the expression in square brackets in (3.26) is a polynomial as well. Hence, by choosing the coefficients b_n in (3.26) in the appropriate way one may obtain $g(z)\equiv 1$.

NECESSARY CONDITIONS OF COMPLETELY REGULAR GROWTH IN THE HALF-PLANE

4 Definition of completely regular growth in the half-plane. List of results on completely regular growth

We begin with a notation introduced in [24]. The relation

$$\lim_{x\to\infty} {}^*\varphi(x) = a$$

means that the limit is taken as x approaches infinity through all positive values except some set E of zero relative measure, i. e. such that

$$\frac{\lim_{x\to\infty} \operatorname{mes}([0,x]\cap E)}{x} = 0.$$

By the *upper relative measure* of a set $E \subset [0,\infty)$ we mean the quantity

$$\overline{\operatorname{mes}}^* E = \limsup_{x\to\infty} x^{-1} \operatorname{mes}([0,x]\cap E).$$

A function $f(z)$ analytic and of finite order $\rho > 0$ in the angle (α, β) is said to be of *finite type* if

$$\sup_{|z|\le r,\ \alpha<\arg z<\beta} \log|f(z)| < Cr^\rho, \quad r > r_0.$$

for some $C > 0$.

Definition 4.1. A function $f(z)$ analytic and of a finite order $\rho > 0$ in the angle $[\alpha, \beta]$ is called a function of *completely regular growth (c.r.g.) in the closed angle* $[\alpha, \beta]$ if the function

$$h_{f,r}(\theta) = r^{-\rho}\log|f(re^{i\theta})|$$

approaches the indicator $h_f(\theta)$ uniformly with respect to θ, $\alpha \le \theta \le \beta$, as $r \to \infty$ and r does not belong to a certain exceptional set E of zero relative measure, the set E being the same for all θ. In other words,

$$\lim_{r\to\infty} {}^*r^{-\rho}\log|f(re^{i\theta})| = \limsup_{r\to\infty} r^{-\rho}\log|f(re^{i\theta})| = h_f(\theta).$$

Definition 4.2. A function $f(z)$ analytic and of order $\rho > 0$ in (α, β) is said to be a function of *completely regular growth* in the open angle (α, β) if it is of finite type in (α,β) and has c.r.g. in each closed angle $[\alpha+\epsilon,\beta-\epsilon]$, $\epsilon>0$.

Denote by $A_{\rho,\alpha,\beta}$ $(\bar{A}_{\rho,\alpha,\beta})$ the class of functions analytic and of order ρ and c.r.g. in the open angle (α,β) (respectively in the closed angle $[\alpha,\beta]$). Denote also

$$A_{\rho,0,\pi}=A_\rho,\quad \bar{A}_{\rho,0,\pi}=\bar{A}_\rho.$$

Definition 4.3. Let ρ_f be the order of a function $f(z)$ analytic in (α,β). Each value $\mu\le\rho_f$ is called the *formal order* of the function $f(z)$.

The following Theorem is a consequence of definition *1.1* and Theorem *1.5.*

Theorem 4.1 *Let $f(z)$ be analytic and of a formal order $\mu>0$ in $[\alpha,\beta]$ and let its formal indicator*

$$h_f^{(\mu)}(\theta)=\limsup_{r\to\infty} r^{-\mu}\log|f(re^{i\theta})|$$

satisfy the inequality

$$-\infty\le h_f^{(\mu)}(\theta)\le H(\theta),\quad \alpha\le\theta\le\beta,$$

where $H(\theta)$ is a continuous function on $[\alpha,\beta]$. Then, for each $\epsilon>0$,

$$\log|f(re^{i\theta})|<(H(\theta)+\epsilon)r^\mu,\quad r>r_{\epsilon,\mu}\quad \alpha\le\theta\le\beta.$$

If $\mu>\rho_f$ we clearly have $h_f^{(\mu)}(\theta)\equiv 0$ and

$$\sup_{\alpha\le\theta\le\beta}\log|f(re^{i\theta})|<\epsilon r^\mu,\ \textit{for } r>r_\mu.$$

Definition 4.4. Denote by $A^*_{\rho,\alpha,\beta}$ $(\bar{A}^*_{\rho,\alpha,\beta})$ the class of functions analytic in (α,β) and of c.r.g. in the open angle (α,β) (respectively, in the closed angle $[\alpha,\beta]$) with respect to the formal order ρ. In what follows we also use the denotations $A^*_\rho=A^*_{\rho,0,\pi}$, $\bar{A}^*_\rho=\bar{A}^*_{\rho,o,\pi}$.

Let us recall the basic results on entire functions of c.r.g. The theory of entire functions of c.r.g.was originated by B. Ya. Levin and A. Pflüger and is presented in [24, Ch.II,III].

Definition 4.5. We say that the point set $\{z_n\}$, $z_n\to\infty$, of the complex plane has an *angular density of order ρ* (or simply *angular density*) if for all values $\theta,\eta\in[0,2\pi]\backslash N$, N is an at most countable set, there exists the finite limit

$$\Delta(\theta,\eta)=\limsup_{r\to\infty} r^{-\rho}n(r,\theta,\eta),$$

where $n(r, \theta, \eta)$ is the number of points z_n within the sector $\{z : |z| \leq r,\ \theta < \arg z \leq \eta\}$.

By the angular density we mean the non-decreasing function $\Delta(\psi) = \Delta(\psi_0, \psi)$ where $\psi_0 \notin N$ is an arbitrary fixed value.

The non-decreasing function determined, to within an additive constant, by the relation $\Delta(\psi) = \Delta(\psi_0, \psi)$, $\psi_0 \notin N$ is some fixed value, is called *angular density.*

Theorem 4.2 *For an entire function $f(z)$ of non-integer order ρ to have c.r.g. it is necessary and sufficient that its zero set has the angular ρ-density.*

Theorem 4.3 *The indicator of the entire function $f(z)$ of c.r.g. and non-integer order ρ admits the representation*

$$h_f(\theta) = \frac{\pi}{\sin \pi\rho} \int_0^{2\pi} \cos \rho\big(|\theta - \psi| - \pi\big) d\Delta(\psi),$$

where $\Delta(\psi)$ is the angular density of zeros of the function $f(z)$.

Theorem 4.4 *For a function $f(z)$ of an integer order $\rho > 0$ to be of c.r.g. it is necessary and sufficient that its zero set $\{z_n\}$ has angular ρ-density and there exists the limit*

$$\delta_0 = \lim_{r \to \infty} \sum_{0 < |z_n| \leq r} z_n^{-\rho}. \tag{4.1}$$

Theorem 4.5 *The indicator of an entire function of c.r.g. of the integer order $\rho > 0$ admits the representation* [7]

$$h_f(\theta) = \int_{\theta - 2\pi}^{\theta} (\theta - \psi) \sin \rho(\psi - \theta) d\Delta(\psi) + \tau_f \cos(\rho\theta + \theta_f),$$

where $\tau_f = |\delta_0/\rho + c_\rho|$, $\theta_f = \arg(\delta_0/\rho + c_\rho)$ and c_ρ is the leading coefficient of the polynomial $P(z)$ in canonical representation (3.1).

Definition 4.6. Let $f(z)$ be an entire function of order $\rho > 0$. We say that its zero set is *regularly distributed* (see ([24, p.119,122]) if it has the angular density and if ρ is an integer, the limit in (4.1) exists.

This definition allows one to combine Theorems 4.2 and 4.4.

Theorem 4.6 *For an entire function $f(z)$ of order $\rho > 0$ to be of c.r.g. it is necessary and sufficient that its zero set is regularly distributed.*

[7] In [24] this relation contains a misprint.

Remark 1. Theorems 4.2–4.6 have been proved by B. Ya. Levin and A. Pflüger, who considered proximate orders, i.e. more general statements. Since we shall not use the notion of proximate order, we formulate these results for a usual order only.

Remark 2. As can easily be seen Theorems 4.2–4.6 are still valid for any formal order and we shall use them in this more general setting. If μ is a formal order and $\mu > \rho$, then both the angular density and the indicator are identically zero.

Remark 3. We formulate Theorems 4.3 and 4.5 in some weakened form. As was proved by B. Ya. Levin and A. Pflüger the relation

$$\lim_{z\to\infty} |z|^{-\rho}\{\log|f(z)| - |z|^\rho h_f(\theta)\} = 0, \quad \theta = \arg z,$$

holds uniformly outside some C^0-set, i.e. the union of countable set of the disks $|z - z_n| < r_n$ having zero linear density, that is

$$\lim_{r\to\infty} \frac{1}{r} \sum_{|z_n|\le r} r_n = 0.$$

B. Ya. Levin and A. Pflüger also obtained other important results on entire functions of c.r.g. (see [24]) that are not formulated here.

The aim of this book is to extend the results of B. Ya.Levin and A. Pflüger to functions of c.r.g. within an angle , i.e. to functions of the classes $A_{\rho,\alpha,\beta}$ and $\bar{A}_{\rho,\alpha,\beta}$. To this end, we introduce an analog of angular density and (for integer ρ) an analog of limit (4.1) which characterizes special symmetry of location of zeros. When doing it we, according to the canonical representation of a function of finite order (Theorem 3.2.), will take into account not only the inner zeros of the function $f(z)$ (that is zeros located inside the angle (α,β)) and the exponential term $\exp(i\sum_0^q a_k z^k)$, but also of the boundary values of the modulus of the function as well as its boundary singular function.

Since any angle (α,β) may be conformaly mapped onto the angle $(0,\pi)$, in what follows we consider functions analytic in the half-plane $\Im z > 0$. If the need be, one may reformulate the results for any angle (α,β). In this chapter we give necessary conditions for a function to belong to the classes A_ρ^* and $\bar{A}_\rho^*$. In the next chapter we establish that these conditions are sufficient in the case of non-integer ρ and also obtain necessary and sufficient conditions for integer $\rho > 0$.

First let us present some known results about functions of c.r.g. in the angle (see [24, Ch.III]).

Theorem 4.7 *Suppose a function $f(z)$ is analytic and of a formal order $\rho > 0$ in the closed angle $[\alpha, \beta]$ and for any $\epsilon > 0$, $\eta > 0$ there exists a set $E_{\epsilon,\eta}$ of real numbers such that $\overline{\text{mes}}^*(E_{\epsilon,\eta}) < \eta$ and*

$$\sup_{\alpha\le\theta\le\beta} |\log|f(re^{i\theta})| - r^\rho h_f(\theta)| < \epsilon r^\rho, \quad r \notin E_{\epsilon,\eta}.$$

Then

$$f(z) \in \bar{A}^*_{\rho,\alpha,\beta}.$$

Theorem 4.8 *Let a function $f(z)$ be analytic and of the formal order $\rho > 0$ in the open angle (α, β) and let $f(z)$ be of c.r.g. on each ray $\{z = \arg\theta\}$ for $\alpha < \theta_1 \le \theta \le \theta_2 < \beta$. Then $f(z)$ has c.r.g. on the whole set $\{z : \theta_1 \le \arg z \le \theta_2\}$.*

Theorem 4.9 *The set of rays $\{\arg z = \theta\}$, $\alpha < \theta_1 \le \theta \le \theta_2 < \beta$, on which $f(z)$ is of c.r.g. is closed.*

Theorem 4.10 *Let $f(z)$ be analytic and of the formal order $\rho > 0$ in the open angle (α, β), and let $f(z)$ be of c.r.g. on the ray $\arg z = \theta$ for some θ, $\alpha < \theta < \beta$. Then there exists the limit*

$$\lim_{r\to\infty} r^{-\rho} \int_\lambda^r \log|f(re^{i\theta})| \frac{dt}{t} = \frac{1}{\rho} h_f(\theta), \quad \lambda > 0.$$

Theorem 4.11 *Let $F(t)$ be a real-valued function defined for $t > 0$ and such that*

$$\limsup_{r\to\infty} r^{-\rho} F(r) \le a = const, \tag{4.2}$$

and

$$\lim_{r\to\infty} r^{-\rho} \int_\lambda^r F(t) \frac{dt}{t} = \frac{a}{\rho}, \quad \lambda > 0. \tag{4.3}$$

Then there exists

$$\lim_{r\to\infty} {}^* r^{-\rho} F(r) = a.$$

It should be noted that Theorem 4.11 is formulated in [24, p.194] with the equality sign in (4.2) instead of that of inequality. But actually this result is proved exactly under assumptions (4.2.), (4.3). Though one may prove that the strict inequality never appeared in (4.2), this formulation is more useful.

The following theorem is important

Theorem 4.12 *If $f(z) \in A^*_{\rho,\alpha,\beta}$, then, for all $\theta \in (\alpha,\beta)\backslash N$ where N is at most countable, there exists the limit*

$$\frac{1}{2\pi\rho}s_f(\theta_1,\theta_2) = \lim_{r\to\infty} r^{-\rho}n(r,\theta_1,\theta_2),$$

where

$$s_f(\theta_1,\theta_2) = h'_f(\theta_2) - h'_f(\theta_1) + \rho^2\int_{\theta_1}^{\theta_2} h_f(\varphi)d\varphi.$$

The exceptional set N contains only those points θ for which $h'_f(\theta+0) \neq h'_f(\theta-0)$.

We shall also use the following theorems.

Theorem 4.13 *Let $f(z)$ and $g(z)$ be analytic and of order $\rho > 0$ in (α,β) and at least one of them be of c.r.g. Then the indicator of their product equals the sum of the indicators of each multiplier.*

Theorem 4.14 *Let $f(z)$ and $g(z)$ be analytic and of order $\rho > 0$ in (α,β), $g(z)$ be of c.r.g. and the function $\psi(z) = f(z)/g(z)$ be analytic in (α,β). Then $h_\psi(\theta) = h_f(\theta) - h_g(\theta)$.*

The following theorem was proved by Hayman [36] .

Theorem 4.15 *Let $f(z)$ be analytic and bounded in the half-plane $\Im z > 0$. Then there exists a set E of finite logarithmic length* [8] *and such that*

$$\lim_{r\to\infty,\ r\notin E} r^{-1}\log|f(re^{i\theta})| = a\sin\theta, \quad -\infty < a \leq 0,$$

and the convergence is uniform with respect to θ.

By using Theorem 1.1 we obtain

Corollary 4.1 *Every function analytic and bounded in the half-plane $\Im z > 0$ belongs to the class A^*_1.*

[8] That is $\int_E d\log g < \infty$. It should be noted that $\int_E d\log r < \infty$ implies $\overline{\text{mes}}^* E = 0$, though the converse statement is false.

5 Relation between completely regular growth in open and closed angles

In this section we prove the following theorem.

Theorem 5.1 *A function $f(z) \in A^*_{\rho,\alpha,\beta}$ belongs to the class $\bar{A}^*_{\rho,\alpha,\beta}$ if and only if its indicator (with respect to the order ρ) is continuous at points α and β, i.e.*

$$h_f(\alpha) = h_f(\alpha + 0), \quad h_f(\beta) = h_f(\beta - 0).$$

Necessity is a direct consequence of Definitions 4.1 and 4.2.

To prove **sufficiency** we need several lemmas.

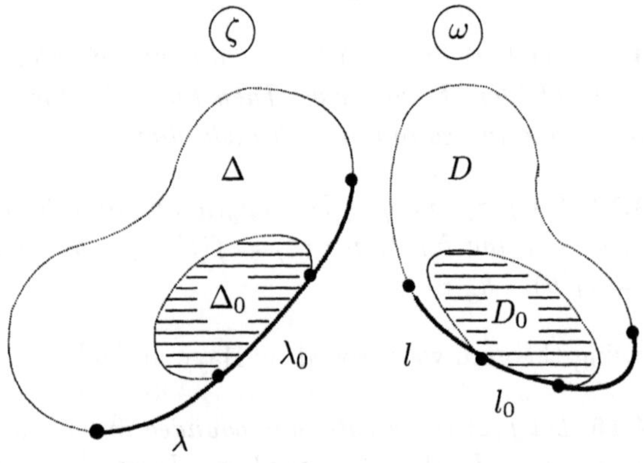

Fig. 1

Recollect that a simple bounded curve L is called a *Lyapunov curve* if the angle $\alpha(t)$ between the tangent at the point $t \in L$ and the real axis satisfies the Hölder condition, i.e.

$$|\alpha(t_1) - \alpha(t_2)| < A|t_1 - t_2|^{\chi}, \quad 0 < \chi \leq 1, \quad A > 0.$$

Lemma 5.1 *Let Δ and D be bounded simply connected domains in the complex planes of variables ζ and w respectively. Let their boundaries contain some Lyapunov arcs λ and l and $\Delta_0 \subset \Delta$ and $D_0 \subset D$ be domains with smooth boundaries such that $\lambda_0 = \partial\Delta \cap \partial\Delta_0$ and $l_0 = \partial D \cap \partial D_0$ are Lyapunov arcs lying strictly inside λ and l respectively. Finally, let $w = w(\zeta)$ be a conformal mapping of Δ onto D such that Δ_0 goes onto D_0. Then the relation*

$$\frac{1}{k} < \left|\frac{w(\zeta_1) - w(\zeta_2)}{\zeta_1 - \zeta_2}\right| < k \tag{5.1}$$

holds for any $\zeta_1, \zeta_2 \in \bar{\Delta}_0$, where k is a positive constant [9].

Proof. We shall prove the right-hand inequality in (5.1) since the left-hand one is proved completely likewise. Suppose the contrary to be true. Then there exist sequences $\{\zeta_{1,n}\}$ and $\{\zeta_{2,n}\}$ such that

$$|w(\zeta_{1,n}) - w(\zeta_{2,n})| > K_n|\zeta_{1,n} - \zeta_{2,n}|, \quad K_n \to \infty. \tag{5.2}$$

Without loss of generality we may assume $\lim_{n\to\infty} \zeta_{1,n} = \lim_{n\to\infty} \zeta_{2,n} = \zeta_0 \in \bar{\Delta}_0$. Since the boundary of Δ is smooth, there exist arcs $\sigma_n \in \bar{\Delta}_0$ with endpoints $\zeta_{1,n}$ and $\zeta_{2,n}$ and such that their lengths ρ_n satisfy

$$\lim_{n\to\infty} \rho_n^{-1}|\zeta_{1,n} - \zeta_{2,n}| = 1.$$

According to the well-known results (see, for example, [18, p.311]) the function $\log w'(\zeta)$ is continuous in $\bar{\Delta}_0$, and hence,

$$\frac{1}{M} < |w'(\zeta)| < M. \tag{5.3}$$

This implies

$$|w(\zeta_{1,n}) - w(\zeta_{2,n})| \leq \int_{\sigma_n} |w'(\zeta)||d\zeta| < M\rho_n <$$

$$< 2M|\zeta_{1,n} - \zeta_{2,n}|, \quad n > n_0,$$

which contradicts to inequality (5.2).

Lemma 5.2 [10] *Let Δ be a simply connected domain, $\zeta_0 \in \partial\Delta$ and the boundary $\partial\Delta_0$ in some neighborhood of ζ_0 be a circular arc. Let the function $z = \varphi(\zeta)$ map conformally the domain Δ onto a domain G. Suppose that the boundary ∂D in the neighborhood of the point $\varphi(\zeta_0)$ consists of two Lyapunov arcs l_1 and l_2 that make an angle of opening α, $0 < \alpha < 2\pi$, with the vertex in this point. Then in some neighborhood of ζ_0*

$$\varphi(\zeta) = \varphi(\zeta_0) + h_1(\zeta)(\zeta - \zeta_0)^{\alpha/\pi}, \tag{5.4}$$

$$|\varphi'(\zeta)| = h_2(\zeta)|\zeta - \zeta_0|^{-1+\alpha/\pi}, \tag{5.5}$$

where

$$\frac{1}{M} < |h_n(\zeta)| < M, \quad n = 1, 2.$$

Proof. Denote $z_0 = \varphi(\zeta_0)$ and take a small domain $G_1 \subset G$ such that G_1

[9] If only one of the arcs l and λ is smooth, then the theorem is false.

[10] The lemma becomes false for smooth curves l_j. On the other hand, one cannot sharpen the lemma even if assuming l_j to be analytic, except some trivial cases, for example, when l_j are linear segments.

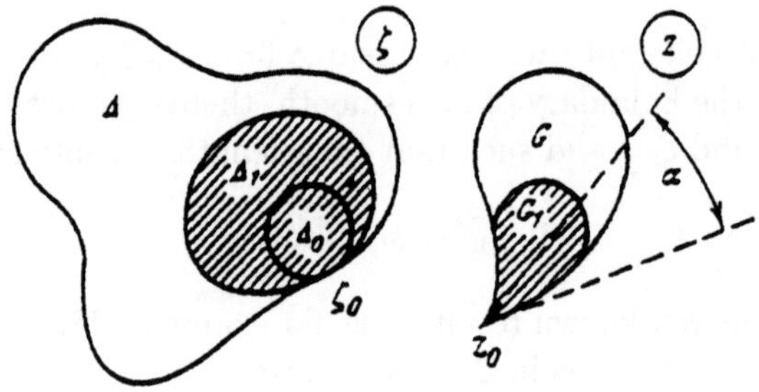

Fig. 2

contains the intersection of G with some neighborhood of z_0 and the boundary ∂G_1 is smooth at all points except z_0 (see, fig. 2).

Let Δ_1 be the pre-image of G_1. The function $w = (z - z_0)^{\pi/\alpha}$ maps the domain G_1 onto a domain B in w-plane and the boundary ∂D in a neighborhood of the point $w = 0$ is a Lyapunov arc. The function

$$w(\zeta) = (\varphi(\zeta) - \varphi(\zeta_0))^{\pi/\alpha} \tag{5.6}$$

maps conformally Δ_1 onto B_1. Applying Lemma 5.1 we obtain

$$\frac{1}{N} < \left|\frac{w(\zeta)}{\zeta - \zeta_0}\right| < N. \tag{5.7}$$

in some domain Δ_0 that contains all points $\zeta \in \Delta_0$, close enough to ζ_0. Now (5.6) and (5.7) imply (5.4). Using (5.6) and (5.3) we obtain

$$\frac{1}{M} < |(\pi/\alpha)(\varphi(\zeta) - \varphi(\zeta_0))^{-1+\pi/\alpha}\varphi'(\zeta)| < M.$$

Together with (5.4) this implies (5.5).

Lemma 5.3 *Let $\varphi(\zeta)$ be a conformal mapping of the disk $|\zeta| < R$ onto the domain*

$$B = \{z : \ R/2 < |z| < 3R/2, \ \Im z > 0\}$$

and let Ω_N, $N > 1$, be the set of all points ζ, $|\zeta| < R$ such that

$$\varphi(\zeta) \in B_N \overset{\text{def}}{=} B \cap \left\{ \left\{ z : \ |z \pm 3R/2| > \frac{R}{8\sqrt{N}} \right\} \cup \left\{ z : \ |z \pm R/2| > \frac{R}{8\sqrt{N}} \right\} \right\}.$$

Then

$$\max_{\zeta \in \Omega_N} |\varphi'(\zeta)| < T\sqrt{N},$$

where T is an absolute constant.

Proof. Without loss of generality we may assume $R = 1$. Let the lemma be false. Then there exist sequences $N_k \to \infty$, $T_k \to \infty$ and $\{\zeta_k\} \subset \Omega_N$ such that ζ_k approaches a point ζ_0, $|\zeta_0| = 1$, which is the preimage of one of the points $z = \pm R/2$, $z = \pm 3R/2$ and

$$|\varphi'(\zeta_n)| > T_n \sqrt{N_n}. \tag{5.8}$$

Lemma 5.2 yields

$$\frac{1}{K} < \frac{|\varphi(\zeta_n) - \varphi(\zeta_0)|}{\sqrt{|\zeta_n - \zeta_0|}} < K, \quad \frac{1}{K} < |\varphi'(\zeta_n)| \sqrt{|\zeta_n - \zeta_0|} < K,$$

as $\zeta_n \to \zeta_0$. Taking into account the definition of Ω_N, we obtain

$$|\varphi'(\zeta_n)| < \frac{K}{\sqrt{|\zeta_n - \zeta_0|}} < \frac{K^2}{|\varphi(\zeta_n) - \varphi(\zeta_0)|} < \frac{8K^2}{\sqrt{N}},$$

which contradicts estimate (5.8).

The following lemma is a generalization of Lemma 1.1.

Lemma 5.4 *Let a function $f(z)$ be analytic and bounded in the domain*

$$B = \{z : \ R/2 < |z| < 3R/2, \ \Im z > 0\}$$

and also $|f(Ri)| \geq 1$. Then there exists some positive absolute constant L such that, for any $N > 0$, the closed domain $\bar{B}$ consisting of all points $z \in B$ satisfying

$$\log |f(z)| \leq -LN \log(\sup_{z \in B} |f(z)|),$$

may be covered by an at most countable union of disks $|z - z_n| < r_n$ such that

$$\sum_n r_n < \frac{2R}{\sqrt{N}}.$$

Proof. The lemma is trivial for $N \leq 1$. So we assume $N > 1$. Let $z = \varphi(\zeta)$ be a conformal mapping of the disk $|\zeta| < R$ onto the domain G and B_N and Ω_N defined as in the previous lemma. According to Theorem 1.8 there exists an at most countable union of disks

$$Q_N = \cup_n \{\zeta : \ |\zeta - \zeta_n| \leq \rho_n\}, \quad \sum_n \rho_n < R/(2TN) \tag{5.9}$$

in the ζ-plane (the constant T is the same as in Lemma 5.3) such that

$$\log|f(\varphi(\zeta))| > -\frac{2AN}{T}\log(\sup_{z\in B}|f(z)|)$$

for $|\zeta| \leq R$ and $\zeta \notin Q_N$.

Let $P_N \subset B$ be the image of Q_N. Divide P_N into two sets $P_N^* = P_N \cap B_N$ and $P_N^{**} = P_N \backslash P_N^*$. That P_N^* is contained in a union of disks with the sum of radii less than $R/(2\sqrt{N})$ is a consequence of (5.9) and Lemma 5.3.

Lemma 5.5 *Let a function $f(z)$ be analytic in $(0, \pi)$ and of finite type with respect to a formal order ρ and also let $f(z)$ have c.r.g. on the ray $\theta = \pi/2$. For $R > 0$, $\eta > 0$ denote by $G_{R,\eta}$ the set of all r, such that $R \leq r \leq 2R$ and*

$$\inf_{0<\theta<\pi} \log|f(re^{i\theta})| < -\eta R^\rho, \quad \eta > 0.$$

Then

$$\limsup_{R\to\infty} R^{-1}\text{mes}G_{R,\eta} \leq 2^\rho C\sqrt{s/\eta}, \tag{5.10)}$$

where $s = \sup_{0<\theta<\pi}|h_f(\theta)|$ and $C > 0$ is an absolute constant.

Proof. Denote

$$K_R = \{z : \ R < |z| < 2R, \ \Im z > 0\},$$

$$D_R = \{z : \ 2R/3 < |z| < 3R, \ \Im z > 0\}.$$

Since $f(z)$ is of c.r.g. on the ray $\theta = \pi/2$, for any $\epsilon > 0$ and $R > R_\epsilon$, there exists an $\tilde{R}$ such that $4R/3 \leq \tilde{R} \leq 2R$ and

$$\log|f(\tilde{R}i)| > (h_f^-(\pi/2) - \epsilon)\tilde{R}^\rho > 2^\rho R^\rho(h_f^-(\pi/2) - \epsilon). \tag{5.11}$$

Here $h_f^-(\pi/2) \equiv \min[0, h_f(\pi/2)]$, and also

$$\sup_{z\in D_R} \log|f(z)| < 3^\rho(s + \epsilon)R^\rho$$

$$s = \sup_{0 \le \theta \le \pi} |h_f(\theta)|.$$

Consider the function

$$f_R(z) = f(z) \exp\{2^\rho R^\rho (\epsilon - h_f^-(\pi/2))\}$$

in the domain

$$\tilde{B}_R = \{z : \ \tilde{R}/2 < |z| < 3\tilde{R}/2, \ \Im z > 0\}.$$

Since $\tilde{B}_R \subset D_R$, we have

$$\sup_{z \in \tilde{B}_R} \log |f_R(z)| \le$$

$$\le \sup_{z \in D_R} \log |f_R(z)| < 2 \cdot 3^\rho (s + \epsilon) R^\rho,$$

and (5.11) yields $|f_R(i\tilde{R})| \ge 1$. For any $N > 1$ consider the set of all points $z \in K_R \subset \tilde{B}_R$ such that

$$\log |f_R(z)| < -LN2 \cdot 3^\rho (s + \epsilon) R^\rho,$$

or, which is the same,

$$\log |f(z)| < -LNR^\rho \{2 \cdot 3^\rho (s + \epsilon)\} + 2^\rho R^\rho (h_f^-(\pi/2) - \epsilon).$$

By virtue of Lemma 5.4 one may cover this set by the disks $|z - z_n| < r_r$ with $\sum r_n \le R/\sqrt{N}$. These disks also cover the set of points $z \in K_R$ for which

$$\log |f(z)| < -LNR^\rho 3^{\rho+1} (s + \epsilon).$$

Setting

$$N = \frac{\eta}{L 3^{\rho+1} (s + \epsilon)}$$

we obtain

$$\limsup_{R \to \infty} R^{-1} \mathrm{mes} G_{R,\eta} \le 4/\sqrt{N} =$$

$$= 4\sqrt{3^{\rho+1} L (s + \epsilon)/\eta} < 2^{\rho+3} \sqrt{L} \sqrt{(s + \epsilon)/\eta}.$$

The left-hand side is independent of ϵ. Letting $\epsilon \to 0$ and denoting $C = 8\sqrt{L}$, we obtain (5.10).

Lemma 5.6 *Let a set $E \subset [0, \infty)$ be such that $\overline{\text{mes}}^* > \omega > 0$, i.e.*

$$\text{mes}\{E \cap [0, r_n]\} > \omega r_n \tag{5.12}$$

for some sequence $r_n \to \infty$. Then, for any δ, $0 < \delta < 1$, there exists a sequence $R_n \to \infty$ such that

$$\text{mes}\{E \cap [R_n, R_n(1+\delta)]\} > \frac{1}{3}\omega\delta R_n.$$

Proof. Let the opposite be true. Then

$$\text{mes}\{E \cap [R, R(1+\delta)]\} \leq \frac{1}{3}\omega\delta R$$

for any $R > R_0$ and, in particular,

$$\text{mes}\{E \cap [(1+\delta)^n, (1+\delta)^{n+1}]\} \leq \frac{1}{3}\omega\delta(1+\delta)^n.$$

for $n \geq n_0$, where $(1+\delta)^{n_0} \geq R_0$.

Taking arbitrary $R \geq (1+\delta)^{n_0}$ and m such that $(1+\delta)^m \leq R \leq (1+\delta)^{m+1}$, we obtain

$$\text{mes}\{E \cap [0, R]\} \leq (1+\delta)^{n_0} + \frac{1}{3}\omega\delta \sum_{n=n_0}^{m} (1+\delta)^n =$$

$$= (1+\delta)^{n_0}\left(1 - \frac{1}{3}\omega\right) + \frac{1}{3}\omega(1+\delta)^{m+1} <$$

$$< (1+\delta)^{n_0}\left(1 - \frac{1}{3}\omega\right) + \frac{2}{3}\omega R,$$

which contradicts (5.12)

Proof of Theorem 5.1 Without loss of generality we may assume $\alpha = 0$, $\beta = \pi$. Then by the hypotheses $f(z) \in A_\rho^*$ and

$$h_f(0) = h_f(+0), \quad h_f(\pi) = h_f(\pi - 0).$$

We restrict ourselves to proving the theorem for some angle $[0, \alpha]$, $\alpha > 0$. One may use the similar arguments to consider the angle $[\pi - \alpha, \pi]$.

We may assume $h_f(0) = 0$, otherwise we would consider the function

$$f(z)\exp\{-h_f(0)z^\rho\}.$$

Suppose that $f(z)$ does not have c.r.g. in a small closed angle $[0, \alpha]$. For any sufficiently small $\delta > 0$ consider the angle $[0, \pi\delta a(\delta)]$ where $a(\delta) =$

$(\delta\log 2)^{-1}\log(1+\delta)$. By the assumption there exist a set $E_\delta\subset[0,\infty)$, $\overline{\text{mes}}^* E_\delta = 2\omega > 0$ and $\epsilon_0 > 0$ such that

$$\sup_{0\le\theta\le\pi\delta a(\delta)} |r^{-\rho}\log|f(re^{i\theta})| - h_f(\theta)| > \epsilon_0, \quad r\in E_\delta, \tag{5.13}$$

and also both ω and ϵ_0 are independent of δ (see Theorem 4.7). By Theorem 4.1 with $H(\theta) = h_f(\theta)$ relation (5.13) takes the form

$$\inf_{0\le\theta\le\pi\delta a(\delta)} \{\log|f(re^{i\theta})| - r^\rho h_f(\theta)\} < -\epsilon_0 r^\rho, \quad r\in E_\delta.$$

Therefore, denoting $s_\delta = \sup_{0\le\theta\le\pi\delta a(\delta)}|h_f(\theta)|$, we have

$$\inf_{0\le\theta\le\pi\delta a(\delta)} \log|f(re^{i\theta})| < -(\epsilon_0 - s_\delta)r^\rho \tag{5.14}$$

for $r\in E(\delta)$. According to Lemma 5.6 there exists a sequence $R_{n,\delta}\to\infty$ such that

$$\text{mes}\{E_\delta\cap[R_{n,\delta}, R_{n,\delta}(1+\delta)]\} > \frac{1}{3}\omega\delta R_{n,\delta}.$$

Consider the function $f(z)$ in the sectors

$$\{z:\ R_{n,\delta} < |z| < R_{n,\delta}(1+\delta),\ 0<\arg z<\pi\delta a(\delta)\}.$$

The transformation $\zeta = z^{1/(\delta a(\delta))}$ gives us the function $\varphi_\delta(\zeta) = f(\zeta^{\delta a(\delta)})$ in the half-annulus

$$\{\zeta:\ R_{n,\delta}^{1/(\delta a(\delta))} < |\zeta| < 2R_{n,\delta}^{1/(\delta a(\delta))},\ \Im\zeta > 0\}.$$

Denote by G_n^δ the image of the set

$$E_\delta\cap[R_{n,\delta}, R_{n,\delta}(1+\delta)].$$

Then

$$\text{mes}G_n^{(\delta)} \ge$$

$$\ge \frac{1}{\delta a(\delta)}(R_{n,\delta}(1+\delta))^{-1+1/(\delta a(\delta))}\frac{1}{3}\omega\delta R_{n,\delta} \ge$$

$$\ge \frac{1}{3}\frac{\omega}{a(\delta)}R_{n,\delta}^{1/(\delta a(\delta))}.$$

Since $a(\delta)\to 1/\log 2$ as $\delta\to 0$, we have

$$\text{mes}G_n^{(\delta)} \ge \frac{1}{6}\omega R_{n,\delta}^{1/(\delta a(\delta))}. \tag{5.15}$$

uniformly for $\delta < \delta_0$. By (5.14) this implies

$$\inf_{0\le\theta\le\pi} \log|\varphi_\delta(re^{i\theta})| < -(\epsilon_0 - s_\delta)^{\rho\delta a(\delta)} <$$

$$< -(\epsilon_0 - s_\delta)(R_{n,\delta}^{1/(\delta a(\delta))})^{\rho\delta a(\delta)}$$

for $r \in G_n^{(\delta)}$ (we assume δ to so be small that $s_\delta < \epsilon_0$). By Lemma 5.5 we have

$$\limsup_{n\to\infty} R_{n,\delta}^{-1/(\delta a(\delta))} \text{mes } G_n^{(\delta)} \le 2^{\rho\delta a(\delta)} C\sqrt{s_\delta/(\epsilon_0 - s_\delta)}. \tag{5.16}$$

Now the relation $h_f(0) = h_f(+0)$ and the assumption $h_f(0) = 0$ imply $\lim_{\delta\to 0} s_\delta = 0$. Therefore, for sufficiently small δ, inequalities (5.15) and (5.16) are incomparable.

Corollary 5.1 *If $f(z) \in A^*_{\rho,\alpha,\beta}$ and, for some $\delta > 0$, $f(z) \notin \bar{A}^*_{\rho,\alpha,\delta}$, then $h_f(\theta)$ has discontinuity at the point $\theta = \alpha$ and*

$$h_f(\alpha + 0) < h_f(\alpha). \tag{5.17}$$

It should be mentioned that the inequality $h_f(\alpha+0) > h_f(\alpha)$ is impossible due to the indicator's properties. So the questions appear, whether there exist either a set $E_{\delta,1}$, $\overline{\text{mes}}^* E_{\delta,1} > 0$ such that

$$\sup_{\alpha<\theta<\alpha+\delta} (r^{-\rho}\log|f(re^{i\theta})| - h_f(\theta)) > \epsilon_1, \quad r \in E_{\delta,1}, \tag{5.18}$$

for some $\epsilon_1 > 0$ or a set $E_{\delta,2}$, $\overline{\text{mes}}^* E_{\delta,2} > 0$, such that

$$\inf_{\alpha<\theta<\alpha+\delta} (r^{-\rho}\log|f(re^{i\theta})| - h_f(\theta)) < \epsilon_2, \quad r \in E_{\delta,2}, \tag{5.19}$$

for some $\epsilon_2 > 0$. The example on p.11 shows that the answers to the both questions are negative. One may also construct a function $f(z)$ of c.r.g. in $(\alpha, \alpha + \delta)$ satisfying (5.17), (5.18) for which (5.19) is false. On the other hand, (5.17) and (5.19) imply (5.18).

6 Asymptotic behavior of the modulus and zero distributions of entire functions of the class A_ρ^*

Theorem 6.1 *If $f(z) \in A_\rho^*$, then for any $\epsilon > 0$ the following estimates hold uniformly with respect to θ , $0 \le \theta \le \pi$,*

$$\int_1^r \log|f(re^{i\theta})|\frac{dt}{t} > \frac{1}{\rho} r^\rho \big(\tilde{h}_f(\theta) - \epsilon\big), \quad r > r_\epsilon.$$

Here [11]

$$\tilde{h}_f(\theta) = \begin{cases} h_f(\theta), & 0 < \theta < \pi, \\ h_f(+0), & \theta = 0, \\ h_f(\pi - 0), & \theta = \pi. \end{cases}$$

To prove the theorem we need the following lemma, which is a direct consequence of Theorem 3.1.

Lemma 6.1 *If a function $F(z)$ is analytic in $\Im z > 0$ and either is bounded or has an order $\rho < 1$, then*

$$\log|F(re^{i\theta})| \le \frac{r\sin\theta}{\pi}\int_{-\infty}^{\infty} \frac{\log|F(t)|dt}{t^2 - 2rt\cos\theta + r^2}. \tag{6.1}$$

Proof of Theorem 6.1 Suppose the statement of the theorem to be false. Then there exist an $\epsilon_0 > 0$ and sequences $\{\theta_n\}$ and $\{r_n\}$, $r_n \to \infty$, such that

$$\int_1^{r_n} \log|f(te^{i\theta_n})|\frac{dt}{t} < \frac{1}{\rho} r_n^\rho \big(\tilde{h}_f(\theta_n) - \epsilon\big). \tag{6.2}$$

Without loss of generality we may assume that $\lim_{n\to\infty}\theta_n = \theta_0$. We restrict ourselves to the case $0 \le \theta_0 \le \pi/2$, since the case $\theta_0 > \pi/2$ is similar. We also suppose

$$|f(z)| \le 1, \quad \text{for} \quad z \in \{z: \theta_n < \arg z \le \theta_n + \pi/(4s)\}, \quad n = 1, 2, \ldots \tag{6.3}$$

Here $s = \max(\rho, 1)$, since otherwise we may consider the function

$$f_*(z) = M^{-1} f(z)\exp(-2Ke^{-i\rho\theta_0} z^\rho) \tag{6.4}$$

where $M > 0$ is sufficiently large and $K > 0$ is the constant from the relation

$$\sup_{0\le\theta\le\pi} |f(re^{i\theta})| < \exp(Kr^\rho); \quad r > r_0. \tag{6.5}$$

[11] If $h(\theta)$ is trigonometrically convex and bounded, then there exist oneside limits $h(+0)$ and $h(\pi - 0)$.

Taking n sufficiently large we have $[\theta_n, \theta_n + \pi/(4s)] \subset [\theta_0 - \pi/(3s), \theta_0 + \pi/(3s)]$, thereby the function $f_*(z)$ is bounded in $[\theta_n, \theta + \pi/(4s)]$. Hence relation (6.3) holds for the function $f_*(z)$ if one takes M in (6.4) sufficiently large and changes the enumeration of θ_n. After proving the theorem for $f_*(z)$ one may easily return to the function $f(z)$.

Consider the sequence of the functions in $\Im\zeta > 0$

$$g_n(\zeta) = f(\zeta^{1/(4s)} e^{i\theta_n}). \tag{6.6}$$

By (6.3) we have, for any $n \geq 1$,

$$|g_n(\zeta)| \leq 1, \quad \text{for } \Im\zeta > 0, \tag{6.7}$$

and by Lemma 6.1

$$\log|g_n(Re^{i\alpha})| \leq \frac{R\sin\alpha}{\pi} \int_1^{r_n^{4s}} \frac{\log|g_n(t)|dt}{t^2 - 2tR\cos\alpha + R^2}, \quad R > 0, \tag{6.8}$$

for any α, $0 < \alpha < \pi$. Now let us choose an arbitrary $\eta > 0$ and keep it fixed through the whole proof. Divide by R the both sides of (6.8) and integrate the result from $R = 1/2$ to $R = (1+\eta)r_n^{4s}$. After changing the order of integration we obtain

$$\int_{1/2}^{(1+\eta)r_n^{4s}} \log|g_n(Re^{i\alpha})| \frac{dR}{R} \leq$$

$$\leq \int_1^{r_n} \log|g_n(t)| \frac{dt}{t} \int_{1/2}^{(1+\eta)r_n^{4s}} \psi(R,t)dR, \tag{6.9}$$

where

$$\pi\psi(R,t) = \frac{t\sin\alpha}{R^2 - 2Rt\cos\alpha + t^2}.$$

Using the relation for the solution of the Dirichlet problem in the half-plane, we have

$$\int_{1/2}^{(1+\eta)r_n^{4s}} \psi(R,t)dR = 1 - \int_{-\infty}^{1/2} \psi(R,t)dR - \int_{(1+\eta)r_n^{4s}}^{\infty} \psi(R,t)dR. \tag{6.10}$$

Since $t \geq 1$, we get

$$\int_{-\infty}^{1/2} \psi(R,t)dR = \frac{1}{\pi}\operatorname{arccotan}\frac{t\cos\alpha - 1/2}{t\sin\alpha} \leq$$

$$\leq \frac{1}{\pi}\operatorname{arccotan}\frac{\cos\alpha - 1/2}{\sin\alpha},$$

and since $t \le r_n^{4s}$

$$\int_{(1+\eta)r_n^{4s}}^{\infty} \psi(R,t)dR = \frac{1}{\pi}\text{arccotan}\frac{(1+\eta)r_n^{4s} - t\cos\alpha}{t\sin\alpha} \le$$

$$\le \frac{1}{\pi}\text{arccotan}\frac{1-\cos\alpha+\eta}{\sin\alpha}$$

Applying the relation

$$\text{arccotan}x + \text{arccotan}y = \text{arccotan}\frac{xy-1}{x+y}, \quad xy > 1,\ x > 0, \tag{6.11}$$

and using (6.7)–(6.10), we have

$$\int_{1/2}^{(1+\eta)r_n^{4s}} \log|g_n(Re^{i\alpha})|\frac{dR}{R} \le$$

$$\le \left(1 - \frac{1}{\pi}\text{arccotan}\frac{(\cos\alpha - 1)(3+2\eta)+\eta}{(1+2\eta)\sin\alpha}\right)\int_1^{r_n^{4s}} \log|g_n(t)|\frac{dt}{t} \tag{6.12}$$

We may use relation (6.11) under the conditions $\cos\alpha > 1/2$, $(\cos\alpha - 1/2)(1-\cos\alpha+\eta)\sin^{-2}(\alpha) > 1$. This, for $0 < \alpha < \pi/2$, is equivalent to

$$0 < \alpha < \arccos\big((3+\eta)/(3+2\eta)\big). \tag{6.13}$$

Let this inequality be fulfilled. Comparing (6.12) and (6.6) we obtain

$$4s\int_{2^{-1/(4s)}}^{(1+\eta)^{1/(4s)}r_n} \log|f(re^{i(\theta_n+\alpha/(4s))})|\frac{dr}{r} \le$$

$$\le 4s\left(1 - \frac{1}{\pi}\text{arccotan}\frac{(\cos\alpha - 1)(3+2\eta)+\eta}{(1+2\eta)\sin\alpha}\right)\int_1^{r_n} \log|f(re^{i\theta_n})|\frac{dr}{r}. \tag{6.14}$$

Inequality (6.14) holds for all n and for α satisfying (6.13). In particular, α may depend on n. Fix a small $\delta > 0$ so that $0 < 4\delta s < \arccos\{(3+\eta)/(3+2\eta)\}$. We have $4s(\theta_0+\delta-\theta_n) < \arccos\{(3+\eta)/(3+2\eta)\}$ for $n > n_\delta$. Setting $\alpha = 4s(\theta_0+\delta-\theta_n)$ in (6.14) and taking into account assumption (6.2) we have

$$\rho r_n^{-\rho}\int_{2^{-1/(4s)}}^{(1+\eta)^{1/(4s)}r_n} \log|f(re^{i(\theta_0+\delta)})|\frac{dr}{r} \le$$

$$\le \left(1 - \frac{1}{\pi}\text{arccotan}\frac{\big(\cos(4s(\theta_0+\delta-\theta_n)) - 1\big)(3+2\eta)+\eta}{(1+2\eta)\sin(4s(\theta_0+\delta-\theta_n))}\right)\times$$

$$\times\big(\tilde{h}_f(\theta_n) - \epsilon_0\big)$$

for $n > n_\delta$. Since $f(z)$ has c.r.g., using Theorem 4.10 and letting $n \to \infty$, we obtain

$$(1+\eta)^{\rho/(4s)}\tilde{h}_f(\theta_0+\delta) \leq$$

$$\leq \left(1 - \frac{1}{\pi}\text{arccotan}\frac{(\cos 4s\delta - 1)(3+2\eta)+\eta}{(1+2\eta)\sin(4s\delta)}\right)(\tilde{h}_f(\theta_0) - \epsilon_0)$$

for each sufficiently small $\delta > 0$. Therefore

$$(1+\eta)^{\rho/(4s)}\tilde{h}_f(\theta_0) \leq \tilde{h}_f(\theta_0) - \epsilon_0,$$

which is impossible if $\eta > 0$ is sufficiently large. The contradiction proves the theorem.

Corollary 6.1 *If $f(z) \in A^*_\rho$, then , for $r > 1$,*

$$\int_1^r |\log|f(te^{i\theta})||\frac{dt}{t} < C_f r^\rho, \quad 0 \leq \theta \leq \pi, \tag{6.15}$$

$$\int_1^r |\log|f(te^{i\theta})||dt < C_f r^{\rho+1}, \quad 0 \leq \theta \leq \pi. \tag{6.16}$$

Here the constant C_f is independent of r and θ.

Proof. Consider the partition of the angle $[0, \pi]$

$$[0,\pi] = \bigcup_{k=0}^{m-1}\left[\frac{k\pi}{4s}, \frac{(k+1)\pi}{4s}\right] \cup \left[\frac{m\pi}{4s}, \pi\right],$$

$$s = \max[\rho, 1], \quad m = [4s-1]. \tag{6.17}$$

Clearly, it suffices to prove (6.15) for every angle from the partition. The function

$$f_k(z) = M_k^{-1}f(z)\exp\{-2Ke^{-i\rho k\pi/(4s)}z^\rho\}, \quad k = 1, \ldots, m-1, \tag{6.18}$$

satisfies the condition $|f_k(z)| < 1$ within the k's angle. Setting

$$N_{k,f} = \sup_{0\leq\theta\leq\pi}\frac{1}{\rho}(1 - h_{f_k}(\theta))$$

and applying Theorem 6.1 with $\epsilon = 1$ to the function $f_k(z)$, we obtain

$$\int_1^r |\log|f_k(te^{i\theta})||\frac{dt}{t} = -\int_1^r \log|f_k(te^{i\theta})|\frac{dt}{t} <$$

$$< N_{k,f}r^{\rho}, \quad k\pi/(4s) \le \theta \le (k+1)\pi/(4s)$$

for $r > r_0$. Together with (6.18) this yields

$$\int_1^r |\log|f_k(te^{i\theta})||\frac{dt}{t} < r^{\rho}(N_{k,f} + 2K\rho^{-1}), \quad r > r_0,$$

and thereby proves (6.15). Denote by $s_\theta(r)$ the left-hand side of (6.15). We have

$$\int_1^r |\log|f_k(te^{i\theta})||dt = \int_1^r x ds_\theta(x) < rs_\theta(r) < C_f r^{1+\rho}.$$

Relation (6.16) is also proved.

Corollary 6.2 *Let $f(z) \in A_\rho^*$. Then the approach to the limit*

$$\lim_{r\to\infty} \rho r^{-\rho} \int_1^r \log|f(te^{i\theta})|\frac{dt}{t} = h_f(\theta) \tag{6.19}$$

is uniform with respect to θ on each segment $0 < \eta \le \theta \le \pi - \eta$.. If $h_f(\theta)$ is continuous on $[0,\pi]$, then the approach to the limit is uniform on the whole segment $0 \le \theta \le \pi$.

Proof. For $0 < \theta < \pi$ relation (6.19) is exactly the statement of Theorem 4.10. That the limit is uniform follows from Theorem 6.1 and the estimate

$$\log|f(re^{i\theta})| < (h_f(\theta) + \epsilon)r^{\rho}, \quad r > r_\epsilon,$$

which holds uniformly with respect to θ either on any segment $\eta \le \theta \le \pi - \eta$ or on $[0,\pi]$ if $h_f(\theta)$ is continuous on $[0,\pi]$. In the latter case Theorem 5.1 implies $f(z) \in \bar{A}_\rho^*$.

Lemma 6.2 *If $f(z) \in A_\rho^*$, then there exists the limit*

$$\lim_{r\to\infty} {}^* r^{-\rho} \int_0^\pi \log|f(te^{i\theta})| \sin\theta d\theta = \int_0^\pi h_f(\theta)\sin\theta d\theta. \tag{6.20}$$

Proof. By the hypothesis

$$\sup_{0\le\theta\le\pi} \log|f(re^{i\theta})| < Cr^{\rho}, \quad r > r_0.$$

Therefore, if r is sufficiently large, the function

$$g_r(\theta) = C - r^{-\rho}\log|f(re^{i\theta})|$$

is non-negative. Setting

$$\psi(r) = \int_0^\pi \log|f(re^{i\theta})| \sin\theta d\theta$$

and applying the Fatou lemma to the integral

$$\int_0^\pi g_r(\theta) \sin\theta d\theta$$

as $r \to \infty$, we obtain

$$\limsup_{r\to\infty} r^{-\rho}\psi(r) \leq \int_0^\pi h_f(\theta) \sin\theta d\theta. \qquad (6.21)$$

Denote $c(\delta) = [0, \delta] \cup [\pi - \delta, \pi]$. We have

$$\int_1^r \psi(t)\frac{dt}{t} = \int_\delta^{\pi-\delta} \sin\theta d\theta \int_1^r \log|f(te^{i\theta})|\frac{dt}{t} +$$

$$+ \int_{c(\delta)} \sin\theta d\theta \int_1^r \log|f(te^{i\theta})|\frac{dt}{t}.$$

Corollaries 1 and 2 of Theorem 6.1, if applied for $r > r_\delta$, yield

$$|r^{-\rho} \int_1^r \psi(t)\frac{dt}{t} - \frac{1}{\rho}\int_\delta^{\pi-\delta} h_f(\theta) \sin\theta d\theta| < \epsilon(\delta),$$

$$\lim_{\delta\to 0} \epsilon(\delta) = 0.$$

Therefore there exists the limit

$$\lim_{r\to\infty} r^{-\rho} \int_1^r \psi(t)\frac{dt}{t} = \frac{1}{\rho}\int_0^\pi h_f(\theta) \sin\theta d\theta. \qquad (6.22)$$

Now Theorem 4.11 taking into account of (6.21) and (6.22) yields (6.20).

Let us now prove a stronger statement.

Theorem 6.2 *Let $f(z) \in A_\rho^*$, $0 < \rho < \infty$. Then there exists the finite limit*

$$\lim_{r\to\infty} r^{-\rho} \int_0^\pi \log|f(re^{i\theta})| \sin\theta d\theta = \int_0^\pi h_f(\theta) \sin\theta d\theta.$$

Proof. In view of (6.21) it suffices to obtain the relation

$$\liminf_{r\to\infty} r^{-\rho}\int_0^\pi \log|f(re^{i\theta})|\sin\theta d\theta \geq \int_0^\pi h_f(\theta)\sin\theta d\theta.$$

Suppose the opposite be true. Then there exist a number $\epsilon > 0$ and a sequence $r_n \to \infty$ such that

$$\int_0^\pi \log|f(r_ne^{i\theta})|\sin\theta d\theta < r_n^\rho\Big\{\int_0^\pi h_f(\theta)\sin\theta d\theta - \epsilon_0\Big\}. \tag{6.23}$$

Applying the generalized Carleman formula (relation (2.15)) we obtain that the left-hand side of (6.23) is continuous with respect to r_n. In addition $f(z)$ has at most a countable number of zeros and $\log|f(t)| > -\infty$ almost everywhere. Therefore, changing r_n if need be, we may assume

$$\sup_{0\leq\theta\leq\pi} |\log|f(r_ne^{i\theta})|| = M_n \neq \infty, \quad n = 1, 2, \ldots. \tag{6.24}$$

Consider the sequence of subharmonic functions defined in the half-disk $B = \{z : \ |z| < 1, \ \Im z > 0\}$

$$u_n(z) = \log|f(zr_n)| - Kr_n^\rho - N. \tag{6.25}$$

Here K is the constant from relation (6.5) and N is chosen sufficiently large so that

$$\sup_{z\in B} u_n(z) \leq 0, \tag{6.26}$$

for all $n \geq 1$. Further, define the harmonic in B functions $u_n^*(z)$ whose boundary values satisfy the relation

$$u_n^*(t) = 0, \quad -1 < t < 1, \qquad u_n^*(e^{i\theta}) = u_n(e^{i\theta}), \quad 0 < \theta < \pi. \tag{6.27}$$

Clearly, the corresponding Dirichlet problem is solvable and in addition $\sup_{z\in B}|u_n^*(z)| \leq M_n$. The definition of $u_n^*(z)$ together with inequality (6.26) yield

$$\limsup_{z\to\zeta}\big(u_n(z) - u_n^*(z)\big) \leq 0$$

for all $\zeta \in \partial B$ except for possibly $\zeta = \pm 1$. Therefore

$$u_n(z) \leq u_n^*(z). \tag{6.28}$$

everywhere in B.

Consider the odd extension of the function $u_n^*(z)$ into the lower half of the disk $|z| < 1$ defined by the relation

$$u_n^*(te^{-i\theta}) = -u^*(te^{i\theta}), \quad 0 \leq t < 1, \quad 0 \leq \theta \leq \pi.$$

The expansion

$$u_n^*(te^{i\theta}) = \sum_{k=1}^{\infty} a_{k,n} t^k \sin k\theta, \quad 0 \le t < 1,$$

together with (6.28) give

$$\int_0^{\pi} u_n(te^{i\theta}) \sin\theta d\theta \le \int_0^{\pi} u_n^*(te^{i\theta}) \sin\theta d\theta =$$

$$= \frac{1}{2}\int_0^{2\pi} u_n^*(te^{i\theta}) \sin\theta d\theta = \frac{\pi}{2} a_{1,n} t. \tag{6.29}$$

The second integral is continuous with respect to t. Therefore one may pass to the limit sending $t \to 1$ in the last equality. Using (6.27) we have

$$\frac{\pi}{2} a_{1,n} = \int_0^{\pi} u_n(e^{i\theta}) \sin\theta d\theta. \tag{6.30}$$

Substituting (6.30) into (6.29) we obtain

$$\int_0^{\pi} u_n(te^{i\theta}) \sin\theta d\theta \le t \int_0^{\pi} u_n(e^{i\theta}) \sin\theta d\theta,$$

and, by (6.25),

$$\int_0^{\pi} \log|f(tr_n e^{i\theta})| \sin\theta d\theta \le$$

$$\le t \int_0^{\pi} \log|f(r_n e^{i\theta})| \sin\theta d\theta - 2(t-1)(Kr_n^{\rho} - N).$$

Assumption (6.23) now yields

$$(tr_n)^{-\rho} \int_0^{\pi} \log|f(tr_n e^{i\theta})| \sin\theta d\theta \le \int_0^{\pi} h_f(\theta) \sin\theta d\theta -$$

$$-\epsilon_0 + (t^{1-\rho} - 1) \int_0^{\pi} h_f(\theta) \sin\theta d\theta - -2t^{-\rho}(t-1)(K - Nr_n^{-\rho}).$$

Now let a number t_0, $0 < t_0 < 1$, be chosen so that, for $t_0 \le t < 1$ the sum in the last line of this expression is less than $\epsilon_0/2$. Then, for any r from the segment $t_0 r_n \le r \le r_n$, we have

$$r^{-\rho} \int_0^{\pi} \log|f(re^{i\theta})| \sin\theta d\theta \le \int_0^{\pi} h_f(\theta) \sin\theta d\theta - \frac{\epsilon_0}{2},$$

which contradicts Lemma 6.2.

Corollary 6.3 *If $f(z) \in A^*_\rho$ and $q > 0$ is an integer, then there exists the finite limit*

$$\lim_{r\to\infty} r^{-\rho}\int_0^\pi \log|f(re^{i\theta})|\sin q\theta d\theta = \int_0^\pi h_f(\theta)\sin q\theta d\theta. \tag{6.31}$$

Proof. Divide $[0,\pi]$ into q equal parts. It suffices to prove that

$$\lim_{r\to\infty} r^{-\rho}\int_{k\pi/q}^{(k+1)\pi/q} \log|f(re^{i\theta})|\sin q\theta d\theta = \int_{k\pi/q}^{(k+1)\pi/q} h_f(\theta)\sin q\theta d\theta \tag{6.32}$$

for any $k \in \{0,1,\dots,q-1\}$. The function $f_k(\zeta) = f(e^{ik\pi/q}\zeta^{1/q})$ has the order ρ/q in the half-plane $\Im\zeta > 0$. By Theorem 6.2 the limit in the left-hand side of (6.32) equals

$$\lim_{r\to\infty}(-1)^k r^{-\rho}\int_0^\pi \log|f(re^{i(\alpha+k\pi)/q})|\sin\alpha d\alpha = \int_0^\pi h_{f_k}(\alpha)\sin\alpha d\alpha.$$

Since $h_{f_k}(\alpha) = h_f\big((\alpha+k\pi)/q\big)$, this implies (6.31).

Let $f(z)$ be a function of order ρ in $\Im z > 0$ and $z_n = r_n e^{i\theta_n}$ be its zeros in the half-plane (we called them the inner zeros) and $\varphi(t)$ be the singular boundary function of $f(z)$. Denote

$$s_1(r) = \frac{1}{2\pi}\int_1^r \log|f(x)f(-x)|dx,$$

$$s_2(r) = \frac{1}{2\pi}\int_1^r d\{\varphi(x)-\varphi(-x)\},$$

$$s_3(r) = \sum_{z_n\in Q_r\setminus Q_1} r_n\sin\theta_n, \quad Q_r = \{z:\ \Im z > 0\}\cap\{z:\ |z|\le r\},$$

$$\sigma(r) = s_3(r) - s_1(r) - s_2(r). \tag{6.33}$$

Without loss of generality we may assume that the arc $\{z:\ |z| = 1,\ \Im z > 0\}$ is free of points z_n, and also $\varphi(1-0) = \varphi(1+0)$, $\varphi(-1-0) = \varphi(-1+0)$. (Otherwise we may begin integration and summation in relation (6.33) from some positive quantity distinct from 1, this will not alter the condition.) So $s_1(1) = s_2(1) = s_3(1) = 0$.

Lemma 6.3 *If $f(z) \in A^*_\rho$, $0 < \rho < 1$* [12]*, and $h_f(\theta)$ is its indicator, then there exists the limit*

$$\lim_{r\to\infty} r^{-1-\rho}\sigma(r) = \frac{\rho-1}{2\pi}\int_0^\pi h_f(\theta)\sin\theta d\theta. \tag{6.34}$$

---

[12] One may prove the lemma as well as Lemma 6.4 below for any $\rho > 0$, but we shall use them for $0 < \rho < 1$ only.

Proof. Suppose first that

$$\sup_{\Im z>0} |f(z)| < 1. \tag{6.35}$$

The generalized Carleman formula (relation (2.15)),being applied to the function $f(z)$, yields

$$\sum_{z_n \in Q_r \setminus Q_1} \left(\frac{1}{r_n} - \frac{r_n}{r^2}\right) \sin\theta_n = \frac{1}{\pi r}\int_0^{\pi} \log|f(re^{i\theta})| \sin\theta d\theta +$$

$$+\frac{1}{2\pi}\int_1^r \left(\frac{1}{x^2} - \frac{1}{r^2}\right) \log|f(x)f(-x)| dx +$$

$$+\frac{1}{2\pi}\int_1^r \left(\frac{1}{x^2} - \frac{1}{r^2}\right) d\{\varphi(x) - \varphi(-x)\} + A_1(f,r), \tag{6.36}$$

where $A_1(r,f)$ is defined by relation (2.16). Using notations (6.33) we have

$$\int_1^r \left(\frac{1}{x^2} - \frac{1}{r^2}\right) d\sigma(x) =$$

$$= \frac{1}{\pi r}\int_0^{\pi} \log|f(re^{i\theta})| \sin\theta d\theta + A_1(f,r),$$

and after integration by parts

$$\int_1^r \frac{\sigma(x)dx}{x^3} = \frac{1}{2\pi r}\int_0^{\pi} \log|f(re^{i\theta})| \sin\theta d\theta + \frac{1}{2}A_1(f,r).$$

Now replace r by kr for some $0 < k < 1$ and subtract the latter relation from the previous one. We obtain

$$\int_{rk}^r \frac{\sigma(x)dx}{x^3} = \frac{1}{2\pi r}\int_0^{\pi} \log|f(re^{i\theta})| \sin\theta d\theta -$$

$$-\frac{1}{2\pi rk}\int_0^{\pi} \log|f(rke^{i\theta})| \sin\theta d\theta + \frac{d(1-k^2)}{4\pi r^2 k^2}, \tag{6.37}$$

where $d = d(1,f)$ is the constant from relation (2.16). According to (6.35), $\sigma(x)$ is an increasing function. Therefore

$$\frac{\sigma(rk)(rk)^{-2}(1-k^2)}{2} \leq$$

$$\leq \int_{kr}^r x^{-3}\sigma(x)dx \leq \frac{\sigma(r)(rk)^{-2}(1-k^2)}{2},$$

and, hence,

$$\sigma(rk) \leq \frac{rk}{(1-k^2)\pi}\Big\{k\int_0^\pi \log|f(re^{i\theta})|\sin\theta d\theta -$$

$$-\int_0^\pi \log|f(rke^{i\theta})|\sin\theta d\theta + \frac{d(1-k^2)}{2kr}\Big\} \leq \sigma(r).$$

From here, using Theorem 6.2, we get

$$\limsup_{r\to\infty}\frac{\sigma(r)}{r^{1+\rho}} = \limsup_{r\to\infty}\frac{\sigma(rk)}{(rk)^{1+\rho}} \leq \frac{k^{1-\rho}-1}{(1-k^2)\pi}\int_0^\pi h_f(\theta)\sin\theta d\theta,$$

$$\liminf_{r\to\infty}\frac{\sigma(r)}{r^{1+\rho}} \geq \frac{k^{1+\rho}(k^{1-\rho}-1)}{(1-k^2)\pi}\int_0^\pi h_f(\theta)\sin\theta d\theta.$$

Sending $k\to\infty$ we obtain (6.34).

To get rid of the additional assumption (6.35) when considering an arbitrary function $f\in A_\rho^*$ one may turn to the function

$$F(z) = f(z)M^{-1}\exp(-Kz^\rho e^{-i\rho\pi/2}),$$

where M and K are sufficiently large.

Now denote

$$l(r) = \frac{1}{2\pi}\int_1^r \log|f(x)f(-x)|\frac{dx}{x} + \frac{1}{2\pi}\int_1^r \frac{1}{x}d\{\varphi(x)-\varphi(-x)\},$$

$$c(r) = \sum_{z_n\in Q_r\setminus Q_1}\sin\theta_n.$$

Lemma 6.4 *If $f(z)\in A_\rho^*$ and $0<\rho<1$, then*

$$\lim_{r\to\infty}\frac{c(r)-l(r)}{r^\rho} = \frac{\rho^2-1}{2\pi\rho}\int_0^\pi h_f(\theta)\sin\theta d\theta. \tag{6.38}$$

Proof. Using (6.33) we have

$$\frac{c(r)-l(r)}{r^\rho} = \frac{1}{r^\rho}\int_1^r\frac{d\sigma(x)}{x} = \frac{\sigma(r)}{r^{\rho+1}} + \frac{1}{r^\rho}\int_1^r\frac{\sigma(x)}{x^2}dx. \tag{6.39}$$

By Lemma 6.3

$$\sigma(r) = ar^{\rho+1} + o(r^{\rho+1}),$$

where

$$a = \frac{\rho-1}{2\pi}\int_0^\pi h_f(\theta)\sin\theta d\theta.$$

Thus letting $r\to\infty$ in (6.39) we obtain (6.38).

7 Existence of argument boundary density for the zero set of a function from the class A_ρ^*

Now introduce the principal function that characterizes the distribution of zeros of a function $f(z) \in A_\rho^*$ with respect to moduli and arguments of these zeros. This function accounts for the inner zeros (lying in the half-plane $\Im z > 0$) and also boundary ones both singular and regular that are characterized by the functions $\varphi(t)$ and $\log|f(t)|$, $-\infty < t < \infty$, respectively.

Let $f(z)$ be a function analytic and of order ρ in $\Im z > 0$, $z_n = r_n e^{i\theta_n}$ being its inner zeros and $\varphi(t)$ being the singular boundary function. Denote

$$(r, \eta_1, \eta_2) = \{z : \ 1 \le |z| \le r\} \cap \{z : \ \eta_1 < \arg z \le \eta_2\}, \tag{7.1}$$

$$c(r, \eta_1, \eta_2) = \sum_{z_n \in (r, \eta_1, \eta_2)} \sin\theta_n, \quad r > 1, \quad \eta_1 < \eta_2, \tag{7.2}$$

$$\tau(t) = \begin{cases} \dfrac{1}{2\pi}\displaystyle\int_1^t \log|f(x)|\frac{dx}{x} + \frac{1}{2\pi}\int_1^t \frac{1}{x}d\varphi(x), & t \ge 1, \\ \dfrac{1}{2\pi}\displaystyle\int_t^{-1} \log|f(x)|\frac{dx}{|x|} + \frac{1}{2\pi}\int_t^{-1} \frac{1}{|x|}d\varphi(x), & t \le -1. \end{cases} \tag{7.3}$$

Also consider the function

$$a(r, \eta_1, \eta_2) = \begin{cases} c(r, \eta_1, \eta_2), & 0 < \eta_1 < \eta_2 < \pi, \\ c(r, 0, \eta_2) - \tau(r), & 0 = \eta_1 < \eta_2 < \pi, \\ c(r, \eta_1, \pi) - \tau(-r), & 0 < \eta_1 < \eta_2 = \pi, \\ c(r, 0, \pi) - \tau(r) - \tau(-r), & \eta_1 = 0, \quad \eta_2 = \pi. \end{cases} \tag{7.4}$$

For $\eta_2 \ge \eta_1$ we additionally define

$$a(r, \eta_1, \eta_2) = -a(r, \eta_2, \eta_1), \quad \eta_1 > \eta_2, \quad a(r, \eta, \eta) \equiv 0.$$

Definition. Let a function $f(z)$ be analytic and of finite order $\rho > 0$ in $\Im z > 0$. We say that the zero set of the function $f(z)$ has *argument boundary density with exponent* ρ (argument bounded ρ-density) if there exists the finite limit

$$\lim_{r\to\infty} r^{-\rho} a(r, \eta_1, \eta_2) = \lambda(\eta_1, \eta_2), \tag{7.5}$$

for all $\eta_1, \eta_2 \in [0, \pi]\backslash N$, where N is some at most countable set and does not contain the points $\eta = 0$ and $\eta = \pi$.

The notion of argument boundary density is closely connected with the notion of angular density (see Definition 4.5), which may be expressed as

$$\Delta(\eta_1, \eta_2) = \lim_{r\to\infty} r^{-\rho} \sum_{z_n \in (r,\eta_1,\eta_2)} 1.$$

By the *argument boundary density* of the function $f(z)$ we mean the function $\lambda(\psi) = \lambda(0, \psi)$, $0 \le \psi \le \pi$. Clearly, this function is non-decreasing in the interval of $(0, \pi)$ and may have arbitrary jumps at the points $0, \pi$. We define $\lambda(\psi)$ to be continuous from the right at the exeptional points $\eta \in N$.

Theorem 7.1 *Let a function $f(z)$ belong to the class A^*_ρ, $0 < \rho < \infty$. Then its zero set has an argument boundary density and*

$$c(r, 0, \pi) < C_1 r^\rho,$$

$$\left| \int_1^r \frac{1}{x} d(\varphi(x) - \varphi(-x)) \right| < C_2 r^\rho, \quad r > r_0, \tag{7.6}$$

where $\varphi(x)$ is the singular boundary function and the value $c(r, 0, \pi)$ is defined by (7.2).

In what follows we shall prove that the existence of argument boundary density specifies functions from the class A^*_ρ for non-integer ρ.

To prove Theorem 7.1 we need several lemmas.

Lemma 7.1 *Let the set $\{z_n\} = \{r_n e^{i\theta_n}\}$ lie within the angle (α, β), $0 < \alpha < \beta < \pi$ and one of the limits*

$$\Delta(\theta, \eta) = \lim_{r\to\infty} r^{-\rho} \sum_{z_n \in (r,\theta,\eta)} 1, \tag{7.7}$$

$$\lambda(\theta, \eta) = \lim_{r\to\infty} r^{-\rho} \sum_{z_n \in (r,\theta,\eta)} \sin\theta_n \tag{7.8}$$

exists for all $\theta, \eta \in (\alpha, \beta)\backslash N$ where N is an at most countable set. Then the other limit exists for the same values θ, η. In addition

$$\lambda(\theta, \eta) = \int_\theta^\eta \sin\psi d\Delta(\psi), \quad \Delta(\theta, \eta) = \int_\theta^\eta (\sin\psi)^{-1} d\lambda(\psi), \tag{7.9}$$

and the relations

$$\int_\theta^\eta g(\psi) d\Delta(\psi) = \int_\theta^\eta \frac{g(\psi)}{\sin\psi} d\lambda(\psi),$$

$$\int_\theta^\eta g(\psi)d\lambda(\psi) = \int_\theta^\eta g(\psi)\sin\psi d\Delta(\psi) \tag{7.10}$$

hold for any function $g(\psi)$ continuous on the segment $[\theta,\eta]$.

Proof. Let limit (7.7) exist for $\theta,\eta \in (\alpha,\beta)\backslash N$. Fixing θ and η, consider the functions

$$\Delta_r(\psi) = r^{-\rho} \sum_{z_n\in(\theta,\psi),\ |z_n|\le r} 1$$

that are non-decreasing on $[\theta,\eta]$. We have $\lim_{r\to\infty}\Delta_r(\psi) = \Delta(\theta,\psi)$ for all $\psi \in [\theta,\eta]\backslash N$ and, in particular, for $\psi=\theta$ and $\psi=\eta$. Applying the Helly theorem we claim that the limit

$$\lim_{r\to\infty}\int_\theta^\eta g(\psi)d\Delta_r(\psi) = \int_\theta^\eta g(\psi)d\Delta(\psi)$$

exists for any function $g(\psi)$ continuous on $[\theta,\eta]$. (Recall, that the function $\Delta(\psi)$ is defined up to a constant by the relation $\Delta(\psi_2)-\Delta(\psi_1) = \Delta(\psi_1,\psi_2)$).

Setting $g(\psi) = \sin\psi$ we claim that limit (7.8) exists and the first of equalities (7.9) holds. Other assertions of the lemma are proved similarly.

Lemma 7.2 *If $f(z)\in A^*_\rho$, $0<\rho<\infty$, then both the angular density $\Delta(\theta,\eta)$ and the argument boundary density $\lambda(\theta,\eta)$ exist for all $\theta,\eta\in(0,\pi)\backslash N$, where N is an at most countable set and $h'_f(\theta+0)\ne h'_f(\theta-0)$ for every $\theta\in N$. In addition*

$$\sup_{\theta,\eta\in(0,\pi)\backslash N}\lambda(\theta,\eta)<\infty,\qquad \int_{+0}^{\pi-0}\sin\psi d\Delta(\psi)<\infty \tag{7.11}$$

and the following estimates are fulfilled

$$\sum_{z_n\in(r,0,\pi)}\sin\theta_n < Cr^\rho,$$

$$\int_1^r \frac{1}{x}\Big|d(\varphi(x)-\varphi(-x))\Big| < Cr^\rho,\quad r\ge 1. \tag{7.12}$$

Here $z_n = r_n e^{i\theta_n}$ are the inner zeros of the function $f(z)$ and $\varphi(x)$ is its singular boundary function.

Proof. The existence of the angular density $\Delta(\theta,\eta)$ and the description of the exceptional set N follow from Theorem 4.12. Now the existence of

$\lambda(\theta,\eta)$ follows from Lemma 7.1. To prove (7.12) we set $k = 1/2$ in (6.37) and apply Theorem 6.2. We obtain

$$\lim_{r\to\infty} r^{1-\rho}\int_{r/2}^{r} x^{-3}\sigma(x)dx = a \neq \infty.$$

Using notation (6.33) we derive

$$\int_{r/2}^{r}\frac{s_3(x)-s_2(x)}{x^3}dx \leq \int_{r/2}^{r}\frac{s_1(x)}{x^3}dx + a_0 r^{\rho-1},$$

$$a_0 > a, \quad r > r_0.$$

Using the definition of order of the function $f(z)$ one can easily see that $s_1(r) < C_1 r^{\rho+1}$. Therefore

$$\int_{r/2}^{r}\frac{s_3(x)-s_2(x)}{x^3}dx < C_2 r^{\rho-1}, \quad r > r_1.$$

Since the functions $s_3(x)$ and $-s_2(x)$ are increasing and positive, we have

$$s_3(r/2) - s_2(r/2) \leq C_3 r^{\rho+1}, \quad r > r_0,$$

$$s_3(r) \leq C_4 r^{\rho+1}, \quad -s_2(r) \leq C_4 r^{\rho+1}, \quad r > r_0.$$

Hence

$$\sum_{z_n\in(r,0,\pi)} \sin\theta_n =$$

$$\int_1^r \frac{ds_3(x)}{x} = \frac{s_3(r)}{r} + \int_1^r \frac{s_3(x)}{x^2}dx < C_5 r^\rho,$$

$$\frac{1}{2\pi}\int_1^r \frac{1}{x}\Big|d(\varphi(x)-\varphi(-x))\Big| = -\frac{s_2(r)}{r} - \int_1^r \frac{s_2(x)}{x^2}dx < C_6 r^\rho,$$

which proves relations (7.12). The first of them yields

$$\sup_{\theta,\eta\in(0,\pi)\setminus N} \lambda(\theta,\eta) < \infty.$$

Now the convergence of the integral in (7.11) follows from (7.9).

Lemma 7.3 *Let $f(z) \in A_\rho^*$, $0<\rho<\infty$, and a direction θ, $0<\theta<\pi$, be such that $h'_f(\theta-0) = h'_f(\theta+0)$. Then there exists the limit*

$$\lim_{r\to\infty}\frac{1}{r^\rho}\sum_{z_n\in(r,0,\theta)}\left(\sin\theta_n - \frac{\theta}{\pi}\sin\frac{\pi\theta_n}{\theta}\right) =$$

$$= \int_{+0}^{\theta}\left(\sin\psi - \frac{\theta}{\pi}\sin\frac{\pi\psi}{\theta}\right)d\Delta(\psi), \tag{7.13}$$

where $\Delta(\psi)$ is the angular density of zeros of $f(z)$.

Proof. Since

$$|\sin\psi - \frac{\theta}{\pi}\sin\frac{\pi\psi}{\theta}| \le P_\theta \sin^3\psi, \quad 0 < \psi \le \pi/2,$$

where P_θ is independent of ψ.Taking an arbitrary δ, $0 < \delta < \theta/2$ and using (7.12), we have

$$\sum_{z_n \in (r,0,\delta)} |\sin\theta_n - \frac{\theta}{\pi}\sin\frac{\pi\theta_n}{\theta}| \le P_\theta \sin^2\delta \sum_{z_n\in(r,0,\delta)} \sin\theta_n \le$$

$$\le P_\theta \sin^2\delta \sum_{z_n\in(r,0,\pi)} \sin\theta_n \le P_\theta C\delta^2 r^\rho. \tag{7.14}$$

We assume that $h'_f(\delta+0) = h'_f(\delta-0)$. Then δ and θ do not belong to the set N and, by Lemma 7.1, there exists the limit

$$\lim_{r\to\infty}\frac{1}{r^\rho}\sum_{z_n\in(r,\delta,\theta)} (\sin\theta_n - \frac{\theta}{\pi}\sin\frac{\pi\theta_n}{\theta}) =$$

$$= \int_\delta^\theta (\sin\psi - \frac{\theta}{\pi}\sin\frac{\pi\psi}{\theta}) d\Delta(\psi). \tag{7.15}$$

Relations (7.14) and (7.15) imply

$$\limsup_{r\to\infty}\frac{1}{r^\rho}\sum_{z_n\in(r,0,\theta)} (\sin\theta_n - \frac{\theta}{\pi}\sin\frac{\pi\theta_n}{\theta}) \le$$

$$\le \int_\delta^\theta (\sin\psi - \frac{\theta}{\pi}\sin\frac{\pi\psi}{\theta}) d\Delta(\psi) + P_\theta C\delta^2,$$

$$\liminf_{r\to\infty}\frac{1}{r^\rho}\sum_{z_n\in(r,0,\theta)} (\sin\theta_n - \frac{\theta}{\pi}\sin\frac{\pi\theta_n}{\theta}) \ge$$

$$\ge \int_\delta^\theta (\sin\psi - \frac{\theta}{\pi}\sin\frac{\pi\psi}{\theta}) d\Delta(\psi) + P_\theta C\delta^2.$$

The value δ may be chosen arbitrarily small. To complete the proof it remains to take account of the convergence of the integral in (7.13).

Lemma 7.4 *If $f(z) \in A^*_\rho$, $0 < \rho < \infty$, and the value θ satisfies*

$$h'_f(\theta-0) = h'_f(\theta+0), \quad 0 < \theta < \pi/s, \quad s = \max(\rho, 1),$$

then there exists the finite limit [13]

$$\lim_{r\to\infty}\frac{1}{r^\rho}\Big(\frac{\theta}{\pi}\sum_{z_n\in(r,0,\theta)}\sin\frac{\pi\theta_n}{\theta}-\tau(r)\Big),$$

where $\tau(r)$ *is defined in (7.3).*

Proof. Define an auxiliary function in the half-plane $\Im z > 0$:

$$g(\zeta)=f(\zeta^{\theta/\pi}),\quad \zeta^{\theta/\pi}>0 \text{ for } \zeta>0.$$

One can easily see that $g(\zeta)\in A^*_{\rho\theta/\pi}$ and $0<\rho\theta/\pi<1$. The zeros $\zeta_k = R_k e^{i\alpha_k}$, $\Im zeta_k > 0$ of the function $g(\zeta)$ and corresponding zeros $z_{n_k} = r_{n_k}e^{i\theta_{n_k}}$, $z_{n_k}\in[0,\theta]$ of the function $f(z)$ are connected by the relations $R_k = r_{n_k}^{\pi/\theta}$, $\alpha_k=(\pi/\theta)\theta_{n_k}$. According to Lemma 6.4 there exists the limit

$$\lim_{R\to\infty}R^{-\rho\theta/\pi}\Big(\sum_{\zeta_k\in(R,0,\pi)}\sin\alpha_k-\frac{1}{2\pi}\int_1^R\log|g(x)g(-x)|\frac{dx}{x}-$$

$$-\frac{1}{2\pi}\int_1^R\frac{1}{x}d\varphi_g(x)\Big)=\nu\neq\infty,$$

where $\varphi_g(x)$ is the singular boundary function for $g(\zeta)$ and $\varphi_g(-x)=0$, $x>0$, since $g(\zeta)$ is analytic for $\zeta<0$. Sincc

$$d\varphi_g(x)=\frac{\pi}{\theta}x^{1-\theta/\pi}d\varphi_f(x^{\theta/\pi}),$$

returning to the function $f(z)$ and its zeros $z_n=r_ne^{i\theta_n}$ and changing R by $r^{\pi/\theta}$, we obtain

$$\lim_{r\to\infty}\Big(\frac{1}{r^\rho}\sum_{z_n\subset(r,0,\theta)}\sin\frac{\pi\theta_n}{\theta}-$$

$$-\frac{1}{2\pi r^\rho}\int_1^{r^{\pi/\theta}}\log|f(x^{\theta/\pi})f(x^{\theta/\pi}e^{i\theta})|\frac{dx}{x}-$$

$$-\frac{1}{2\theta r^\rho}\int_1^r\frac{d\varphi_f(x^{\theta/\pi})}{x^{\theta/\pi}}\Big)=\nu.$$

Let us change the variable $x=t^{\pi/\theta}$ and recall that, by Theorem 4.10, there exists the limit

$$\lim_{r\to\infty}\frac{1}{2\pi r^\rho}\int_1^r\log|f(te^{i\theta})|\frac{dt}{t}=\frac{1}{2\pi\rho}h_f(\theta),$$

[13] One may get rid of the restriction $0<\theta<\pi/s$.

Therefore

$$\lim_{r\to\infty}\frac{1}{r_\rho}\Big(\sum_{z_n\in(r,0,\theta)}\sin\frac{\pi\theta_n}{\theta}-\frac{\pi}{\theta}\tau(r)\Big)=\nu+\frac{1}{2\pi\rho}h_f(\theta).$$

Proof of Theorem 7.1. That the argument boundary density $\lambda(\theta,\eta)$ exists for $\theta,\eta\in[0,\pi]\backslash N$ where N is an at most countable set as well as estimates (7.6) are valid is proved in Lemma 7.2. It remains to prove that one may remove the points 0 and π from the set N. Consider the point 0. Choosing θ, $0<\theta<\pi/\rho$, such that $h'_f(\theta+0)=h'_f(\theta-0)$ and using notations (7.2), (7.4) we have

$$\frac{a(r,0,\theta)}{r^\rho}=\frac{1}{r^\rho}\sum_{z_n\in(r,0,\theta)}\left(\sin\theta_n-\frac{\theta}{\pi}\sin\frac{\pi\theta_n}{\theta}\right)+$$

$$+\frac{1}{r_\rho}\Big(\sum_{z_n\in(r,0,\theta)}\frac{\theta}{\pi}\sin\frac{\pi\theta_n}{\theta}-\tau(r)\Big).$$

By Lemmas 7.3 and 7.4 the both terms in the right-hand side have the limits as $r\to\infty$. Therefore the limit $\lim_{p\to\infty}r^{-\rho}a(r,0,\theta)$ exists.

8 Existence of boundary and argument densities for the zero set of a function from A^*_ρ

The notion of argument boundary density describes the joint behavior of the inner zeros of function and of the boundary values of its modulus. Now we introduce notions that describe separate behavior of each entries.

Definition 8.1 Let $\{z_n\}$ be a set in the half-plane $\Im z > 0$ with only real limit points. We say that the set $\{z_n\}$ has *argument density with exponent* ρ *(argument* ρ*-density)* if there exists the finite limit

$$\lim_{r\to\infty} r^{-\rho} c(r,\eta_1,\eta_2) = \mu(\eta_1,\eta_2), \tag{8.1}$$

for all $\eta_1,\eta_2 \in [0,\pi]\backslash N$. Here N is an at most countable set, $0,\pi \notin N$, the value $c(r,\eta_1,\eta_2)$ is defined in (7.2) and $\rho > 0$.

By changing lim by lim sup in the left-hand side of (8.1) we obtain the upper argument density $\mu^*(\eta_1,\eta_2)$.

It should be mentioned that, if $\{z_n\}$ has the argument density, then there exists the limit

$$\lim_{r\to\infty} r^{-\rho} c(r,0,\pi) = \mu(0,\pi) < \infty.$$

Definition 8.2. Let $f(z)$ be analytic and of order $\rho > 0$ in $\Im z > 0$ and $\tau(t)$ be defined by (7.3). If the limits

$$\lim_{r\to\infty} r^{-\rho}\tau(r) = l_1, \quad \lim_{r\to\infty} r^{-\rho}\tau(-r) = l_2 \tag{8.2}$$

exist and are finite, then they are called *right-hand and left-hand boundary densities* of the zero set of the function $f(z)$. By taking the upper limits in (8.2) we obtain the upper right-hand and left-hand densities l_1^* and l_2^* respectively. If both the right-hand and left-hand densities exist we say that the zero set of $f(z)$ *possesses double-sided boundary density.*

It follows from Definitions 7.1, 8.1, 8.2 that if $f(z)$ is a function of order $\rho > 0$ and of finite type in $\Im z > 0$, and its zero set has the argument boundary density, then the upper argument density of its zeros is bounded, i.e. $c(r,0,\pi) < Cr^\rho$, $r \geq r_0$, and also the upper densities l_1^* and l_2^* are finite.

Lemma 8.1 *If $f(z)$ is a function from the class A^*_ρ, $\rho > 0$, then both the left-hand and right-hand upper boundary densities l_1^*, l_2^* are finite and the* *upper argument density of zeros is bounded.* [14] *In addition*

$$\lambda(\psi) = \begin{cases} 0, & \psi = 0, \\ \mu^*(\psi) - l_1^*, & 0 < \psi < \pi, \\ \mu^*(\pi) - l_1^* - l_2^*, & \psi = \pi, \end{cases} \tag{8.3}$$

for all $\psi \notin N$, *where* N *is an at most countable set,* $0, \pi \notin N$. *Here* $\lambda(\psi) = \lambda(0, \psi)$, $\mu^*(\psi) = \mu^*(0, \psi)$.

Proof. By Theorem 7.1 (see relation (7.6)) and Corollary 1 of Theorem 6.1 we obtain

$$\sup_{0 \le \eta_1, \eta_2 \le \pi} \mu^*(\eta_1, \eta_2) < \infty, \quad l_1^* \neq \infty, \quad l_2^* \neq \infty.$$

Now relation (8.3) is a consequence of Theorem 7.1, relations (7.4), (7.5) and the equality $\lim(x - y) = \limsup x - \limsup y$, which is true if all the limits exists and are finite.

Corollary 8.1 *For the zero set of a function of the class* A_ρ^* *to have the argument density it is necessary and sufficient that the function has the left-hand and right-hand boundary densities.*

Lemma 8.2 *Let a function* $f(z)$ *be analytic and of order* $\rho > 0$ *in* $\Im z > 0$, *and its zero set have the argument boundary density. Then the argument and boundary densities* $\mu(\psi)$ *and* $\Delta(\psi)$ *exist within every angle* $[\delta, \pi - \delta]$, $\delta > 0$ *and* $\delta \notin N$, N *is at most countable, and also*

$$\mu(\psi_1, \psi_2) = \int_{\psi_1}^{\psi_2} \sin \psi d\Delta(\psi),$$

$$\Delta(\psi_1, \psi_2) = \int_{\psi_1}^{\psi_2} (\sin \psi)^{-1} d\mu(\psi).$$

The lemma is a direct consequence of Lemma 7.1.

Let us obtain a sharpening of Theorem 7.1 for functions from the class $\bar{A}_\rho^*$.

Theorem 8.1 *Let a function* $f(z)$ *belong to the class* $\bar{A}_\rho^*$, $0 < \rho < \infty$. *Then*

1) its zero set has the argument density;

2) the argument density $\mu(\psi)$ *is continuous at* $\psi = 0$ *and* $\psi = \pi$;

---

[14] One can prove that $\sup \mu^*(\eta_1, \eta_2) < \infty$ and $l_1^*, l_2^* \neq \infty$ for every function of finite order $\rho > 1$ and of finite type. This statement ceases to be true if $\rho \leq 1$.

3) the singular boundary function $\varphi(t)$ satisfies the relations

$$\lim_{r\to\infty} \frac{1}{r^\rho} \int_1^r \frac{d\varphi(x)}{x} = \lim_{r\to\infty} \frac{1}{r^\rho} \int_1^r \frac{d\varphi(-x)}{x} = 0. \tag{8.4}$$

Proof. Taking an arbitrary δ, $0 < \delta < \min(\pi, \pi/\rho)$, consider the auxiliary function $g_\delta(\zeta) = f(\zeta^{(\delta/\pi)})$ in the half-plane $\Im\zeta > 0$. One can easily see that

a) $g_\rho(\zeta) \in \bar{A}^*_{\rho\delta/\pi}$ and also $\rho\delta/\pi < 1$;

b) the indicators of the functions $g_\rho(\zeta)$ and $f(z)$ satisfy

$$h_{g_\delta}(\alpha) = h_f(\alpha\delta/\pi), \quad 0 \le \alpha \le \pi;$$

c) the zeros $\zeta_k = R_k e^{i\alpha_k}$ of the function $g_\delta(\zeta)$ and the zeros $r_{n_k} e^{i\theta_{n_k}}$ of the function $f(z)$, lying in $(0, \delta)$ are connected by the relations $R_k = r_{n_k}^{\pi/\delta}$, $\alpha_k = (\pi/\delta)\theta_{n_k}$.

According to Lemma 6.4 there exists the limit

$$\lim_{R\to\infty} R^{-\rho\delta/\pi}\Big(\sum_{\zeta_k\in(R,0,\pi)} \sin\alpha_k -$$

$$-\frac{1}{2\pi}\int_1^R \log|g_\delta(x)g_\delta(-x)|\frac{dx}{x} - \frac{1}{2\pi}\int_1^R \frac{1}{x} d\varphi_{g_\delta}(x)\Big) =$$

$$= \frac{(\rho\delta/\pi)^2 - 1}{2\rho\delta}\int_0^\pi h_{g_\delta}(\alpha)\sin(\alpha)d\alpha.$$

And by Theorem 4.10 there exists the limit

$$\lim_{R\to\infty} R^{-\rho\delta/\pi}\Big(\sum_{\zeta_k\in(R,0,\pi)} \sin\alpha_k - \frac{1}{2\pi}\int_1^{r^{\pi/\delta}} \frac{1}{x} d\varphi_{g_\delta}(x)\Big) =$$

$$= \frac{\rho^2\delta^2 - \pi^2}{2\rho\delta\pi^2}\int_0^\pi h_{g_\delta}(\alpha)\sin\alpha d\alpha + \frac{1}{2\rho\delta}\big(h_{g_\delta}(0) + h_{g_\delta}(\pi)\big).$$

Returning to the function $f(z)$ and using the substitution $x^{\delta/\pi} = t$, we obtain

$$\lim_{r\to\infty} \frac{1}{r^\rho}\Big(\sum_{\zeta_k\in(r,0,\delta)} \sin\frac{\pi\theta_n}{\delta} + \frac{1}{2\delta}\int_1^r \frac{-1}{x} d\varphi_f(x)\Big) =$$

$$= \frac{\rho^2\delta^2 - \pi^2}{2\rho\delta\pi^2}\int_0^\pi h_f\big(\frac{\alpha\delta}{\pi}\big)\sin\alpha d\alpha + \frac{1}{2\rho\delta}\big(h_f(0) + h_f(\delta)\big). \tag{8.5}$$

Now we apply the inequality $\sin\theta < \theta < \frac{\delta}{2}\sin(\pi\theta/\delta)$, $0 < \theta \le \delta/2$, and observe that both the summands and the integral in the square brackets in (8.5) are non-negative. We obtain

$$\limsup_{r\to\infty} r^{-\rho} \sum_{z_n\in(r,0,\delta/2)} \sin\theta_n \le \omega_f(\delta), \tag{8.6}$$

$$\limsup_{r\to\infty} r^{-\rho} \int_1^r x^{-1}|d\varphi_f(x)| \le 4\omega_f(\delta), \tag{8.7}$$

where

$$\omega_f(\delta) = \frac{1}{4\rho}\Big\{\Big[\Big(\frac{\rho\delta}{\pi}\Big)^2 - 1\Big]\int_0^\pi h_f\Big(\frac{\alpha\delta}{\pi}\Big)\sin\alpha d\alpha + h_f(0) + h_f(\delta)\Big\}.$$

Since the function $h_f(\theta)$ is continuous on $[0,\pi]$, we obtain

$$\lim_{\delta\to 0}\omega_f(\delta) = 0,$$

and thereby (8.6) and (8.7) yield

$$\lim_{\delta\to 0}\mu^*(0,\delta/2) = 0, \quad \lim_{r\to\infty} r^{-\rho}\int_1^r x^{-1}d\varphi(x) = 0. \tag{8.8}$$

Likewise,

$$\lim_{\delta\to 0}\mu^*(\pi-\delta/2,\pi) = 0, \quad \lim_{r\to\infty} r^{-\rho}\int_1^r x^{-1}d\varphi(-x) = 0. \tag{8.9}$$

Relation (8.4) is proved. Now taking $\delta,\theta \in (0,\pi)$ so that the derivatives $h_f'(\delta)$, $h_f'(\theta)$ exist, and using relations (8.8), (8.9) and Lemma 7.1, we get

$$\limsup_{r\to\infty} r^{-\rho}c(r,0,\theta) \le \lim_{r\to\infty} r^{-\rho}c(r,\delta,\theta) + \omega_f(\delta),$$

$$\liminf_{r\to\infty} r^{-\rho}c(r,0,\theta) \ge \lim_{r\to\infty} r^{-\rho}c(r,\delta,\theta),$$

whence

$$\limsup_{r\to\infty} r^{-\rho}c(r,0,\theta) - \liminf_{r\to\infty} r^{-\rho}c(r,0,\theta) \le \omega_f(\delta).$$

So we have proved that there exists the limit

$$\mu(0,\theta) = \lim_{r\to\infty} c(r,0,\theta).$$

One can similarly prove that there exists the limit

$$\mu(\theta,\pi) = \lim_{r\to\infty} r^{-\rho}c(r,\theta,\pi).$$

That the argument density $\mu(\psi)$ is continuous at $\psi = 0$ and $\psi = \pi$ is a consequence of the first relations in (8.8) and (8.9) respectively.

Corollary 8.2 *Every function* $f(z) \in \bar{A}_\rho^*$ *has the right- and left-hand boundary densities* l_1 *and* l_2. *In addition*

$$l_1 = \lim_{r\to\infty} \frac{1}{2\pi r^\rho} \int_1^r \log|f(t)|\frac{dt}{t} = \frac{1}{2\pi\rho} h_f(0),$$

$$l_2 = \lim_{r\to\infty} \frac{1}{2\pi r^\rho} \int_1^r \log|f(-t)|\frac{dt}{t} = \frac{1}{2\pi\rho} h_f(\pi). \tag{8.10}$$

Applying Theorem 5.1 we obtain

Corollary 8.3 *If* $f(z) \in A_\rho^*$ *and the indicator* $h_f(\theta)$ *is continuous on* $[0, \pi]$, *then the statement of the previous corollary is true for the function* $f(z)$.

SUFFICIENT CONDITIONS OF COMPLETELY REGULAR GROWTH IN THE HALF-PLANE AND FORMULAS FOR INDICATORS

9 The growth of some auxiliary functions of non-integer order

Let us prove that the existence of argument-boundary density characterizes the functions from the class A_ρ^* for non-integer ρ. First we need some auxiliary statements.

Lemma 9.1 *Let a set* $\{z_n\} \subset \{z : \Im z > 0\}$ *have the argument-boundary* ρ*-density* $\lambda(\psi)$ *with a countable exceptional set* N*,* ρ *is non-integer. Let* $q = [\rho]$ *and* D_q *define the canonical Nevanlinna factor (see Sect.3). Then the function*

$$B_\delta(z) = \prod_{z_n \in (\delta, \pi-\delta)} D_q(z, z_n), \quad 0 < \delta < \frac{\pi}{2}, \quad \delta, \pi - \delta \notin N,$$

has c.r.g. within the angle $[0, \pi]$ *and its indicator admits the representation*

$$h_{B_\delta}(\theta) = \frac{\pi}{\sin \pi\rho} \int_\delta^{\pi-\delta} (\cos \rho(|\theta - \psi| - \pi) - \cos \rho(\theta + \psi - \pi)) \times$$

$$\times \frac{d\lambda(\psi)}{\sin \psi}. \tag{9.1}$$

Proof. By Lemma 7.1 the set $z_n \cap (\delta, \pi - \delta)$ has the angular ρ density. Consider the infinite products

$$B_\delta^+(z) = \prod_{z_n \in (\delta, \pi-\delta)} E(z/z_n, q),$$

$$B_\delta^-(z) = \prod_{z_n \in (\delta, \pi-\delta)} E(z/\bar{z}_n, q),$$

where $E(u, q)$ is the canonical Weierstrass factor. By Theorem 4.2 both $B_\delta^+(z)$ and $-B_\delta^-(z)$ are entire functions of c.r.g. Since $B_\delta^-(z)$ does not vanish in $\Im z > 0$, then $B_\delta(z) \in \bar{A}_\rho^*$ and by Theorem 4.14 we have

$$h_{B_\delta}(\theta) = h_{B_\delta^+}(\theta) - h_{B_\delta^-}(\theta).$$

Clearly

$$h_{B_\delta^-}(\theta) = h_{B_\delta^+}(-\theta),$$

and Theorem 4.3 yields

$$h_{B_\delta}(\theta) = \frac{\pi}{\sin \pi\rho} \int_\delta^{\pi-\delta} (\cos \rho(|\theta - \psi| - \pi) - \cos \rho(\theta + \psi - \pi))d\Delta(\psi).$$

Now relation (7.10) gives (9.1).

Lemma 9.2 *Let $\nu_1(x)$, $\nu_2(x)$ be real functions on $[1, \infty]$ having bounded variation on $[1, R]$ for any $R > 1$. If, for some non-integer $\rho > 0$, the conditions*

$$\lim_{t\to+\infty} t^{-\rho}\nu_1(t) = a_1, \quad \lim_{t\to+\infty} t^{-\rho}\nu_2(t) = a_2 \tag{9.2}$$

hold, then the functions

$$F_1(z) = \exp\Big\{2iz^{q+1} \int_1^\infty \frac{d\nu_1(x)}{x^q(x-z)}\Big\},$$

$$F_2(z) = \exp\Big\{(-1)^q 2iz^{q+1} \int_1^\infty \frac{d\nu_2(x)}{x^q(x+z)}\Big\},$$

with $q = [\rho]$, belong to the class A_ρ^* and have the following indicators

$$h_{F_1}(\theta) = \lim_{r\to\infty} r^{-\rho} \log |F_1(re^{i\theta})| = \frac{2\pi\rho a_1}{\sin \pi\rho} \sin \rho(\theta - \pi), \quad 0 < \theta < \pi,$$

$$h_{F_2}(\theta) = \lim_{r\to\infty} r^{-\rho} \log |F_2(re^{i\theta})| = -\frac{2\pi\rho a_2}{\sin \pi\rho} \sin \rho\theta, \quad 0 < \theta < \pi, \tag{9.3}$$

in addition the convergence is uniform with respect to θ on every segment $0 < \eta \leq \theta \leq \pi - \eta$.

Proof. Restrict ourselves to considering the function $F_1(z)$, the function $F_2(z)$ admits the similar reasoning. By (9.2) we have $\nu_1(t) = a_1 t^\rho + \epsilon_1(t)t^\rho$, where $\epsilon_1(t) \to 0$ as $t \to +\infty$. The integration by parts gives

$$\log F_1(z) = 2iz^{q+1} a_1\rho \int_1^\infty \frac{dx}{x^{q+1-\rho}(x-z)} +$$

$$+2iz^{q+1}\Big\{\frac{\epsilon_1(1)}{z-1} + \int_1^\infty \frac{(x(1+q) - qz)\epsilon_1(x)}{x^{q+1-\rho}(x-z)^2} dx\Big\} \equiv$$

$$\equiv \Omega_1(z) + \omega_1(z).$$

Using the residue theorem, we obtain

$$\Omega_1(z) = 2a_1\rho i z^{q+1}\Big(\int_0^\infty - \int_0^1\Big)\frac{dx}{x^{q-\rho+1}(x-z)} =$$

$$= \frac{2\pi\rho 1 e^{-i(\rho-q)\pi}z^\rho}{i\sin\pi(\rho-q)} + d_1(z) = \frac{2\pi\rho a_1 e^{-i\rho\pi}}{i\sin\pi\rho}z^\rho + d_1(z),$$

$$\max_{|z|=r}|d_1(z)| = O(r^q), \quad r\to\infty.$$

Therefore the limit

$$\lim_{r\to\infty} r^{-\rho}\Re\Omega_1(re^{i\theta}) = \frac{2\pi\rho a_1}{\sin\pi\rho}\sin\rho(\theta-\pi).$$

exists uniformly with respect to θ, $0\le\theta\le\pi$.

Let $z = re^{i\theta}$, $0<\eta\le\theta\le\pi-\eta$. Then

$$|x-z|^2 = x^2 - 2xr\cos\theta + r^2 \le$$

$$\le x^2 - 2xr\cos\eta + r^2 = |x - re^{i\eta}|^2, \tag{9.4}$$

In addition $|\epsilon_1(x)|<\delta$ for $x>r_\delta$. Therefore

$$\limsup_{r\to\infty} r^{-\rho}\max_{\eta\le\theta\le\pi-\eta}|\omega_1(re^{i\theta})| \le$$

$$\le 2\delta\limsup_{r\to\infty} r^{q+1-\rho}\int_{r_\delta}^\infty \frac{(x(1+q)+qr)dx}{x^{q+1-\rho}|x-re^{i\eta}|^2} =$$

$$= 2\delta\int_0^\infty \frac{(t(1+q)+q)dt}{t^{q-\rho+1}|t-e^{i\eta}|^2}.$$

The limit in the left-hand side is independent of δ and, hence, vanishes. Therefore $F_1(z)\in A_\rho^*$ and relation (9.3) holds.

Lemma 9.3 *Let $f(z)$ be a function analytic in the half-plane $\Im z>0$ and of finite type with respect to some non-integer order $\rho>0$. Suppose that its zero set $\{z_n\}$ has the argument-boundary density $\lambda(\psi)$ with an at most countable exceptional set N and denote*

$$F_{1,\delta}^*(z) = \exp\Big\{\frac{z^{q+1}}{\pi i}\Big(\int_1^\infty \frac{\log|f(t)|dt}{t^{q+1}(t-z)} +$$

$$+\int_1^\infty \frac{d\varphi(t)}{t^{q+1}(t-z)}\Big)\Big\}\prod_{z_n\in(0,\delta)} D_q(z,z_n),$$

$$F^*_{2,\delta}(z) = \exp\Big\{\frac{z^{q+1}}{\pi i}\Big(\int_{-\infty}^{-1} \frac{\log|f(t)|dt}{t^{q+1}(t-z)} +$$

$$+\int_{-\infty}^{-1} \frac{d\varphi(t)}{t^{q+1}(t-z)}\Big)\Big\} \prod_{z_n \in (\pi-\delta,\pi)} D_q(z, z_n),$$

where $\varphi(t)$ is the singular boundary function for $f(z)$ and δ satisfies

$$\delta \notin N, \quad \pi - \delta \notin N, \quad 0 < \delta < \eta/2, \quad \eta < \pi/2.$$

Then the functions $F^*_{k,\delta}(z)$ admit the representation

$$F^*_{k,\delta}(z) = F_{k,\delta}(z)\alpha_{k,\delta}(z), \quad k = 1, 2,$$

In addition, the following relations hold uniformly with respect to θ, $\eta \leq \theta \leq \pi - \eta$,

$$\lim_{r\to\infty} r^{-\rho} \log|F_{1,\delta}(re^{i\theta})| = \frac{2\pi\rho}{\sin \pi\rho}\lambda(\delta) \sin \rho(\theta - \pi),$$

$$\lim_{r\to\infty} r^{-\rho} \log|F_{2,\delta}(re^{i\theta})| =$$

$$= -\frac{2\pi\rho}{\sin \pi\rho}\big(\lambda(\pi) - \lambda(\pi - \delta)\big) \sin \rho\theta,$$

and for $r \geq 1$

$$\max_{\eta\leq\theta\leq\pi-\eta} |\log|\alpha_{k,\delta}(re^{i\theta})|| < N_\eta \delta r^\rho, \quad k = 1, 2,$$

where the constant N_η is independent of r and δ.

Proof. Using notations (7.2)–(7.4) we put $c_\delta(r) = c(r, 0, \delta)$, $a_\delta(r) = a(r, 0, \delta)$. By Lemma 9.3 we have

$$\log \prod_{z_n\in(0,\delta)} D_q(z, z_n) = 2iz^{q+1} \int_1^\infty \frac{dc_\delta(t)}{t^q(t-z)} +$$

$$+2iz^{q+1} \sum_{z_n\in(0,\delta)} \frac{\beta_q(z, z_n)}{r_n^q(r_n - z)} \sin \theta_n.$$

Since, by (7.2)–(7.4)

$$c_\delta(r) - a_\delta(r) = \tau(r) = \frac{1}{2\pi}\int_1^r \frac{\log|f(x)|}{x}dx + \frac{1}{2\pi}\int_1^r \frac{d\varphi(x)}{x}, \quad r \geq 1,$$

we obtain the representation

$$\log F^*_{1,\delta}(z) = 2iz^{q+1}\int_1^\infty \frac{da_\delta(t)}{t^q(t-z)}+$$

$$+2iz^{q+1}\sum_{z_n\in(0,\delta)}\frac{\beta_q(z,z_n)}{r_n^q(r_n-z)}\sin\theta_n \equiv I_1(z)+I_2(z).$$

We assume $\delta \notin N$. Therefore, by the definition of the argument-boundary density, $\lim_{t\to\infty} t^{-\rho}a_\delta(t) - \lambda(\delta)$. By Lemma 9.2 we have

$$\lim_{r\to\infty} r^{-\rho}I_1(re^{i\theta}) = \frac{2\pi\rho}{\sin\pi\rho}\lambda(\delta)\sin\rho(\theta-\pi). \tag{9.5}$$

uniformly with respect to θ, $\eta \le \theta \le \pi-\eta$.

Now,taking into account (9.7) and (9.4), we get

$$\max_{\eta\le\theta\le\pi-\eta}|I_2(re^{i\theta})| \le \frac{4(q+1)\delta}{\sin(\eta/2)}r^{q+1}\int_1^\infty \frac{dc_\delta(t)}{t^q|t-re^{i\eta}|}.$$

Since the existence of the argument-boundary density for a function of finite type implies $c(r,0,\pi) < Cr^\rho$, $r \ge r_0$, then we have $c_\delta(t) < Ct^\rho$, $t \ge 1$. The integration by parts gives

$$\max_{\eta\le\theta\le\pi-\eta}|I_2(re^{i\theta})| \le$$

$$\le -\frac{4(q+1)C\delta}{\sin(\eta/2)}r^{q+1}\int_1^\infty t^\rho\frac{d}{dt}\frac{1}{t^q|t-re^{i\eta}|}dt <$$

$$< \frac{4(q+1)C\delta}{\sin(\eta/2)}\Big\{\frac{1}{|1-re^{i\eta}|}+\rho\int_0^\infty \frac{dt}{t^{q-\rho+1}|t_re^{i\eta}|}\Big\}r^{q+1} \le$$

$$\le N_\eta\delta r^\rho, \tag{9.6}$$

where N_η is a positive constant. Now if setting $F_{1,\delta}(z) = \exp\{I_1(z)\}$, $\alpha_{1,\delta}(z)\exp\{I_2(z)\}$, relations (9.5) and (9.6) yield the desired representation of the function $F^*_{1,\delta}(z)$.

The representation of the function $F^*_{2,\delta}(z)$ may be obtained in the same way.

10 A criterion for a function to belong to the class A_ρ^*, ρ being non-integer

Consider the function $g(\psi,\theta)$ defined on the square $D = \{(\psi,\theta) : 0 \le \psi,\theta \le \pi\}$ by the relation

$$g(\psi,\theta) = \begin{cases} (\cos\rho(|\theta-\psi|-\pi)- & \\ -\cos\rho(\theta-\psi-\pi)/\sin\psi, & 0<\psi<\pi, \quad 0\le\theta\le\pi \\ 2\rho\sin\rho(\theta-\pi), & \psi=0, \quad 0<\theta\le\pi, \\ -2\rho\sin\rho\theta, & \psi=\pi, \quad 0\le\theta<\pi, \\ 0 & \psi=\theta=0, \quad \psi=\theta=\pi. \end{cases} \tag{10.1}$$

Lemma 10.1 *The function $g(\psi,\theta)$ possesses the following properties:*

1. $g(\psi,\theta)$ is continuous with respect to ψ on the segment $0\le\psi\le\pi$ for any θ, $0\le\theta\le\pi$.

2. $g(\psi,\theta)$ is continuous with respect to θ on the segment $0\le\theta\le\pi$ for any ψ, $0<\psi<\pi$.

3. $g(\psi,\theta)$ is bounded in D and $\sup_{(\psi,\theta)\in D}|g(\psi,\theta)|\le\pi\rho$.

Proof. Omitted as obvious.

The following theorems are the main results of the present section.

Theorem 10.1 *For a function $f(z)$ analytic in $\Im z>0$ and of non-integer formal order ρ and finite type to belong to the class A_ρ^* it is necessary and sufficient that its zero set has the argument-boundary density.*

Theorem 10.2 *The indicator of the function $f(z)\in A_\rho^*$, ρ being non-integer, may be expressed by the relation*

$$h_f(\theta) = \frac{\pi}{\sin\pi\rho}\int_0^\pi g(\psi,\theta)d\lambda(\psi), \quad 0<\theta<\pi, \tag{10.2}$$

where $\lambda(\psi)$ is the argument-boundary density of the zero set of $f(z)$ and $g(\psi,\theta)$ is defined by (10.1).

Proof of Theorem 10.1 and 10.2. The necessity in Theorem 10.1 follows from Theorem 7.1. It remains to prove the sufficiency and then to obtain Theorem 10.2.

Let $f(z)$ have the formal non-integer order ρ and a finite type in the half-plane $\Im z>0$ and let its zero set have the argument-boundary density with the exceptional set N. We shall prove that $f(z)$ has c.r.g. in the angle $[\eta,\pi-\eta]$, i.e., that $f(z)\in A^*_{\rho,\eta,\pi-\eta}$ for any η, $0<\eta<\pi/2$. Denote the right-hand side in (10.2) by $h_f^*(\theta)$. In view of Theorem 4.7 to prove the two statements under consideration one has, for any $\epsilon > 0$, to find a set $E_\epsilon \subset [0,\infty)$ and r_ϵ such that $\operatorname{mes}^* E_\epsilon = 0$,

$$\sup_{\eta\le\theta\le\pi-\eta} |\log|f(re^{i\theta})| - r^\rho h_f^*(\theta)| \le \epsilon r^\rho, \quad r \notin E_\epsilon, \tag{10.3}$$

and

$$\log|f(re^{i\theta})| < \big(h_f^*(\theta)+\epsilon\big)r^\rho \quad \forall\theta \quad \eta \le \theta \le \pi-\eta. \tag{10.4}$$

for $r > r_\epsilon > 0$.

Fix an arbitrary δ such that $0 < \delta < \eta/2$, $\delta \notin N$, $\pi - \delta \notin N$, and represent $f(z)$ in the form

$$f(z) = f_0(z) f_\delta(z) \alpha_\delta(z),$$

where

$$f_0(z) = \exp\Big\{i\sum_{k=0}^{q} b_k z^k +$$

$$+\frac{1}{\pi i}\Big(\int_{-1}^{1}\frac{\log|f(t)|}{(t-z)}dt + \int_{-1}^{1}\frac{d\varphi(t)}{(t-z)}\Big)\Big\}\prod_{|z_n|\le 1}\frac{z-z_n}{z-\bar z_n},$$

$$f_\delta(z) = B_\delta(z)F_{1,\delta}(z)F_{2,\delta}(z), \quad \alpha_\delta(z) = \alpha_{1,\delta}(z)\alpha_{2,\delta}(z), \tag{10.5}$$

(in the right-hand sides of (10.5) we use the same notations as in Lemmas 9.1,9.3). Since ρ is non-integer, we have

$$\lim_{r\to\infty} r^\rho |\log|f_0(re^{i\theta})|| = 0, \quad 0 \le \theta \le 2\pi, \tag{10.6}$$

uniformly with respect to θ. Lemmas 9.1 and 9.3 yield $f_\delta(z) \in \bar A^*_{\rho,\eta,\pi-\eta}$, in addition using notation (10.1) we obtain the following representation of the indicator

$$h_{f_\delta}(\theta) = \frac{\pi}{\sin\pi\rho}\Big\{\int_\delta^{\pi-\delta} g(\psi,\theta)d\lambda(\psi) +$$

$$+\lambda(\delta)g(0,\theta) + \big(\lambda(\pi)-\lambda(\pi-\delta)\big)g(\pi,\theta)\Big\}, \quad \eta \le \theta \le \pi-\eta. \tag{10.7}$$

Now by Lemma 9.3 we have

$$\sup_{\eta\le\theta\le\pi-\delta} |\log|\alpha_\delta(re^{i\theta})|| \le N_\eta \delta r^\rho, \tag{10.2}$$

for $r \ge 1$, where the constant N_η is independent of δ. Using the notation $h_f^*(\theta)$, we obtain from (10.7) that

$$h_f^*(\theta) - h_{f_\delta}(\theta) = \frac{\pi}{\sin\pi\rho}\Big\{\int_{+0}^{\delta} g(\psi,\theta)d\lambda(\psi) + \int_{\pi-\delta}^{\pi-0} g(\psi,\theta)d\lambda(\psi) +$$

$$+\big(\lambda(+0)-\lambda(\delta)\big)g(0,\theta)+\big(\lambda(\pi-0)-\lambda(\pi-\delta)\big)g(\pi,\theta)\Big\}, \quad \eta\le\theta\le\pi-\eta.$$

Here, when writing $+0$ and $\pi-0$ as the limits of the integration, we mean that the endpoints $\psi=0$ and $\psi=\pi$ are excluded, that is we do not account for the possible jumps of the function $\lambda(\psi)$ at these points. By Lemma 10.1 we have [15]

$$\max_{\eta\le\theta\le\pi-\eta}|h_f^*(\theta)-h_{f_\delta}(\theta)|\le\frac{\pi^2\rho}{\sin\pi\rho}\{\mathrm{Var}_{0<\psi\le\delta}\lambda(\psi)+$$

$$+\mathrm{Var}_{\pi-\delta\le\psi<\pi}\lambda(\psi)+|\lambda(+0)-\lambda(\delta)|+|\lambda(\pi-0)-\lambda(\pi-\delta)|\} \tag{10.9}$$

Denoting the right-hand side of this inequality by $\omega(\delta)$, we see that

$$\lim_{\delta\to 0}\omega(\delta)=0,$$

therefore

$$|\omega(\delta)|\le\epsilon/4, \quad \delta<\epsilon/(4N_\eta), \tag{10.10}$$

for sufficiently small δ, $\delta<\eta/2$. Since $f_\delta(z)\in\bar{A}^*_{\rho,\eta,\pi-\eta}$, then by relations (10.9) and (10.10) one may choose a set $\tilde{E}_\epsilon\subset[0,\infty)$ of zero relative measure such that

$$\max_{\eta\le\theta\le\pi-\eta}\left|\log|f_\delta(re^{i\theta})|-h_f^*(\theta)r^\rho\right|\le\frac{1}{2}\epsilon r^\rho, \quad r\notin\tilde{E}_\epsilon,$$

and the number $r_{1,\epsilon}$ such that

$$\log|f_\delta(re^{i\theta})|\le\{h_f^*(\theta)+\frac{1}{2}\epsilon\}r^\rho, \quad \eta\le\theta\le\pi-\eta,$$

for $r>r_{1,\epsilon}$. By (10.6) we have

$$\max_{0\le\theta\le 2\pi}|\log|f_0(re^{i\theta})||<\frac{1}{4}\epsilon r^\rho$$

for $r>r_{2,\epsilon}$. By (10.8) and (10.10) we have

$$\max_{\eta\le\theta\le\pi-\eta}|\log|\alpha_\delta(re^{i\theta})||\le\frac{1}{4}\epsilon r^\rho.$$

Therefore setting $E_\epsilon=\tilde{E}_\epsilon\cap(r_\epsilon,\infty)$ for sufficiently large r, we obtain the set E_ϵ with $\mathrm{mes}^*E_\epsilon=0$ and such that inequalities (10.3) and (10.4) hold for $r>r_\epsilon$ and $r\notin E_\epsilon$. We have proved Theorems 10.1 and 10.2.

Using Lemmas 8.1 and 10.1, one may transform the expression (10.8) for the indicator as follows.

[15] Here, by definition, $\mathrm{Var}_{a<\psi\le b}\lambda(\psi)=\lim_{\epsilon\to 0}\mathrm{Var}_{a+\epsilon\le\psi\le b}\lambda(\psi)$.

Corollary 10.1 *If $f(z) \in A^*_\rho$ and ρ is a non-integer, then the indicator $h_f(\theta)$ admits the representation*

$$h_f(\theta) =$$

$$= \frac{\pi}{\sin \pi\rho}\Big\{\int_{+0}^{\pi-0} g(\psi,\theta)d\mu^*(\psi) - 2\rho l_1^* \sin\rho(\theta-\pi) + 2\rho l_2^* \sin\rho\theta\Big\}, \quad (10.11)$$

where $\mu^*(\psi)$ is the upper argument density of the zero set of $f(z)$, l_1^* and l_2^* are the left-hand and right-hand boundary densities respectively, and $g(\psi,\theta)$ is defined by relation (10.1).

The advantage of relation (10.11) to that of (10.2) is that it represents the indicator via the values $\mu^*(\psi)$ and l_k^*, the first depending only on the asymptotic behavior of the inner zeros, while the values l_k^* are defined by the asymptotic behavior of the modulus of the function $f(z)$ on the boundary. It should be noted that the right-hand side in (10.11) may be also considered for functions of the order ρ, that do not belong to the class A^*_ρ, though one cannot guarantee that it coincides with the indicator. Here are two examples of applying relation (10.2).

Example 1. Let ρ be non-integer and $0 < \nu < 1$. Denote

$$S = \bigcup_{n=1}^{\infty}(2^{2n}, 2^{2n+1}), \quad P = \bigcup_{n=1}^{\infty}(2^{2n-1}, 2^{2n}),$$

$$u(x) = \frac{2\pi\rho}{\nu-1}x^\rho, \quad (x>0), \quad z_k = k^{(1-\nu)/\rho}e^{i\theta_k},$$

$$\theta_k = \arcsin k^{-\nu}, \quad k = 1,2,\dots. \quad (10.12)$$

Set

$$f_1(z) = \exp\Big\{\frac{z^{q+1}}{\pi i}\int_P \frac{u(x)dx}{x^{q+1}(x-z)}\Big\}\prod_{|z_k|\in S} D_q(z,z_k), \quad (10.13)$$

where $q = [\rho]$. One can easily see that $\log|f_1(t)| = u(t)$ for $t \in P$ and $\log|f_1(t)| = 0$ for $t \in (-\infty,\infty)\backslash P$, and also the singular boundary function of the function $f_1(z)$ is a constant. Therefore, $\tau(t) = 0$ for $t < -1$ and, for $t \geq 1$, we obtain

$$\tau(t) = \frac{1}{2\pi}\int_1^t \frac{\log|f_1(x)|}{x}dx =$$

$$= \frac{1}{2\pi}\int_{P\cap(1,t)} \frac{u(x)}{x}dx = -\frac{\rho}{1-\nu}\int_{P\cap(1,t)} x^{\rho-1}dx. \quad (10.14)$$

Further, we have

$$c(t,0,\pi) = \sum_{|z_k|\in S\cap(1,t)} \sin\theta_k = \sum_{k^{(1-\nu)/\rho}\in S\cap(1,t)} k^{-\nu} =$$

$$= \int_{y^{(1-\nu)/\rho}\in S\cap(1,t)} y^{-\nu}dy + O(1) = \frac{\rho}{1-\nu}\int_{S\cap(1,t)} x^{\rho-1}dx + O(1). \quad (10.15)$$

Now relations (10.14) and (10.15) yield

$$a(t,0,\pi) = \frac{\rho}{1-\nu}\int_1^t x^{\rho-1}dx + O(1) = \frac{t^\rho}{1-\nu} + O(1), \quad t\to\infty.$$

Since any angle $[\delta,\pi]$, $\delta>0$, contains only a finite number of the zeros of the function $f_1(z)$, the latter asymptotic expression still holds for $a(t,0,\delta)$, and the argument-boundary density exists and has the form

$$\lambda(\psi) = \begin{cases} 0, & \psi = 0, \\ 1/(1-\nu), & 0<\psi\le\pi. \end{cases} \quad (10.16)$$

One can easily see that the function $f_1(z)$ has a finite type with respect to the formal order ρ and by Theorem 10.1 it belongs to A_ρ^*. Therefore $f_1(z)$ and relations (10.1),(10.16) and (10.2) yield

$$h_{f_1}(\theta) = \frac{2\pi\rho}{(1-\nu)\sin\pi\rho}\sin\rho(\theta-\pi), \quad 0<\theta<\pi. \quad (10.17)$$

Since $\log|f_1(t)| = 0$ for $t<0$, this expression is still valid for $\theta=\pi$. But it fails for $\theta=0$, since $\log|f_1(t)| = u(t) < 0$ for $t\in P$, and $\log|f_1(t)| = 0$ for $t\in[1,\infty)\backslash P$, and, hence, $h_{f_1}(0)=0$. We obtain that $f(z)\notin \bar{A}_\rho^*$ and, moreover, $f_1(z)$ does not have c.r.g. on the ray $\theta=0$. Relations (10.14) and (10.15) show that neither argument nor right-hand boundary densities exist, so one cannot sharpen Theorem 7.1.

Example 2. Denote

$$f_2(z) = \prod_{k=1}^\infty D_q(z,z_k), \quad (10.18)$$

where z_k are defined as in (10.12). For the function $f_2(z)$ we have $\tau(t)=0$ both for $t\ge1$ and $t\le-1$ and

$$c(t,0,\pi) = \sum_{|z_k|<t} \sin\theta_k =$$

$$= \frac{\rho}{1-\nu}\int_1^t x^{\rho-1}dx + O(1) = \frac{t^\rho}{1-\nu} + O(1).$$

The function $f_2(z)$ belongs to A_ρ^* and, in contrast to $f_1(z)$, both the argument and two-sided boundary densities exist for this function. Since relation (10.16) holds for the function $f_2(z)$, then, according to (10.2), for $0 < \theta < \pi$ its indicator is given by the right-hand side of (10.17). Clearly, $h_{f_2}(0) = h_{f_2}(\pi) = 0$, hence, the indicator is discontinuous at the point $\theta = 0$ and thereby $f_2(z) \notin \bar{A}_\rho^*$. One can draw the same conclusion by using Theorem 8.1 since the argument density of the zero set of the function $f_2(z)$ coincides with the argument-boundary density (10.16) and is discontinuous at $\psi = 0$. It should be noted that the function $f_2(z)$ possesses c.r.g. along the rays $\theta = 0$ and $\theta = \pi$.

11 A criterion for a function to belong to the class $\bar{A}_\rho^*$, ρ being non-integer.

In Section 8 (Theorem 8.1 and Corollary 1) we have proved that a function $f(z)$ belongs to the class $\bar{A}_\rho^*$, then its zero set possesses both the angular and two-sided boundary densities (and the angular density is continuous at the points $\psi = 0$ and $\psi = \pi$) and also relations (8.4) and (8.10) hold. These conditions appear to be sufficient for a function to belong to the class $\bar{A}_\rho^*$. The following theorems hold.

Theorem 11.1 *For a function $f(z)$, which is analytic and of non-integer finite order ρ and of finite type in the half-plane $\Im z > 0$, to belong to the class $\bar{A}_\rho^*$ it is necessary and sufficient that*

1) The angular density $\mu(\psi)$ exists and is continuous at the points $\psi = 0$ and $\psi = \pi$.

2) The right-hand and left-hand densities l_1 and l_2 exist.

3) The singular boundary function satisfies

$$\lim_{r\to\infty} r^{-\rho}\int_1^r \frac{d\varphi(x)}{x} = \lim_{r\to\infty} r^{-\rho}\int_1^r \frac{d\varphi(-x)}{x} = 0, \tag{11.1}$$

4) The indicator $h_f(\theta)$ satisfies the conditions

$$h_f(0) = 2\pi\rho l_1, \quad h_f(\pi) = 2\pi\rho l_2 \tag{11.2}$$

Theorem 11.2 *The indicator of a function $f(z) \in \bar{A}_\rho^*$, ρ non-integer, has the representation by the relation*

$$h_f(\theta) = \frac{\pi}{\sin\pi\rho}\Big\{\int_{+0}^{\pi-0} g(\psi,\theta)d\mu(\psi) -$$

$$-2\rho l_1 \sin\rho(\theta-\pi) + 2\rho l_2 \sin\rho\theta\Big\}, \quad 0 \le \theta \le \pi, \tag{11.3}$$

where $g(\theta,\psi)$ is defined by equality (10.1).

Proof of Theorem 11.2. Let $f \in \bar{A}_\rho^*$. According to Theorem 8.1 and its Corollary 1 there exist the argument and two-sided boundary densities of the zero set of the function $f(z)$. Using (10.2) we see that equality (11.3) holds for $0 < \theta < \pi$, it remains to prove this equality for $\theta = 0$ and $\theta = \pi$. Since $h_f(\theta)$ is continuous on $[0,\pi]$ it suffices to prove that the integral in the right-hand side of (11.3) is continuous as well. According to Lemma 10.1, the family of functions $g_\theta(\psi) = g(\psi,\theta)$ is uniformly bounded and, for any $\psi \in (0,\pi)$, we have $g_\theta(\psi) \to g_{\theta_0}(\psi)$ as $\theta \to \theta_0$, $\theta, \theta_0 \in [0,\pi]$. Since each function $g_\theta(\psi)$ is continuous on the interval $0 < \psi < \theta$, the continuity of the integral in (11.3) follows from the Lebesgue bounded convergence theorem. This completes the proof of Theorem 11.2.

It should be mentioned that, generally speaking, relation (11.3) for $\theta = 0$ and $\theta = \pi$ fails if $f(z) \in A^*_\rho$ but $f(z) \notin \bar{A}^*_\rho$, and relation (10.2) fails at these points even for functions from the class $\bar{A}^*_\rho$. It is caused by discontinuity of the function $g(0,\theta)$ at $\theta = 0$ and $g(\pi,\theta)$ at $\theta = \pi$.

Lemmas 7.1 and 7.2 together with relation (11.3) give us the following result

Corollary 11.1 *The indicator of the function $f(z) \in \bar{A}^*_\rho$, ρ non-integer, admits the representation*

$$h_f(\theta) =$$

$$\frac{\pi}{\sin \pi\rho}\Big\{\int_{+0}^{\pi-0} \big(\cos\rho(|\theta-\psi|-\pi) - \cos\rho(\theta+\psi-\pi)\big)d\Delta(\psi)-$$

$$-2\rho l_1 \sin\rho(\theta-\pi) + 2\rho l_2 \sin\rho\theta\Big\}, \quad 0 \le \theta \le \pi, \tag{11.4}$$

where $\Delta(\psi)$ is the angular density of the zero set of the function $f(z)$ and l_1, l_2 are the right-hand and left-hand boundary densities.

Thus one may calculate the indicator of a function from the class $\bar{A}^*_\rho$ using the boundary and angular densities only. Example 2 in the previous section shows that, generally speaking, this fact ceases to be true for functions from the class A^*_ρ.

Proof of Theorem 11.1. The necessity of the conditions 1–4 is proved in Theorem 8.1 and its Corollary 1, so we have to prove the sufficiency. Let these conditions be true. According to Theorem 10.1, we have $f(z) \in A^*_\rho$, hence from (10.11) we see that representation (11.3) holds for $0 < \theta < \pi$. For $\theta = 0$ and $\theta = \pi$ this representation follows from (11.2) and taking into account that the integral in the right-hand side is continuous, we prove $h_f(\theta)$ to be continuous on $[0,\pi]$. Now Theorem 5.1 yields $f(z) \in \bar{A}^*_\rho$.

Remark. One may see that the necessity still remains true if condition 3 is omitted. Such a form of this theorem is formulated in article [8].

Combining Theorems 10.1 and 5.1, one may obtain one more condition for a function to belong to the class $\bar{A}^*_\rho$.

Theorem 11.3 *Let $f(z)$ be an analytic function of non-integer formal order ρ and of finite type in the half-plane $\Im z > 0$. For this function to belong to the class $\bar{A}^*_\rho$ it is necessary and sufficient that the zero set of $f(z)$ has the argument-boundary density and the indicator of $h_f(\theta)$ is continuous at the endpoints of the segment $[0,\pi]$.*

Theorem 11.3 seems to be rather simple in comparison with Theorem 11.1. Nevertheless this simplicity is tricky, since checking the continuity of the indicator at the points $\theta = 0$ and $\theta = \pi$ may appear more complicated than checking the conditions of Theorem 11.1.

12 The argument-boundary symmetry of the zero set of a function of the class A_ρ^*, ρ being integer

As when considering entire functions, the case of functions analytic and of integer order in the half-plane has its own peculiarities. Let $f(z)$ be such a function. By Theorem 7.1 the existence of the argument-boundary density of the zero set is a necessary condition for $f(z)$ to belong to the class A_ρ^* (as well as a non-integer ρ). But now this condition ceases to be sufficient.

Definition 12.1. Let $f(z)$ be a function analytic and of an integer formal order $\rho > 0$ in the half-plane $\Im z > 0$ and $z_n = r_n e^{i\theta_n}$ be its inner zeros, $\varphi(t)$ be its singular boundary function. We say that the zero set of the function $f(z)$ has an *argument-boundary symmetry with exponent ρ (argument-boundary ρ-symmetry)*, if there exists the finite limit [16]

$$\sigma = \lim_{r\to\infty}\left\{\frac{1}{\rho}\sum_{1\le r_n\le r}\frac{\sin\rho\theta_n}{r_n^\rho}-\right.$$

$$\left.-\frac{1}{2\pi}\int_{1\le|x|\le r}\frac{\log|f(x)|}{x^{\rho+1}}dx-\frac{1}{2\pi}\int_{1\le|x|\le r}\frac{d\varphi(x)}{x^{\rho+1}}\right\}. \tag{12.1}$$

The value σ is called the *coefficient of ρ-symmetry.*

Theorem 12.1 *If a function $f(z)$ belongs to the class A_ρ^*, ρ integer, then the zero set of this function possesses the argument-boundary symmetry.*

Proof. Divide the angle $[0,\pi]$ into ρ equal parts and denote by l_k, $k = 0,1\dots,\rho-1$, the union of the two rays

$$l_k' = \{z:\quad \arg z = k\pi/\rho,\quad |z|\ge 1\},$$

$$l_k'' = \{z:\quad \arg z = (k+1)\pi/\rho,\quad |z|\ge 1\},$$

directed in such a way that the interior of the angle $[k\pi/\rho,(k+1)\pi/\rho]$ lies from the left. Set also

$$s_k(r) = \{z:\quad k\pi/\rho < \arg z < (k+1)\pi/\rho,\quad 1\le|z|\le r\}.$$

Let us extend the domain of definition of $\varphi(t)$ by including all rays l_k and defining $\varphi(\zeta) = 0$ if ζ does not belong to the real axis.

[16]We remind that, by definition, $\int_{1\le|x|\le r} = \int_{-r}^{-1}+\int_1^r$.

We shall prove that there exist the finite limits

$$\lim_{r\to\infty} r^{-2\rho}\Big\{\frac{1}{\rho}\sum_{z_n\in s_k(r)} r_n^\rho \sin\rho\theta_n -$$

$$-\frac{(-1)^k}{2\pi}\int_{l_k} |\zeta|^{\rho-1}\log|f(\zeta)||d\zeta| -$$

$$-\frac{(-1)^k}{2\pi}\int_{l_k}|\zeta|^{\rho-1}d\varphi(\zeta)\Big\} = \mu_k, \quad k=0,1\dots,\rho-1. \tag{12.2}$$

The proofs are similar for all k, so we assume $k=0$. Consider the function $f_0(\zeta) = f(\zeta^{1/\rho})$ in the half-plane $\Im\zeta > 0$. Since $f_0(\zeta)\in A_1^*$, then, by Theorem 7.1, there exists the finite limit

$$\lim_{R\to\infty}\frac{1}{R}\Big\{\sum_{\zeta_n\in(R,0,\pi)}\sin\alpha_n -$$

$$-\frac{1}{2\pi}\int_{1\le|t|\le R}\frac{\log|f_0(t)|}{|t|}dt - \frac{1}{2\pi}\int_{1\le|t|\le R}\frac{d\varphi_0(t)|}{|t|}\Big\} = \nu, \tag{12.3}$$

where $\varphi_0(t)$ is the singular boundary function of the function $f_0(z)$. It may be related with the function $\varphi(t)$ by the expression $d\varphi_0(t) = \rho t^{1-1/\rho}d\varphi(t^{1/\rho})$. The zeros $\zeta_k = R_k e^{i\alpha_k}$ of the function $f_0(\zeta)$ lying in $\Im\zeta > 0$ correspond to appropriate zeros $z_{n_k} = r_{n_k}e^{i\theta_{n_k}}$ of the function $f(z)$ lying in the angle $[0,\pi/\rho]$ in such a way $R_k = r_{n_k}^\rho$, $\alpha_k = \rho\theta_{n_k}$. Returning to $f(z)$ in (12.3) we obtain

$$\lim_{r\to\infty} r^{-\rho}\Big\{\sum_{z_n\in(r,0,\pi/\rho)}\sin\rho\theta_n - \frac{1}{2\pi}\int_1^{r^\rho}\log|f(t^{1/\rho})|\frac{dt}{t} -$$

$$-\frac{1}{2\pi}\int_1^{r^\rho}\log|f(s^{1/\rho}e^{\pi i/\rho})|\frac{ds}{s} -$$

$$\frac{\rho}{2\pi}\int_1^{r^\rho}\frac{d\varphi(t^{1/\rho})}{t^{1/\rho}} - \frac{\rho}{2\pi}\int_1^{r^\rho}\frac{d\varphi(s^{1/\rho}e^{\pi i/\rho})}{s^{1/\rho}}\Big\} = \nu,$$

and after the change of variables

$$\lim_{r\to\infty} r^{-\rho}\Big\{\sum_{z_n\in(r,0,\pi/\rho)}\sin\rho\theta_n - \frac{\rho}{2\pi}\int_1^r\frac{\log|f(x)|}{x}dx -$$

$$-\frac{\rho}{2\pi}\int_1^r\log|f(xe^{i\pi/\rho})|\frac{dx}{x} -$$

$$-\frac{\rho}{2\pi}\int_1^r \frac{d\varphi(x)}{x} - \frac{\rho}{2\pi}\int_1^r \frac{d\varphi(xe^{i\pi/\rho})}{x}\Big\} = \nu. \tag{12.4}$$

Denote the expression in the braces by $s(r)$. It follows from (12.4) that there exists the limit

$$\lim_{r\to\infty} r^{-2\rho}\int_1^r x^\rho ds(x) = \lim_{r\to\infty}\Big\{\frac{s(r)}{r^\rho} - \frac{\rho}{r^{2\rho}}\int_1^r s(x)x^{\rho-1}dx\Big\} = \frac{\nu}{2}. \tag{12.5}$$

The left-hand side of this equality coincides with that of (12.2) for $k = 0$, so we obtain relation (12.2) for $k = 0$.

Now adding together relations (12.2) for all k from 0 until $\rho - 1$ we obtain

$$\lim_{r\to\infty} r^{-2\rho}\Big\{\frac{1}{\rho}\sum_{z_n\in(r,0,\pi)} r_n^\rho \sin\rho\theta_n -$$

$$-\frac{1}{2\pi}\int_{1\le|x|\le r} x^{\rho-1}\log|f(x)|dx - \frac{1}{2\pi}\int_{1\le|x|\le r} x^{\rho-1}d\varphi(x)\Big\} = \mu. \tag{12.6}$$

Now use Theorem 2.2. Setting $q = \rho$, $\lambda = 1$, $R = r$, we rewrite relation (2.13) as

$$\frac{1}{\rho}\sum_{1<r_n\le r}\frac{\sin\rho\theta_n}{r_n^\rho} - \frac{1}{2\pi}\int_{1\le|x|\le r}\frac{\log|f(x)|}{x^{\rho+1}}dx -$$

$$-\frac{1}{2\pi}\int_{1\le|x|\le r}\frac{d\varphi(x)}{x^{\rho+1}} = \frac{1}{r^{2\rho}}\Big\{\sum_{1<r_n\le r} r_n^\rho\sin\rho\theta_n -$$

$$-\frac{1}{2\pi}\int_{1\le|x|\le r} x^{\rho-1}\log|f(x)|dx - \frac{1}{2\pi}\int_{1\le|x|\le r} x^{\rho-1}d\varphi(x)\Big\} +$$

$$+\frac{1}{\pi r^\rho}\int_0^\pi \log|f(re^{i\theta})|\sin\rho\theta d\theta + A_1^{(\rho)}(f,r),$$

and (2.14) provides the existence of $\lim_{r\to\infty} A_1^{(\rho)}(f,r)$. Therefore (12.6) yields the theorem.

Thus for a function of integer order ρ to belong to the class A_ρ^* it is necessary that its zeros possess the argument-boundary density and also the argument-boundary symmetry. Below we prove that these conditions are also sufficient.

First we obtain a consequence of Theorem 12.1 which shows that the existence of the argument-boundary symmetry affects properties of the argument-boundary density.

Corollary 12.1 *The argument-boundary density $\lambda(\psi)$ of a function $f(z) \in A_\rho^*$, with $\rho > 0$ is integer and satisfies the condition*

$$\int_0^\pi \frac{\sin \rho\psi}{\sin \psi} d\lambda(\psi) = 0 \tag{12.7}$$

(the function $\lambda(\psi)$ is defined at the endpoints by continuity).

Proof. Consider the family of the functions $\lambda_r(\psi) = r^{-\rho} a(r, 0, \psi)$, that are non-decreasing on the segment $[0, \pi]$. By Theorem 7.1 we have $\lim_{r\to\infty} \lambda_r(\psi) = \lambda(\psi)$ for all $\psi \in [0, \pi] \backslash N$, where N is an at most countable set, $0, \pi \notin N$. In addition, Theorem 7.1 and Corollary 1 to Theorem 6.1 yield that the functions $\lambda_r(\psi)$ are uniformly bounded. According to the Helly theorem, there exists the finite limit

$$a = \lim_{r\to\infty} \int_0^\pi \frac{\sin \rho\psi}{\sin \psi} d\lambda_r(\psi) = \int_0^\pi \frac{\sin \rho\psi}{\sin \psi} d\lambda(\psi).$$

Let us prove that $a = 0$. Setting

$$\omega(r) = \frac{r^\rho}{\rho} \int_0^\pi \frac{\sin \rho\psi}{\sin \psi} d\lambda_r(\psi) = \frac{1}{\rho} \sum_{1<r_n\leq r} \sin \rho\theta_n -$$

$$-\frac{1}{2\pi} \int_{1\leq|x|\leq r} \frac{\log|f(x)|}{x(\operatorname{sign} x)^\rho} dx - \frac{1}{2\pi} \int_{1\leq|x|\leq r} \frac{d\varphi(x)}{x(\operatorname{sign} x)^\rho}, \tag{12.8}$$

we see that

$$\frac{1}{\rho} \sum_{1<r_n\leq r} \frac{\sin \rho\theta_n}{r_n^\rho} - \frac{1}{2\pi} \int_{1\leq|x|\leq r} \frac{\log|f(x)|}{x^{\rho+1}} dx -$$

$$-\frac{1}{2\pi} \int_{1\leq|x|\leq r} \frac{d\varphi(x)}{x^{\rho+1}} = \int_1^r \frac{d\omega(x)}{x^\rho} = \frac{\omega(r)}{r^\rho} + \rho \int_1^r \frac{\omega(x)}{x^{\rho+1}} dx.$$

In view of relation (12.1) and Theorem 12.1 this value must have a finite limit, which is possible only if $a = 0$, since $\omega(r) = \frac{a}{\rho} r^\rho + o(r^\rho)$, $r \to \infty$.

13 The growth of some auxiliary functions of integer order

In what follows we use the notations

$$(r,\alpha,\beta)=\{z:\ |z|\le r\}\cap\{z:\ \alpha<\arg z\le\beta\},$$

and

$$(\alpha,\beta,r)=\{z:\ |z|>r\}\cap\{z:\ \alpha<\arg z\le\beta\}.$$

The auxiliary functions that are considered in this section differ from those of Section 9, as they are not analytic in the half-plane $\Im z>0$. Nevertheless for these functions we may use the notion of c.r.g. in the sense of Definitions 4.1 and 4.2, preserving the usual notion of indicator.

Lemma 13.1 *Let a function $f(z)$ be analytic and of integer order $\rho>0$ in the half-plane $\Im z>0$ and let its zero set $\{z_n\}$ possess the argument-boundary density $\lambda(\psi)$ with the exceptional set N. Denote*

$$B_\delta(z)=\prod_{z_n\in(r,\delta,\pi-\delta)}D_{\rho-1}(z,z_n)\prod_{z_n\in(\delta,\pi-\delta,r)}D_\rho(z,z_n),$$

where $r=|z|$, δ, $\pi-\delta\notin N$, $0<\delta<\pi/2$. Then the function $B_\delta(z)$ possesses c.r.g. in $[0,\pi]$ with the indicator

$$h_{B_\delta}(\theta)=2\sin\rho\theta\Big\{\int_\delta^\theta\frac{\psi\cos\rho\psi}{\sin\psi}d\lambda(\psi)+\int_\theta^{\pi-\delta}\frac{(\psi-\pi)\cos\rho\psi}{\sin\psi}d\lambda(\psi)\Big\}-$$

$$-2\pi\cos\rho\theta\int_\delta^\theta\frac{\sin\rho\psi}{\sin\psi}d\lambda(\psi)+2(\theta\cos\rho\theta-$$

$$-\frac{1}{\rho}\sin\rho\theta)\int_\delta^{\pi-\delta}\frac{\sin\rho\psi}{\sin\psi}d\lambda(\psi),\quad \delta\le\theta\le\pi-\delta. \tag{13.1}$$

Proof. Represent $B_\delta(z)$ as the ratio of the functions

$$B_\delta^+(z)=\prod_{(r,\delta,\pi-\delta)}E(z/z_n,\rho-1)\prod_{(\delta,\pi-\delta,r)}E(z/z_n,\rho)$$

and

$$B_\delta^-(z)=\prod_{(r,\delta,\pi-\delta)}E(z/\bar z_n,\rho-1)\prod_{(\delta,\pi-\delta,r)}E(z/\bar z_n,\rho).$$

As is shown in [24, p.146], both these functions have c.r.g. in the whole complex plane with the indicators [17]

$$h^+(\theta) = \int_{\theta-2\pi}^{\theta} \{(\theta - \psi - \pi)\sin\rho(\psi - \theta) -$$

$$-\frac{1}{\rho}\cos\rho(\theta - \psi)\}d\Delta(\psi), \quad 0 < \theta < 2\pi,$$

$$h^-(\theta) = h^+(2\pi - \theta) = -\int_{-\theta}^{2\pi-\theta} \{(\theta + \psi - \pi)\sin\rho(\psi + \theta) +$$

$$+\frac{1}{\rho}\cos\rho(\theta + \psi)\}d\Delta(\psi), \quad 0 < \theta < 2\pi, \tag{13.2}$$

respectively. Here $\Delta(\psi)$ is the angular density of the zero set of the function $B_\delta^+(z)$. Since this set is located in the angles $[\delta, \pi - \delta]$, then, for $\delta \le \theta \le \pi - \delta$, we may rewrite relations (13.2) in the form

$$h^+(\theta) = \int_{\delta}^{\pi-\theta} \{(\theta - \psi - \pi)\sin\rho(\psi - \theta) -$$

$$-\frac{1}{\rho}\cos\rho(\theta - \psi)\}d\Delta(\psi) + 2\pi\int_{\theta}^{\pi-\delta} \sin\rho(\psi - \theta)d\Delta(\psi),$$

$$h^-(\theta) = -\int_{\delta}^{\pi-\delta} \{(\theta + \psi - \pi)\sin\rho(\psi + \theta) + \frac{1}{\rho}\cos\rho(\theta + \psi)\}d\Delta(\psi).$$

The function $B_\delta^-(z)$ does not vanish in $\Im z > 0$. Therefore the function $B_\delta(z)$ also has c.r.g. in $\Im z > 0$ with the indicator $h^+(\theta) - h^-(\theta)$. After simplifications we have

$$h_{B_\delta}(\theta) = 2(\theta - \pi)\cos\rho\theta \int_{\delta}^{\pi-\delta} \sin\rho\psi d\Delta(\psi) +$$

$$+2\sin\rho\theta \int_{\delta}^{\pi-\delta} \psi\cos\rho\psi d\Delta(\psi) -$$

$$-\frac{2}{\rho}\sin\rho\theta \int_{\delta}^{\pi-\delta} \sin\rho\psi d\Delta(\psi) + 2\pi\int_{\theta}^{\pi-\delta} \sin\rho(\psi - \theta)d\Delta(\psi).$$

From here, by rearranging the summands and using Lemma 7.1 we obtain relation (13.1)

[17] We recall that at the points of discontinuity the angular density $\Delta(\psi)$ is defined as the right-hand limit.

Lemma 13.2 *Let $\nu_1(x)$ and $\nu_2(x)$ be real functions on $[1,\infty)$ having bounded variation on $[1,R]$ for any $R>1$ and such that*

$$\lim_{t\to\infty} t^{-\rho}\nu_1(t) = a_1, \quad \lim_{t\to\infty} t^{-\rho}\nu_2(t) = a_2 \tag{13.3}$$

for some integer $\rho > 0$.

Then the functions

$$F_1(z) = \exp\Big\{2iz^\rho\Big(\int_1^r \frac{d\nu_1(x)}{x^{\rho-1}(x-z)} + z\int_r^\infty \frac{d\nu_1(x)}{x^\rho(x-z)}\Big)\Big\},$$

and

$$F_2(z) = \exp\Big\{2iz^\rho(-1)^{\rho-1}\Big(\int_1^r \frac{d\nu_2(x)}{x^{\rho-1}(x+z)} - z\int_r^\infty \frac{d\nu_2(x)}{x^\rho(x+z)}\Big)\Big\}$$

possess c.r.g. with respect to the formal order ρ in the open angle $(0,\pi)$ with the indicators

$$h_{F_1}(\theta) = \lim_{r\to\infty} r^{-\rho}\log|F_1(re^{i\theta})| =$$

$$= 2a_1\rho(\theta-\pi)\cos\rho\theta, \quad 0<\theta<\pi,$$

$$h_{F_2}(\theta) = \lim_{r\to\infty} r^{-\rho}\log|F_2(re^{i\theta})| =$$

$$= (-1)^{\rho-1}2a_2\rho\theta\cos\rho\theta, \quad 0<\theta<\pi. \tag{13.4}$$

Here the convergence is uniform with respect to θ on every segment $0 < \eta\le\theta\le\pi-\eta$.

Proof. We consider the function $F_1(z)$ only, since the arguments for the function $F_2(z)$ are similar. According to (13.3) we have

$$\nu_1(t) = a_1t^\rho + \epsilon_1(t)t^\rho, \quad \lim_{t\to\infty}\epsilon_1(t) = 0.$$

Therefore

$$\log F_1(z) = 2ia_1\rho z^\rho\Big\{\int_1^r \frac{dx}{x-z} + z\int_r^\infty \frac{dx}{x(x-z)}\Big\} +$$

$$+2iz^\rho\Big\{\int_1^r \frac{d\{\epsilon_1(x)x^\rho\}}{x^{\rho-1}(x-z)} + z\int_r^\infty \frac{d\{\epsilon_1(x)x^\rho\}}{x^\rho(x-z)}\Big\} =$$

$$2ia_1\rho z^\rho\log\frac{r}{1-z} + 2iz^\rho\Big\{\epsilon_1(r) - \frac{\epsilon_1(1)}{1-z} +$$

$$+\int_1^r \frac{x\rho-(\rho-1)z}{(x-z)^2}\epsilon_1(x)dx + z\int_r^\infty \frac{x(\rho+1)-\rho z}{x(x-z)^2}\epsilon_1(x)dx\Big\} \equiv$$

$$\equiv \Omega_1(z)+\omega_1(z).$$

As can be easily seen

$$\lim_{r\to\infty} r^{-\rho}\Re\Omega_1(re^{i\theta}) =$$

$$= 2a_1\rho(\theta-\pi)\cos\rho\theta, \quad 0\le\theta\le\pi,$$

uniformly with respect to θ. Using inequality (9.3) and arguing as when proving Lemma 9.2, we obtain

$$\limsup_{r\to\infty} r^{-\rho}\max_{\eta\le\theta\le\pi-\eta}|\omega_1(re^{i\theta})| \le$$

$$\le 4\delta\Big\{\int_0^1 \frac{t\rho+\rho-1}{|t-e^{i\eta}|^2}dt + \int_1^\infty \frac{t(1+\rho)+\rho}{t|t-e^{i\eta}|^2}dt\Big\}$$

for every arbitrary small $\delta>0$. This yields (13.4)

Lemma 13.3 *Let a function $f(z)$ be analytic and of integer order $\rho>0$ and finite type in $\Im z>0$. Let its zero set have the argument-boundary density $\lambda(\psi)$ with an exceptional set N. Denote*

$$F^*_{1,\delta}(z) = \exp\Big\{\frac{z^\rho}{\pi i}\Big(\int_1^r \frac{\log|f(t)|}{t^\rho(t-z)}dt + z\int_r^\infty \frac{\log|f(t)|}{t^{\rho+1}(t-z)}dt +$$

$$+\int_1^r \frac{d\varphi(t)}{t^\rho(t-z)} + z\int_r^\infty \frac{d\varphi(t)}{t^{\rho+1}(t-z)}\Big)\Big\}\times$$

$$\times \prod_{z_n\in(r,0,\delta)} D_{\rho-1}(z,z_n) \prod_{z_n\in(0,\delta,r)} D_\rho(z,z_n),$$

$$F^*_{2,\delta}(z) = \exp\Big\{\frac{z^\rho}{\pi i}\Big(\int_{-r}^{-1} \frac{\log|f(t)|}{t^\rho(t-z)}dt + z\int_{-\infty}^{-r} \frac{\log|f(t)|}{t^{\rho+1}(t-z)}dt +$$

$$+\int_{-r}^{-1} \frac{d\varphi(t)}{t^\rho(t-z)} + z\int_{-\infty}^{-r} \frac{d\varphi(t)}{t^{\rho+1}(t-z)}\Big)\Big\}\times$$

$$\times \prod_{z_n\in(r,\pi-\delta,\pi)} D_{\rho-1}(z,z_n) \prod_{z_n\in(\pi-\delta,\pi,r)} D_\rho(z,z_n),$$

where $\varphi(t)$ is the singular boundary function of $f(z)$ and δ satisfies the condition

$$\delta\notin N \quad \pi-\delta\notin N, \quad 0<\delta<\eta/2, \quad \eta<\pi/2 \qquad (13.5)$$

Then the functions $F^*_{k,\delta}(z)$ *admit the representation*

$$F^*_{k,\delta}(z) = F_{k,\delta}(z)\alpha_{k,\delta}(z), \quad k = 1, 2,$$

in $[\eta, \pi - \eta]$ *and the relations*

$$\lim_{r\to\infty} r^{-\rho} \log |F_{1,\delta}(re^{i\theta})| = 2\rho\lambda(\delta)(\theta - \pi)\cos\rho\theta,$$

$$\lim_{r\to\infty} r^{-\rho} \log |F_{2,\delta}(re^{i\theta})| = (-1)^{\rho-1} 2\rho\big(\lambda(\pi) - \lambda(\pi - \delta)\big)\theta \cos\rho\theta$$

hold uniformly with respect to θ, $\eta \le \theta \le \pi - \eta$, *and also*

$$\max_{\eta\le\theta\le\pi-\eta} |\log |\alpha_{k,\delta}(re^{i\theta})|| < N_\eta \delta r^\rho, \quad k = 1, 2,$$

for $r > 1$, *where the constant* N_η *is independent of* r *and* δ.

Proof. Using the notations from the proof of Lemma 9.3 and applying Lemma 3.4 we have

$$\log F^*_{1,\delta}(z) = 2iz^\rho \Big\{ \int_1^r \frac{da_\delta(t)}{t^{\rho-1}(t - z)} + z \int_r^\infty \frac{da_\delta(t)}{t^\rho(t - z)} \Big\} +$$

$$+2iz^\rho \Big\{ \sum_{z_n \in (r,0,\delta)} \frac{\beta_{\rho-1}(z, z_n)\sin\theta_n}{r_n^{\rho-1}(r_n - z)} +$$

$$+z \sum_{z_n \in (0,\delta,r)} \frac{\beta_\rho(z, z_n)\sin\theta_n}{r_n^\rho(r_n - z)} \Big\} \equiv I_1(z) + I_2(z).$$

Since $\delta \notin N$, then, by the definition of the argument-boundary density, $\lim_{t\to\infty} t^{-\rho} a_\delta(t) = \lambda(\delta)$. By Lemma 13.2 we obtain

$$\lim_{r\to\infty} r^{-\rho} I_1(re^{i\theta}) = 2\rho\lambda(\delta)(\theta - \pi)\cos\rho\theta \tag{13.6}$$

uniformly with respect to θ, $\eta \le \theta \le \pi - \eta$.

Inequalities (9.7) and (9.4) yield

$$\max_{\eta\le\theta\le\pi-\eta} |I_2(re^{i\theta})| \le$$

$$\le \frac{4(\rho+1)\delta}{\sin(\eta/2)} r^\rho \Big\{ \int_1^r \frac{dc_\delta(t)}{t^{\rho-1}|t - re^{i\eta}|} + r \int_r^\infty \frac{dc_\delta(t)}{t^\rho |t - re^{i\eta}|} \Big\}$$

for $r > 1$.

Taking into account that $c_\delta(t) < Ct^\rho$, $t \ge 1$, and integrating by parts, we obtain

$$\max_{\eta\le\theta\le\pi-\eta} |I_2(re^{i\theta})| \le$$

$$\leq -\frac{4(\rho+1)C\delta}{\sin(\eta/2)}r^{\rho}\left\{\int_1^r t^{\rho}\frac{d}{dt}\frac{1}{t^{\rho-1}|t-re^{i\eta}|}dt+\right.$$

$$\left.+r\int_r^{\infty} t^{\rho}\frac{d}{dt}\frac{1}{t^{\rho}|t-re^{i\eta}|}dt\right\} \leq \frac{4(\rho+1)C\delta}{\sin(\eta/2)}r^{\rho}\times$$

$$\times\left\{\frac{1}{|1-re^{i\eta}|}+\rho\int_1^r \frac{dt}{|t-re^{i\eta}|}+\rho r\int_r^{\infty}\frac{dt}{t|t-re^{i\eta}|}\right\} \leq$$

$$\leq N_{\eta}\delta r^{\rho}, \tag{13.7}$$

where $N_\eta > 0$ is a constant. Setting $F_{1,\delta}(z) = \exp I_1(z)$ and $\alpha_{1,\delta}(z) = \exp I_2(z)$ we obtain the assertion of the lemma for the function $F^*_{1,\delta}(z)$. One may obtain the assertion for the function $F^*_{2,\delta}(z)$ in a similar way.

14 A criterion for a function to belong to the class A^*_ρ for integer ρ

By Theorem 3.3 every function of an integer order $\rho > 0$ in the half-plane $\Im z > 0$ admits the canonical representation

$$f(z) = \exp\Big\{i\sum_{k=0}^{\rho} a_k z^k +$$

$$+\frac{1}{\pi i}\Big(\int_{-1}^{1}\frac{\log|f(t)|}{t-z}dt + \int_{-1}^{1}\frac{d\varphi(t)}{t-z}\Big)\Big\}\prod_{|z_n|<1}\frac{z-z_n}{z-\bar{z}_n}\times$$

$$\times\exp\Big\{\frac{z^{\rho+1}}{\pi i}\Big(\int_{|t|\geq 1}\frac{log|f(t)|}{t^{\rho+1}(t-z)}dt+$$

$$\int_{|t|\geq 1}\frac{d\varphi(t)}{t^{\rho+1}(t-z)}\Big)\Big\}\prod_{|z_n|\geq 1}D_\rho(z,z_n). \tag{14.1}$$

It should be mentioned that, for integer ρ, the indicator depends on the coefficient a_ρ. The values a_k in representations (3.9) and (14.1) are, generally speaking, different. To avoid misunderstanding in what follows we use representation (14.1). If need be, one can easily return to representation (3.9).

Theorem 14.1 *For a function $f(z)$ analytic and of integer formal order $\rho > 0$ and finite type in the half-plane $\Im z > 0$ to belong to the class A^*_ρ, it is necessary and sufficient that its zero set possesses both the argument-boundary ρ-density and the argument-boundary ρ-symmetry.*

Theorem 14.2 *The indicator of a function $f(z) \in A^*_\rho$, ρ being integer, admits the representation*

$$h_f(\theta) = 2\pi\cos\rho\theta\int_0^\pi\frac{\sin\rho\psi}{\sin\psi}d\lambda(\psi)+$$

$$+2\sin\rho\theta\Big\{\sigma - \frac{1}{2}a_\rho + \int_0^\theta\frac{\psi\cos\rho\psi}{\sin\psi}d\lambda(\psi)+$$

$$+\int_\theta^\pi\frac{(\psi-\pi)\cos\rho\psi}{\sin\psi}d\lambda(\psi)\Big\},\quad 0<\theta<\pi, \tag{14.2}$$

where $\lambda(\psi)$ is the argument-boundary density of the zero set of $f(z)$, and σ is the coefficient of the argument-boundary symmetry, and a_ρ is the coefficient in the canonical representation (14.1).

Proof of Theorems 14.1 and 14.2. The necessity of the condition of Theorem 14.1 follows from Theorems 7.1 and 12.1. What remains is the sufficiency and Theorem 14.2.

Let $f(z)$ have an integer formal order $\rho > 0$ and a finite type in $\Im z > 0$ and let its zero set $\{z_n\} = \{re^{i\theta_n}\}$ have both the argument-boundary density and the argument-boundary symmetry. In view of Theorem 4.7, it suffices to prove that for any $\epsilon > 0$ there exists a set $E_\epsilon \subset (0, \infty)$, $\text{mes}^* E_\epsilon = 0$ and $r_\epsilon > 0$ such that

$$\sup_{\eta\le\theta\le\pi-\eta} |\log|f(re^{i\theta})| - r^\rho h_f^*(\theta)| \le \epsilon r^\rho, \quad r \notin E_\epsilon, \tag{14.3}$$

and

$$\log|f(re^{i\theta})| < \{h_f^*(\theta) + \epsilon\} r^\rho, \quad \eta \le \theta \le \pi - \eta, \tag{14.4}$$

for $r > r_\epsilon$, here $h_f^*(\theta)$ is the right-hand side of (14.2).

Let $r = |z|$. Denote

$$f_0(z) = \exp\Big\{\frac{z^\rho}{i}\Big(\frac{2}{\rho}\sum_{1<r_n\le r}\frac{\sin\rho\theta_n}{r_n^\rho} -$$

$$-\frac{1}{\pi}\int_{1\le|t|\le r}\frac{\log|f(t)|}{t^{\rho+1}}dt - \frac{1}{\pi}\int_{1\le|t|\le r}\frac{d\varphi(t)}{t^{\rho+1}}\Big)\Big\}.$$

By Theorem 12.1 the function $f_0(z)$ possesses c.r.g. in $\Im z > 0$ (though this function is not analytic), and also

$$\lim_{r\to\infty} r^{-\rho}\log|f_0(re^{i\theta})| = 2\sigma\sin\rho\theta, \quad 0 \le \theta \le \pi, \tag{14.5}$$

uniformly with respect to θ.

Now let us prove that $f(z) \in A^*_{\rho,\eta,\pi-\eta}$ for each $\eta \in (0, \pi/2)$. Choose δ satisfying condition (13.5) and represent $f(z)$ in the form

$$f(z) = f_\delta(z)\alpha_\delta(z), \tag{14.6}$$

where

$$\alpha_\delta(z) = \alpha_{1,\delta}(z)\alpha_{2,\delta}(z),$$

$$f_\delta(z) = f_0(z)B_\delta(z)F_{1,\delta}(z)F_{2,\delta}(z)\exp\{i\sum_{k=1}^{\rho}a_k z^k\}\times$$

$$\times\exp\Big\{\frac{1}{\pi i}\Big(\int_{-1}^{1}\frac{\log|f(t)|}{t-z}dt + \int_{-1}^{1}\frac{d\varphi(t)}{t-z}\Big)\Big\}\prod_{|z_n|\le 1}\frac{z-z_n}{z-\bar z_n} \tag{14.7}$$

(we use the notations of Lemmas 13.1 and 13.3 representation (14.1)). The logarithm of the function in the last line of (14.7) is analytic at infinity. Using (14.5) and Lemmas 13.1 and 13.3, we see that $f_\delta(z)$ has c.r.g. in $[\eta, \pi - \eta]$ with the indicator

$$h_{f_\delta}(\theta) = 2\sin\rho\theta\Big\{\int_\delta^\theta \frac{\psi\cos\rho\psi}{\sin\psi}d\lambda(\psi)+$$

$$+\int_\theta^{\pi-\delta} \frac{(\psi-\pi)\cos\rho\psi}{\sin\psi}d\lambda(\psi) + \sigma - \frac{1}{2}a_\rho\Big\}-$$

$$-2\pi\cos\rho\theta\int_\delta^\theta \frac{\sin\rho\psi}{\sin\psi}d\lambda(\psi) + 2(\theta\cos\rho\theta - \frac{1}{\rho}\sin\rho\theta)\times$$

$$\times\int_\delta^{\pi-\delta} \frac{\sin\rho\psi}{\sin\psi}d\lambda(\psi)+$$

$$+2\rho\lambda(\delta)(\theta-\pi)\cos\rho\theta + (-1)^{\rho-1}2\rho(\lambda(\pi)-\lambda(\pi-\delta))\theta\cos\rho\theta. \qquad (14.8)$$

By Lemma 13.3 we have

$$\max_{\eta\le\theta\le\pi-\eta} |\log|\alpha_\delta(re^{i\theta})|| < N_\eta\delta r^\rho, \quad r \ge 1, \qquad (14.9)$$

where the constant N_η is independent of δ.

Using equality (12.7), one may rewrite the representation of $h_f^*(\theta)$ in the form

$$h_f^*(\theta) = 2\sin\rho\theta\Big\{\int_{+0}^\theta \frac{\psi\cos\rho\psi}{\sin\psi}d\lambda(\psi)+$$

$$+\int_\theta^{\pi-0} \frac{(\psi-\pi)\cos\rho\psi}{\sin\psi}d\lambda(\psi) + \sigma - \frac{1}{2}a_\rho\Big\}-$$

$$-2\pi\cos\rho\theta\int_0^\theta \frac{\sin\rho\psi}{\sin\psi}d\lambda(\psi) + 2\theta\cos\rho\theta\int_0^\pi \frac{\sin\rho\psi}{\sin\psi}d\lambda(\psi)-$$

$$-\frac{2}{\rho}\sin\rho\theta\int_{+0}^{\pi-0} \frac{\sin\rho\psi}{\sin\psi}d\lambda(\psi) = 2\sin\rho\theta\Big\{\int_{+0}^\theta \frac{\psi\cos\rho\psi}{\sin\psi}d\lambda(\psi)+$$

$$+\int_\theta^{\pi-0} \frac{(\psi-\pi)\cos\rho\psi}{\sin\psi}d\lambda(\psi) + \sigma - \frac{1}{2}a_\rho\Big\}-$$

$$-2\pi\cos\rho\theta\int_{+0}^\theta \frac{\sin\rho\psi}{\sin\psi}d\lambda(\psi) + (2\theta\cos\rho\theta - \frac{2}{\rho}\sin\rho\theta)\times$$

$$\times\int_{+0}^{\pi-0} \frac{\sin\rho\psi}{\sin\psi}d\lambda(\psi) + 2\rho\lambda(+0)(\theta-\pi)\cos\rho\theta+$$

$$+(-1)^{\rho-1}2\rho(\lambda(\pi)-\lambda(\pi-0))\theta\cos\rho\theta, \tag{14.10}$$

where $\eta \le \theta \le \pi-\eta$. Now (14.8) and (14.10) yield

$$h_f^*(\theta)-h_{f_\delta}(\theta)=$$

$$=2\sin\rho\theta\Big\{\int_{+0}^{\delta}\frac{\psi\cos\rho\psi}{\sin\psi}d\lambda(\psi)+\int_{\pi-\delta}^{\pi-0}\frac{(\psi-\pi)\cos\rho\psi}{\sin\psi}d\lambda(\psi)\Big\}-$$

$$-2\pi\cos\rho\theta\int_{+0}^{\delta}\frac{\sin\rho\psi}{\sin\psi}d\lambda(\psi)+\big(2\theta\cos\rho\theta-\frac{2}{\rho}\sin\rho\theta\big)\times$$

$$\times\Big\{\int_{+0}^{\delta}\frac{\sin\rho\psi}{\sin\psi}d\lambda(\psi)+\int_{\pi-\delta}^{\pi-0}\frac{\sin\rho\psi}{\sin\psi}d\lambda(\psi)\Big\}+$$

$$+2\rho\cos\rho\theta\{(\theta-\pi)(\lambda(+0)-\lambda(\delta))+(-1)^{\rho-1}\theta(\lambda(\pi)-\lambda(\pi-\delta))\}.$$

Therefore for an arbitrary $\epsilon>0$ there exists $\delta_\epsilon>0$ such that

$$\max_{\eta\le\theta\le\pi-\eta}|h_f^*(\theta)-h_{f_\delta}(\theta)|<\frac{\epsilon}{3}$$

for $0<\delta<\delta_\epsilon$. Hence, there exist a set E_ϵ with $\mathrm{mes}^*E_\epsilon=0$ and $r_\epsilon>0$ such that

$$\max_{\eta\le\theta\le\pi-\eta}|\log|f_\delta(re^{i\theta})|-r^\rho h_f^*(\theta)|<\frac{\epsilon}{3}r^\rho,$$

for $r\notin E_\epsilon$ and

$$\log|f_\delta(re^{i\theta})|<\big(h_f^*(\theta)+\frac{1}{3}\epsilon\big)r^\rho$$

for $r\ge r_\epsilon$.

Assuming $\delta<\frac{\epsilon}{3N_\eta}$ and using (14.6) and (14.9), we obtain inequalities (14.3) and (14.4) for $r\notin E_\epsilon$, $r\ge r_\epsilon$, and thereby we prove Theorems 14.1 and 14.2

Remark 1. In contrast to the argument-boundary density $\lambda(\psi)$, the coefficient of the argument-boundary symmetry σ cannot be defined only by the asymptotic properties of the function; even elimination of a single inner zero alters the value σ, though leaves the indicator unchanged. Therefore, when applying formula (14.2), it is convenient to know the argument-boundary density and the value of the indicator at least at one point $\theta=\theta_0$, $0<\theta_0<\pi$; with this in hand, one can easily find the value $\sigma-\frac{1}{2}a_\rho$ from relation (14.2).

Remark 2. Relation (12.7) allows one to explain the absence of symmetry with respect to the points $\psi=0$ and $\psi=\pi$ in the first summand of (14.2) since

$$\int_\theta^\pi\frac{\sin\rho\psi}{\sin\psi}d\lambda(\psi)=-\int_0^\theta\frac{\sin\rho\psi}{\sin\psi}d\lambda(\psi).$$

Representation (14.2) may be rewritten in other forms. For example

$$h_f(\theta) = (2\sigma - a_\rho)\sin\rho\theta + 2\cos\rho\theta\int_0^\pi \frac{\psi\sin\rho\psi}{\sin\psi}d\lambda(\psi)+$$

$$+2\int_0^\theta \frac{\psi\sin\rho(\theta-\psi)}{\sin\psi}d\lambda(\psi) + 2\int_\theta^\pi \frac{(\psi-\pi)\sin\rho(\theta-\psi)}{\sin\psi}d\lambda(\psi).$$

This form is given in [10].

In just the same way as it was done in Section 10, we may split the argument-boundary density into the upper argument densities $\mu^*(\psi)$ and the upper boundary densities l_1^* and l_2^*. Then we obtain

Corollary 14.1 *The indicator of a function $f(z) \in A_\rho^*$, with $\rho > 0$ integer, admits the representation*

$$h_f(\theta) = 2\pi\cos\rho\theta\Big\{\int_0^\pi \frac{\sin\rho\psi}{\sin\psi}d\mu^*(\psi) + \rho(-1)^\rho l_2^*\Big\}+$$

$$+2\sin\rho\theta\Big\{\sigma - \frac{1}{2}a_\rho - l_1^* + (-1)^\rho l_2^*+$$

$$+\int_0^\theta \frac{\psi\cos\rho\psi}{\sin\psi}d\mu^*(\psi) + \int_0^\pi \frac{(\psi-\pi)\cos\rho\psi}{\sin\psi}d\mu^*(\psi)\Big\}. \qquad (14.11)$$

15 A criterion for a function to belong to the class $\bar{A}_\rho^*$ for integer ρ

Theorem 15.1 *A function $f(z)$ analytic and of an integer formal order $\rho > 0$ and finite type in $\Im z > 0$ belongs to the class $\bar{A}_\rho^*$ if and only if*

1) The argument density $\mu(\psi)$ exists and is continuous at $\psi = 0$ and $\psi = \pi$.

2) The right-hand and left-hand boundary densities l_1 and l_2 exist.

3) The singular boundary function $\varphi(x)$ satisfies the conditions

$$\lim_{r\to\infty} r^{-\rho}\int_1^r \frac{d\varphi(x)}{x} = \lim_{r\to\infty} r^{-\rho}\int_1^r \frac{d\varphi(-x)}{x} = 0.$$

4) The indicator $h_f(\theta)$ satisfies the conditions

$$h_f(0) = 2\pi\rho l_1, \quad h_f(\pi) = 2\pi\rho l_2. \tag{15.1}$$

5) The zero set of $f(z)$ possesses the argument-boundary symmetry.

Theorem 15.2 *The indicator of a function $f(z) \in \bar{A}_\rho^*$, ρ integer, admits the representation*

$$h_f(\theta) = 2\pi\cos\rho\theta\Big\{\int_0^\pi \frac{\sin\rho\psi}{\sin\psi}d\mu(\psi) + \rho(-1)^\rho l_2\Big\}+$$
$$+2\sin\rho\theta\Big\{\sigma - \frac{1}{2}a_\rho - l_1 + (-1)^\rho l_2+$$
$$+\int_0^\theta \frac{\psi\cos\rho\psi}{\sin\psi}d\mu(\psi) + \int_\theta^\pi \frac{(\psi-\pi)\cos\rho\psi}{\sin\psi}d\mu(\psi)\Big\},$$
$$0 \le \theta \le \pi. \tag{15.2}$$

Proof of Theorem 15.2 is simple. Equality (15.2) is proved in the Corollary to Theorem 14.2 for $0 < \theta < \pi$. The right-hand side of (15.2) is continuous at the point $\theta = 0$ and $\theta = \pi$ since, by Theorem 8.1, $\mu(\psi)$ is continuous at $\psi = 0$ and $\psi = \pi$. On the other hand, $h_f(\theta)$ also is continuous at $\theta = 0$ and $\theta = \pi$ (see Theorem 5.1). Therefore (15.2) holds for all θ, $0 \le \theta \le \pi$.

Proof of Theorem 15.1. The necessity of conditions 1)–5) follows from Theorem 8.1 and 14.1 and relations (8.10). Let us prove the sufficiency. Let conditions 1)–5) hold. By Theorem 14.1 we have $f(z) \in A_\rho^*$ and, hence, the indicator $h_f(\theta)$ admits, for $0 < \theta < \pi$, representation (14.11) and also representation (15.2) since $l_k^* = l_k$, $\mu^*(\psi) = \mu(\psi)$. The right-hand side of (15.2) is continuous on $[0, \pi]$ and takes the values $2\pi\rho l_1$ and $2\pi\rho l_2$ at the points $\theta = 0$ and $\theta = \pi$ respectively. To prove this statement one may use the relation

$$\int_0^\pi \frac{\sin\rho\psi}{\sin\psi} d\mu(\psi) = \rho l_1 + (-1)^{\rho-1}\rho l_2, \tag{15.3}$$

which follows from (12.7). Therefore condition 4) of Theorem 15.1 implies that the indicator $h_f(\theta)$ is continuous on $[0,\pi]$, and using Theorem 5.1 we obtain $f(z) \in \bar{A}^*_\rho$.

The following criterion of belonging to the class $\bar{A}^*_\rho$ is a direct consequence of Theorems 15.1 and 5.1.

Theorem 15.3 *For a function $f(z)$ analytic and of an integer formal order $\rho > 0$ and finite type in $\Im z > 0$ to belongs to the class $\bar{A}^*_\rho$ it is necessary and sufficient that its zero set possesses both the argument-boundary density and the argument-boundary symmetry and the indicator $h_f(\theta)$ is continuous at the endpoints of the segment $[0,\pi]$.*

Corollary 15.1 *The indicator of a function $f(z) \in \bar{A}^*_\rho$, with $\rho > 0$ integer, admits the representation* [18]

$$h_f(\theta) = 2\pi\cos\rho\theta\Big\{\int_\theta^{\pi-0} \sin\rho\psi d\Delta(\psi) + \rho(-1)^\rho l_2\Big\} +$$

$$+2\sin\rho\theta\Big\{\sigma - \frac{1}{2}a_\rho - l_1 + (-1)^\rho l_2 +$$

$$+\int_{+0}^\theta \psi\cos\rho\psi d\Delta(\psi) + \int_\theta^{\pi-0} (\psi-\pi)\cos\rho\psi d\Delta(\psi)\}, \quad 0 \le \theta \le \pi, \tag{15.4}$$

where $\Delta(\psi)$ is the angular density of the zero set of the function $f(z)$.

This corollary may be obtained in the same way as the corollary to Theorem 11.2.

Thus we see, that, both for integer and non-integer ρ, one may find the indicator of a function $f(z)$ without using the argument density (though we have used the notion of the argument density when proving this fact). In the same way as in Section 10 one may construct functions of the class $\bar{A}^*_\rho$ whose indicator cannot be expressed by the angular and boundary densities only.

[18] In this representation the first integral vanishes when $\theta = \pi$ and when $\theta = 0$ its lower limit equals to $+0$.

16 Functions of the class $\bar{A}^*_\rho$ for even and for odd ρ

Consider some differences of behavior of functions of the class $\bar{A}^*_\rho$ for even and odd values of ρ. First let us mention the following lemma.

Lemma 16.1 *If $f(z) \in \bar{A}^*_\rho$ and $\rho > 0$ is an integer, then the indicator and the argument density satisfy the relation*

$$h_f(0) + (-1)^{\rho-1} h_f(\pi) = 2\pi \int_0^\pi \frac{\sin \rho\psi}{\sin \psi} d\mu(\psi). \tag{16.1}$$

This lemma directly follows from (15.3) which in turn is a consequence of relation (12.7) and equation (15.1).

Theorem 16.1 *For a function $f(z)$ analytic and of an odd formal order ρ and finite type in $\Im z > 0$ to belong to the class $\bar{A}^*_\rho$ it is necessary and sufficient that*

1) the zero set of $f(z)$ has the angular density $\Delta(\psi)$ in the open angle $(0, \pi)$,

2) the zero set of $f(z)$ has the angular-boundary symmetry,

3) the indicator $z_f(\theta)$ satisfies the condition

$$h_f(0) + h_f(\pi) = 2\pi \int_{+0}^{\pi-0} \sin \rho\psi d\Delta(\psi). \tag{16.2}$$

Proof. The necessity is a direct consequence of Theorem 15.1 and Lemmas 7.1 and 16.1.To prove the sufficiency, suppose that $f(z)$ has an odd order ρ and a finite type and conditions 1)–3) are fulfilled. Denote

$$\omega(r) = \frac{1}{\rho} \sum_{1<r_n\le r} \sin \rho\theta_n - \frac{1}{2\pi} \int_1^r \frac{\log |f(x)f(-x)|}{x} dx -$$

$$-\frac{1}{2\pi} \int_1^r \frac{d(\varphi(x) - \varphi(-x))}{x}$$

(since ρ is odd, the value $\omega(r)$ coincides with that of (12.8)). For any δ, $0 < \delta < \pi/2$, we have

$$r^{-\rho}\omega(r) = r^{-\rho}\Big\{\frac{1}{\rho} \sum_{z_n\in(r,0,\delta)} \sin \rho\theta_n +$$

$$+\frac{1}{\rho} \sum_{z_n\in(r,\pi-\delta,\pi)} \sin \rho\theta_n - \frac{1}{2\pi} \int_1^r \frac{\log |f(x)f(-x)|}{x} dx -$$

$$-\frac{1}{2\pi}\int_1^r \frac{d(\varphi(x)-\varphi(-x))}{x}\Big\} + \frac{1}{\rho}\int_\delta^{\pi-\delta} \sin\rho\psi d\Delta_r(\psi),$$

where $\Delta_r(\psi) = r^{-\rho}n(r,0,\psi)$. Let N be the set of points ψ for which the limit $\lim_{r\to\infty}\Delta_r(\psi) = \Delta(\psi)$ does not exist. This set is at most countable. Taking δ such that δ, $\pi-\delta \notin N$ and letting $r\to\infty$ and then applying the Helly theorem, we obtain

$$\limsup_{r\to\infty} r^{-\rho}\Big\{\frac{1}{\rho}\sum_{z_n\in(r,0,\delta)} \sin\rho\theta_n + \frac{1}{\rho}\sum_{z_n\in(r,\pi-\delta,\pi)} \sin\rho\theta_n -$$

$$-\frac{1}{2\pi}\int_1^r \frac{d(\varphi(x)-\varphi(-x))}{x}\Big\} \le \limsup_{r\to\infty} r^{-\rho}\omega(r) +$$

$$+\limsup_{r\to\infty}\frac{r^{-\rho}}{2\pi}\int_1^r \frac{\log|f(x)f(-x)|}{x}dx - \frac{1}{\rho}\int_\delta^{\pi-\delta}\sin\rho\psi d\Delta(\psi).$$

Now, using condition (16.2) we get

$$\limsup_{r\to\infty}\frac{r^{-\rho}}{2\pi}\int_1^r \frac{\log|f(x)f(-x)|}{x}dx \le$$

$$\le \frac{1}{2\pi\rho}(h_f(0)+h_f(\pi)) = \frac{1}{\rho}\int_{+0}^{\pi-0}\sin\rho\psi d\Delta(\psi).$$

Let δ be sufficiently small so that $\sin\rho\theta_n > \frac{1}{2}\sin\rho\theta_n$ for $0<\sin\theta_n<\sin\delta_0$ (it may be done since ρ is odd). Then, the latter inequality yields

$$\limsup_{r\to\infty} r^{-\rho}\Big\{\frac{1}{2}\sum_{z_n\in(r,0,\delta)} \sin\rho\theta_n + \frac{1}{2}\sum_{z_n\in(r,\pi-\delta,\pi)} \sin\theta_n -$$

$$-\frac{1}{2\pi}\int_1^r \frac{d(\varphi(x)-\varphi(-x))}{x}\Big\} \le$$

$$\le \limsup_{r\to\infty} r^{-\rho}\omega(r) + \eta(\delta), \quad \lim_{\delta\to 0}\eta(\delta) = 0 \tag{16.3}$$

for $\delta<\delta_0$. Set

$$\sigma(r) = \frac{1}{\rho}\sum_{1<r_n\le r}\frac{\sin\rho\theta_n}{r_n^\rho} -$$

$$-\frac{1}{2\pi}\int_1^r \frac{\log|f(x)f(-x)|}{x^{\rho+1}}dx - \frac{1}{2\pi}\int_1^r \frac{d(\varphi(x)-\varphi(-x))}{x^{\rho+1}}.$$

Since ρ is an odd integer, this value coincides with that of (12.1) and, by condition 2), there exists the finite limit $\lim_{r\to\infty}\sigma(r) = \sigma < \infty$. Since

$$\omega(r) = \int_1^r x^\rho d\sigma(x) = r^\rho\sigma(r) - \rho\int_1^r \sigma(x)x^{\rho-1}dx =$$

$$= r^\rho(\sigma(r) - \sigma) - \rho \int_1^r (\sigma(x) - \sigma)x^{\rho-1}dx + \sigma = o(r^\rho),$$

we get $\lim_{r\to\infty} r^{-\rho}\omega(r) = 0$ and (16.3) implies

$$\limsup_{r\to\infty} r^{-\rho}\Big\{ \sum_{z_n\in(r,0,\delta)} \sin\rho\theta_n + \sum_{z_n\in(r,\pi-\delta,\pi)} \sin\theta_n -$$

$$-\frac{1}{\pi}\int_1^r \frac{d(\varphi(x) - \varphi(-x))}{x}\Big\} \le \eta(\delta).$$

Since $d\varphi(x) \le 0$, we have

$$\limsup_{r\to\infty} r^{-\rho}\Big\{ \sum_{z_n\in(r,0,\delta)} \sin\rho\theta_n + \sum_{z_n\in(r,\pi-\delta,\pi)} \sin\theta_n\Big\} = 0 \tag{16.4}$$

and

$$\limsup_{r\to\infty} r^{-\rho} \int_1^r \frac{d(\varphi(x) - \varphi(-x))}{x} = 0. \tag{16.5}$$

That the zero set of $f(z)$ has the argument density $\mu(\psi)$ in $[0,\pi]$, continuous at the points $\theta = 0$ and $\theta = \pi$, is proved in the same way as in Theorem 8.1 (see relations (8.6) and (8.7)) if using (16.4) and taking account of the existence of the angular density of the zero-set. We use this and relation (16.5) and let $r \to \infty$ in the relation

$$\frac{\omega(r)}{r^\rho} = \frac{1}{\rho}\int_0^\pi \frac{\sin\rho\psi}{\sin\psi} d_\psi \frac{c(r,0,\psi)}{r^\rho} -$$

$$-\frac{1}{2\pi r^\rho}\int_1^r \frac{\log|f(x)f(-x)|}{x}dx - \frac{1}{2\pi r^\rho}\int_1^r \frac{d(\varphi(x) - \varphi(-x))}{x}.$$

We obtain

$$\lim_{r\to\infty} \frac{1}{2\pi r^\rho}\Big\{\int_1^r \frac{\log|f(x)|}{x}dx + \int_1^r \frac{\log|f(-x)|}{x}dx\Big\} =$$

$$= \frac{1}{\rho}\int_0^\pi \frac{\sin\rho\psi}{\sin\psi}d\mu(\psi) = \frac{1}{\rho}\int_{+0}^{\pi-0} \sin\rho\psi d\Delta(\psi),$$

whence, by (16.2),

$$\lim_{r\to\infty} \frac{1}{2\pi r^\rho}\Big\{\int_1^r \frac{\log|f(x)|}{x}dx + \int_1^r \frac{\log|f(-x)|}{x}dx\Big\} =$$

$$= \frac{h_f(0) + h_f(\pi)}{2\pi\rho}.$$

Since

$$\limsup_{r\to\infty}\frac{1}{2\pi r^\rho}\int_1^r\frac{\log|f(x)|}{x}dx\le\frac{h_f(0)}{2\pi\rho},$$

$$\limsup_{r\to\infty}\frac{1}{2\pi r^\rho}\int_1^r\frac{\log|f(-x)|}{x}dx\le\frac{h_f(\pi)}{2\pi\rho},$$

there exist the limits

$$l_1=\lim_{r\to\infty}\frac{1}{2\pi r^\rho}\int_1^r\frac{\log|f(x)|}{x}dx=\frac{h_f(0)}{2\pi\rho},$$

$$l_2=\lim_{r\to\infty}\frac{1}{2\pi r^\rho}\int_1^r\frac{\log|f(-x)|}{x}dx=\frac{h_f(\pi)}{2\pi\rho}.$$

Thus we have proved that the function $f(z)$ satisfies the conditions of Theorem 15.1.

Relation (16.1) holds for all integer $\rho>0$. Therefore the question appears,whether Theorem 16.1 still holds for an even $\rho>0$ if change condition (16.2) is replaced by

$$h_f(0)-h_f(\pi)=2\pi\int_{+0}^{\pi-0}\sin\rho\psi d\Delta(\psi). \tag{16.6}$$

One may prove the necessity of conditions 1),2) and (16.6) in the same way as for an odd ρ. But the sufficiency ceases to be true. Below we construct an example of a function $f(z)$ of an even order ρ such that conditions 1) and 2) and (16.6) hold, but the function fails to belong to the class A_ρ^*.

Example. Consider the function

$$u(x)=\begin{cases} x^\rho, & 2^{2n}<|x|\le 2^{2n+1},\\ -x^\rho, & 2^{2n-1}<|x|\le 2^{2n}, \quad n=1,2,\dots\\ 0 & |x|\le 2,\end{cases}$$

where $\rho>0$ is an arbitrary even integer. Let

$$f(z)=\exp\left\{\frac{z^{\rho+1}}{\pi i}\int_{-\infty}^{\infty}\frac{u(x)dx}{x^{\rho+1}(x-z)}\right\}.$$

Denote $r=|z|$. We have

$$\log f(z)=\frac{z}{\pi i}\int_{-2r}^{2r}\frac{u(x)dx}{x(x-z)}-\sum_{k+1}^{\rho}\frac{z^k}{\pi i}\int_{-2r}^{2r}\frac{u(x)}{x^{k+1}}dx+$$

$$+\frac{z^{\rho+1}}{\pi i}\int_{|x|\geq 2r}\frac{u(x)dx}{x^{\rho+1}(x-z)}\equiv I_1(z)+I_2(z)+\cdots+I_{\rho+2}(z),$$

$$\Re I_1(re^{i\theta})=\frac{r\sin\theta}{\pi}\int_{-2r}^{2r}\frac{u(x)dx}{r^2-2rx\cos\theta+x^2}\leq$$

$$\leq\max_{|x|\leq 2r}|u(x)|\leq 2^\rho r^\rho,\quad I_{\rho+1}(z)\equiv 0,$$

$$|I_2(z)+\cdots+I_\rho(z)|\leq\sum_{k=1}^{\rho-1}\frac{r^k}{\pi}\int_0^{2r}x^{\rho-k-1}dx<C_1r^\rho,$$

$$|I_{\rho+2}(z)|\leq\frac{r^{\rho+1}}{\pi}\int_{2r}^{\infty}\frac{dx}{x(x-r)}=\frac{r^\rho}{\pi}\int_2^\infty\frac{du}{u(u-1)}.$$

Combining the obtained estimates, we get

$$\sup_{0<\theta<\pi}\log|f(re^{i\theta})|<C_2r^\rho,\quad r>r_0,$$

hence, the order of the function does not exceed ρ. Since

$$h_f(0)=\limsup_{x\to\infty}x^{-\rho}u(x)=1,\quad h_f(\pi)=\limsup_{x\to-\infty}|x|^{-\rho}u(x)=1,\qquad(16.7)$$

this order equals ρ. Since the function $f(z)$ does not vanish in $\Im z\geq 0$, relation (16.7) provides the fulfillment of condition (16.6) which takes the form $h_f(0)-h_f(\pi)=0$. Clearly, conditions 1) and 2) are fulfilled as well. Nevertheless the function $f(z)$ does not belong to the class A_ρ^*. Indeed, since $f(z)$ has no inner zeros, then, due to the Corollary to Lemma 8.1, the inclusion $f(z)\in A_\rho^*$ provides the existence of the right-hand and left-hand boundary densities, which is impossible because of

$$\limsup_{r\to\infty}r^{-\rho}\int_1^r\frac{u(x)}{x}dx\neq\lim_{r\to\infty}r^{-\rho}\int_1^r\frac{u(x)}{x}dx.$$

One may also construct the example of a function $f(z)\in A_\rho^*$, ρ even, such that $f(z)\notin\bar{A}_\rho^*$, though conditions 1), 2) and (16.6) are fulfilled.

17 Functions of a finite degree in the half-plane

By a function of finite degree we mean a function whose growth is not faster than of the first order a and finite type. Since such functions are often used in various applications (see, e.g.,[24, Ch.5]), we will formulate the obtained results espesially for this class of functions.

The following theorem is a special case of Theorem 3.3.

Theorem 17.1 *Every function $f(z)$ analytic and of a finite degree in $\Im z > 0$ admits the representation*

$$f(z) = \exp\Big\{i(a_0 + a_1 z) + \frac{1}{\pi i}\Big(\int_{-1}^{1} \frac{\log|f(t)|}{t-z}dt +$$

$$+ \int_{-1}^{1} \frac{d\varphi(t)}{t-z}\Big)\Big\} \prod_{|z_n|\le 1} \frac{z-z_n}{z-\bar{z}_n} \times \exp\Big\{\frac{z^2}{\pi i}\Big(\int_{|t|\ge 1} \frac{\log|f(t)|}{t^2(t-z)}dt +$$

$$+ \int_{|t|\ge 1} \frac{d\varphi(t)}{t^2(t-z)}\Big)\Big\} \prod_{|z_n|>1} \frac{(1-z/z_n)e^{z/z_n}}{(1-z/\bar{z}_n)e^{z/\bar{z}_n}}, \tag{17.1}$$

where a_0, a_1 are real constants, $z_n = r_n e^{i\theta_n}$ are the inner zeros of $f(z)$ and $\varphi(t)$ is the singular boundary function. In addition

$$\sum_n \frac{\sin\theta_n}{1+r_n^{1+\epsilon}} + \int_{-\infty}^{\infty} \frac{|\log|f(t)||}{1+|t|^{2+\epsilon}}dt + \int_{-\infty}^{\infty} \frac{|d\varphi(t)|}{1+|t|^{2+\epsilon}} < \infty$$

for any $\epsilon > 0$.

Now consider functions of a finite degree and c.r.g. Theorem 14.1 yields

Theorem 17.2 *For a function $f(z)$ analytic and of a finite degree in $\Im z > 0$ to belong to the class A_1^* it is necessary and sufficient that its zero set possesses the argument-boundary density and the argument-boundary symmetry with the exponent $\rho = 1$. It should be noted that, for $\rho = 1$, the coefficient of the argument-boundary symmetry admits the representation*

$$\sigma = \lim_{r\to\infty}\Big\{\sum_{1<r_n<r} \frac{\sin\theta_n}{r_n} - \frac{1}{2\pi}\int_1^r \frac{\log|f(x)f(-x)|}{x^2}dx -$$

$$-\frac{1}{2\pi}\int_1^r \frac{d\big(\varphi(x)-\varphi(-x)\big)}{x^2}\Big\}. \tag{17.2}$$

According to relation (12.7), we always have $\lambda(\pi) - \lambda(0) = 0$ for a function of the class A_1^* and since $\lambda(0) = 0$, we obtain $\lambda(\pi) = 0$.

Using Theorem 17.2 one may transform representation (17.1).

Theorem 17.3 *Every function from the class A_1^* admits the representation*

$$f(z) = \exp\Big\{i(b_0 + b_1 z) + \frac{1}{\pi i}\Big(\int_{-1}^{1} \frac{\log|f(t)|}{t-z} dt +$$

$$+ \int_{-1}^{1} \frac{d\varphi(t)}{t-z}\Big)\Big\} \prod_{|z_n|<1} \frac{z-z_n}{z-\bar{z}_n} \times$$

$$\times \lim_{R\to\infty} \Big\{\exp\Big\{\frac{z}{\pi i}\Big(\int_{1\le|t|\le R} \frac{\log|f(t)|}{t(t-z)} dt +$$

$$+ \int_{1\le|t|\le R} \frac{d\varphi(t)}{t(t-z)}\Big)\Big\} \prod_{1\le|z_n|\le R} \frac{1-z/z_n}{1-z/\bar{z}_n}\Big\},$$

where b_0, b_1 are real constants.

This formula is an analog of a known formula for entire functions of c.r.g. (see [24,p.206]).

For $\rho = 1$ Theorems 14.2,15.2 and 15.1 may be formulated as follows.

Theorem 17.4 *If $f(z)$ is of finite degree in the half-plane $\Im z > 0$ and belongs to the class A_1^*, then its indicator admits the representation*

$$h_f(\theta) = 2\sin\theta\Big\{\sigma - \frac{1}{2}a_1 + \int_0^\theta \psi\cot\psi d\lambda(\psi) +$$

$$+ \int_\theta^\pi (\psi - \pi)\cot\psi d\lambda(\psi)\Big\} - 2\pi\lambda(\theta)\cos\theta, \quad 0 < \theta < \pi.$$

Theorem 17.5 *If $f(z)$ is of finite degree in the half-plane $\Im z > 0$ and belongs to the class $\bar{A}_1^*$, then its indicator admits the representation*

$$h_f(\theta) = 2\pi\big(l_1 - \mu(\theta)\big) + 2\sin\theta\Big\{\sigma - \frac{1}{2}a_1 - l_1 - l_2 +$$

$$+ \int_0^\theta \psi\cot\psi d\mu(\psi) + \int_\theta^\pi (\psi - \pi)\cot\psi d\mu(\psi)\Big\}, \quad 0 \le \theta \le \pi.$$

Theorem 17.6 *For a function $f(z)$ analytic and of finite degree in $\Im z > 0$ to belong to the class $\bar{A}_1^*$ it is necessary and sufficient that*

1) the zero set of $f(z)$ possesses the argument-boundary density $\mu(\psi)$ continuous at the points $\psi = 0$ and $\psi - \pi$,

2) there exists the two-sided boundary density,

3) there exists the limit in the right-hand side of (17.2),

4) the singular boundary function $\varphi(t)$ satisfies the condition

$$\lim_{r\to\infty}\frac{1}{r}\int_{1\le|t|\le r}\frac{d\varphi(t)}{|t|}=0,$$

5) the indicator $h_f(\theta)$ satisfies

$$h_f(0)=2\pi l_1,\quad h_f(\pi)=2\pi l_2.$$

By virtue of (15.4) the indicator of a function $f(z)\in\bar{A}_1^*$ may be read as

$$h_f(\theta)=2\pi\cos\theta\Big\{l_1-\int_{+0}^{\theta}\sin\psi d\Delta(\psi)\Big\}+2\sin\theta\Big\{\sigma-\frac{1}{2}a_1-$$

$$-l_1-l_2+\int_{+0}^{\theta}\psi\cos\psi d\Delta(\psi)+\int_{\theta}^{\pi-0}(\psi-\pi)\cos\psi d\Delta(\psi)\Big\},\quad 0\le\theta\le\pi.$$

In the case $\rho=1$ Theorem 16.1 gives

Theorem 17.7 *For a function $f(z)$ analytic and of a finite degree in $\Im z>0$ to belong to the class $\bar{A}_1^*$ it is necessary and sufficient that*

1) the zero set of $f(z)$ possesses the angular density $\Delta(\psi)$ in the open angle $(0,\pi)$,

2) there exists the limit in the right-hand side of (17.2),

3) the indicator satisfies the condition

$$h_f(0)+h_f(\pi)=2\pi\int_{+0}^{\pi-0}\sin\psi d\Delta(\psi).$$

Now consider the applications of the obtained results to studying functions analytic in the closed half-plane $\Im z\ge 0$, whose zeros $z_n=r_ne^{i\theta_n}$ satisfies the condition

$$\sum_n\frac{\sin\theta_n}{r_n}<\infty. \tag{17.3}$$

Following B. Levin ([24, Ch.5]) we call such functions *functions from the class A.*

Theorem 17.8 *For a function $f(z)$ analytic and of finite degree to belong both to the class A and A_1^* it is necessary and sufficient that there exist the limits*

$$\lim_{r\to\infty}\frac{1}{r}\int_1^r\frac{\log|f(x)|}{x}dx=l_1,$$

$$\lim_{r\to\infty}\frac{1}{r}\int_1^r\frac{\log|f(-x)|}{x}dx=l_2 \tag{17.4}$$

and the integral

$$\int_1^\infty \frac{\log|f(x)f(-x)|}{x^2}dx \tag{17.5}$$

converges.

Theorem 17.9 *For a function $f(z)$ analytic and of finite degree in $\Im z \geq 0$ to belong both to the classes A and $\bar{A}_1^*$ it is necessary and sufficient that integral (17.5) converges and in addition $h_f(0) + h_f(\pi) = 0$.*

Proof of Theorem 17.8. Since $f(z)$ is analytic in the closed half-plane, we may assume that the singular boundary function is identically zero. Assume $f(z) \in A \cap A_1^*$. Then, by (17.3), the argument density $\mu(\psi)$ vanishes. Actually, denote

$$\lambda(r) = \sum_{1<r_n\leq r} \frac{\sin\theta_n}{r_n}, \quad c(r) = \sum_{1<r_n\leq r} \sin\theta_n.$$

We have

$$c(r) = \int_1^r x d\lambda(x) = \lambda(r) + \int_1^r (\lambda(r) - \lambda(x))dx,$$

whence, taking into account that the limit $\lim_{r\to\infty}\lambda(r) = \lambda \neq \infty$ exists, we obtain

$$\lim_{r\to\infty} r^{-1}c(r) = 0.$$

Since $f(z) \in A_1^*$, by Theorem 17.2, the zero set of $f(z)$ possesses the argument-boundary density. Since the argument density also exists, Corollary to Lemma 8.1 yields the existence of the limits in relation (17.4). The convergence of the integral in (17.5) follows from the limit in (17.2), taking into account that $\varphi(x) \equiv 0$.

Now let the limits in (17.4) exist and the integral in (17.5) converge. Then, as is shown in [24, p.295–296], $f(z)$ is a function from the class A. Convergence of the integral in (17.5) and the series in (17.3) yields the existence of the argument-boundary symmetry (17.2). At last, as has been just shown, the convergence of the series in (17.3) implies the existence of the argument density $\mu(\psi) \equiv 0$. Since both the boundary and argument densities exist, the argument-boundary density exists as well and, by Theorem 17.2, we have $f(z) \in A_1^*$.

Theorem 17.9 may be proved similarly by using Theorem 17.6. It should be noted that in a weakened form the sufficiency in Theorem 17.9 was proved earlier in ([24, p.317]); it was proved there that $f(z) \in A_1^*$. Under the assumption that $f(z)$ is an entire function, Theorem 17.9 was proved by A. Pflüger [37].

PART II

RIEMANN BOUNDARY PROBLEM WITH AN INFINITE INDEX WHEN THE VERTICITY INDEX IS LESS THAN 1/2

18 Statement of the homogeneous problem

Let D be a domain in the complex z-plane such that the boundary ∂D is a simple smooth curve L which begins at the point $z = t_0$ and ends at $z = \infty$. In what follows the letters t and τ will denote the points of the curve L. Denote by $\psi(t)$ the angle between the tangent to the contour L at the point t and the positive real axis. Since L is smooth, the function $\psi(t)$ is continuous at all points $t \in L$ including the point $t = \infty$ (the latter means that $\psi(t)$ approaches a limit as $t \to \infty$, $t \in L$). Everywhere below we assume that $\lim_{t\to\infty} \psi(t) = 0$, $\Re t_0 > 0$ and $0 \notin L$. These assumptions do not restrict generality, since they may be satisfied by applying an appropriate rotation and shift.

Let $G(t)$ be a continuous function on L non-vanishing at all points $t \in L$, $t \neq \infty$. We say that the index of the function $G(t)$ equals $+\infty$ (respectively $-\infty$) if $\lim_{t\to\infty} \arg G(t) = +\infty$. In this case we write

$$\mathrm{Ind}_L G(t) = +\infty \ (-\infty),$$

and say that the point $t = \infty$ is a point of *positive (negative) verticity* of the function $G(t)$.

Definition 18.1 We say that a function $f(t)$ on the curve L *satisfies the Hölder condition with the exponent* μ, $0 < \mu \leq 1$, if the following inequality holds

$$|f(t_1) - f(t_2)| \leq C\left|\frac{1}{t_1} - \frac{1}{t_2}\right|^{\mu} \tag{18.1}$$

for any $t_1, t_2 \in L$, and some positive constant C depending on f. For a finite curve L this definition is equivalent to the traditional one:

$$|f(t_1) - f(t_2)| \leq C|t_1 - t_2|^{\mu}. \tag{18.2}$$

We write $f(t) \in H_L(\mu)$, if f satisfies the Hölder condition with the exponent μ and $f(t) \in H_L$ or simply $f(t) \in H$if $f \in H(\mu)$ for some μ, $0 < \mu \leq 1$.

Let a function $G(t)$ on the curve L satisfy the conditions

1)
$$\arg G(t) = 2\pi\varphi(t)|t|^{\rho}, \tag{18.3}$$

where

$$\rho > 0, \quad \varphi(t) \in H_L, \quad \varphi(\infty) = \lambda > 0, \quad -1 < \varphi(t_0) \le 0, \tag{18.4}$$

2)
$$\log|G(t)| \in H_L. \tag{18.5}$$

Denote $\tilde{L} = L \backslash \{t_0, \infty\}$. By the *homogeneous Riemann boundary problem with index* $+\infty$, *verticity order* ρ, *and verticity coefficient* λ we mean the following problem:

Find a function $\Phi(z)$ analytic and bounded in D whose boundary values on $\tilde{L}$ satisfy the relation

$$\Phi^+(t) = G(t)\Phi^-(t), \quad t \in \tilde{L}. \tag{18.6}$$

In this chapter we consider this problem in the case $0 < \rho < 1/2$.

19 Canonical function

Let us study the problem posed in the previous section. Denote

$$X(z) = \exp\left\{\frac{z}{2\pi i}\int_L \frac{\log G(\tau)}{\tau(\tau - z)}d\tau\right\}. \tag{19.1}$$

The integral in the right-hand side is convergent, since $0\bar{\in}L$ and relations (18.3)–(18.5) imply $|\log G(t)| = O(|t|^\rho)$, $t \to \infty$. We shall prove that the function $X(z)$ possesses the following properties:

1) *$X(z)$ is analytic in D, continuous up to $\tilde{L}$ and does not vanish* in $D \cup \tilde{L}$,

2) *$X(z)$ satisfies the boundary condition* (18.6), i.e.

$$X^+(t) = G(t)X^-(t), \quad t \in \tilde{L},$$

3) *$X(z)$ is normalized by the condition $X(0) = 1$,*

4) *in a sufficiently small neighborhood of the beginning of the curve L the following estimate holds*

$$C_1|z - t_0|^\alpha < |X(z)| < C_2, \quad 0 \le \alpha < 1, \tag{19.2}$$

5) *$X(z)$ is bounded in D.*

Since $\rho < 1/2$, we have

$$||t_1|^\rho - |t_2|^\rho| \le |t_1 - t_2|^\rho.$$

Therefore by (18.3) and (18.5) we obtain $\log G(t) \in H_{L'}$, on any finite arc $L' \subset L$. The boundary value of the Caushy integral in the relation (19.1) also belongs to $H_{L'}$ (see [2, p.53]), which proves property 1). Property 2) is a direct consequence of the Sohotski formula and property 3) is evident.

To prove property 4) divide the curve L into two parts l and L^*, where l is a finite curve starting at $z = t_0$. We obtain

$$\log X(z) = \frac{1}{2\pi i}\int_l \frac{\log G(\tau)}{\tau - z}d\tau -$$

$$-\frac{1}{2\pi i}\int_l \frac{\log G(\tau)}{\tau}d\tau + \frac{z}{2\pi i}\int_{L^*} \frac{\log G(\tau)}{\tau(\tau - z)}d\tau.$$

According to [2, p.67], we have

$$\frac{1}{2\pi i}\int_l \frac{\log G(\tau)}{\tau - z}d\tau =$$

$$= \frac{1}{2\pi}\log|z - t_0|\{i\log|G(t_0)| - \arg G(t_0)\} + O(1), \quad z \to t_0,$$

Therefore setting

$$\alpha = -\frac{1}{2\pi}\arg G(t_0),$$

and taking into account (18.4), we obtain (as $z \to t_0$)

$$\log|X(z)| = \alpha L|z - t_0| + O(1), \quad 0 \le \alpha < 1,$$

which yields 4).

We shall prove a statement which is stronger than 5), using the following theorem.

Theorem 19.1 *Let a function $\varphi(t) \in H_L$ on the curve L be such that $\varphi(\infty) = \lambda \neq 0$. Consider the function*

$$f(z) = \exp\Big\{z\int_L \frac{\varphi(\tau)|\tau|^\rho}{\tau(\tau - z)}d\tau\Big\}, \quad 0 < \rho < 1.$$

This function is of order ρ in the domain D and has the indicator

$$h_f(\theta) = \lim_{r\to\infty} r^{-\rho}\log|f(re^{i\theta})| =$$

$$= -\frac{\pi\lambda}{\sin\pi\rho}\cos\rho(\theta - \pi), \quad 0 \le \theta \le 2\pi.$$

Here the convergence is uniform with respect to θ, $0 \le \theta \le 2\pi$.

To prove the theorem we need the following three lemmas.

Lemma 19.1 *If a function $\omega(r)$ is continuous on the segment $[0, r_0]$ and $\rho > 0$, $a > 0$, then*

$$\int_{aR}^{r_0} |\omega(r) - \omega(R)|\frac{dr}{r^{\rho+1}} = o\Big(\frac{1}{R^\rho}\Big), \quad R \to 0.$$

Proof. For any $k > 0$ we have

$$\limsup_{R\to 0} R^\rho \int_{aR}^{r_0} |\omega(r) - \omega(R)|\frac{dr}{r^{\rho+1}} \le$$

$$\le \limsup_{R\to 0} R^\rho \int_{k}^{r_0} |\omega(r) - \omega(R)|\frac{dr}{r^{\rho+1}} +$$

$$+ \limsup_{R\to 0} R^\rho \max_{0\le r\le k}\{|\omega(r) - \omega(R)|\}\int_{aR}^{k} \frac{dr}{r^{\rho+1}} =$$

$$= \frac{a^{-\rho}}{\rho}\max_{0\le r\le k}|\omega(r) - \omega(0)|.$$

This implies the desired estimate since k is an arbitrary positive number.

Lemma 19.2 *If a complex-valued function $\omega(r)$ is continuous on the segment $[0, r_0]$ and, for $0 < r \leq r_0$, has the derivative $\omega'(r)$ satisfying the condition*

$$|\omega'(r)| < \frac{\eta(r)}{r},$$

for some $\eta(r)$, such that $\lim_{r\to 0}\eta(r) = 0$, then the following estimate

$$I(R) = \int_0^{r_0} \frac{|\omega(r) - \omega(R)|}{r^\rho |r - R|} dr = o\Big(\frac{1}{R^\rho}\Big),$$

holds for any ρ, $0 < \rho < 1$.

Proof. Assuming $R < r_0/2$, we represent $I(R)$ in the form

$$I(R) = \Big(\int_0^{R/2} + \int_{R/2}^{2R} + \int_{2R}^{r_0}\Big)\psi(r, R)dr, \quad \psi(r, R) = \frac{|\omega(r) - \omega(R)|}{r^\rho |r - R|}.$$

Clearly

$$\int_0^{R/2} \psi(r, R)dr \leq \max_{0\leq r\leq R}\{|\omega(r) - \omega(R)|\}\frac{1}{R^\rho}\int_0^{1/2} \frac{dx}{x^\rho(1-x)} =$$

$$= o\Big(\frac{1}{R^\rho}\Big).$$

Applying the Lagrange theorem we have

$$\int_{R/2}^{R} \psi(r, R)dr \leq 2 \max_{R/2\leq r\leq R}\{\eta(r)\}\int_{R/2}^{2R} \frac{dr}{r^{\rho+1}} = o\Big(\frac{1}{R^\rho}\Big).$$

Since $r - R \geq r/2$ for $r \geq 2R$, Lemma 19.1 yields

$$\int_{2R}^{r_0} \psi(r, R)dr \leq 2\int_{2R}^{r_0} |\omega(r) - \omega(R)|\frac{dr}{r^{\rho+1}} = o\Big(\frac{1}{R^\rho}\Big).$$

Lemma 19.3 *Let l be a smooth open curve in the ζ-plane which is tangent to the positive real axis at its endpoint $u = 0$ and let a point $v \in l$ be such that $|v| = |\zeta|$ (this point is uniquely defined for a sufficiently small $|\zeta|$). Then there is a neighborhood of $\zeta = 0$ in wich the estimate*

$$J(\zeta) = \int_l \frac{e^{i\rho \arg u} - e^{i\rho \arg v}}{u^\rho(u - \zeta)} du = o\Big(\frac{1}{|\zeta|^\rho}\Big), \quad \zeta \to 0, \tag{19.3}$$

holds for $0 < \rho < 1$.

Proof. Without loss of generality we may assume that l is so small that

1) any circle $|\zeta| = r$ intersects l no more than at one point.

2)

$$|\arg(du/ds) - \arg u| < \alpha_0 < \pi/2, \tag{19.4}$$

where s is the arc abscissa of the point u.

Thus, we may use $r = |u|$ as a parameter on the curve l and the function $\omega(r) = e^{i\rho \arg u(r)}$ satisfies the conditions of Lemma 19.2. Indeed,

$$|\omega'(r)| = \rho \left| \frac{d \arg u}{ds} \cdot \frac{ds}{dr} \right| =$$

$$= \frac{\rho}{\cos \alpha(u)} \left| \Im\left(\frac{1}{u}\frac{du}{ds}\right) \right| = \frac{\eta(r)}{r}.$$

So using condition (19.4) and taking into account that l is a smooth curve, we obtain

$$\eta(r) \leq \frac{\rho}{\cos \alpha} \sin \left| \arg \frac{du}{ds} - \arg u \right| \to 0, \quad r \to 0.$$

Since $|du| = ds \leq dr/\cos\alpha_0$, $|u - \zeta| \geq |r - R|$ where $|\zeta| = R = |v|$, we get

$$|J(\zeta)| \leq \frac{1}{\cos \alpha_0} \int_0^{r_0} \frac{|\omega(r) - \omega(R)|}{r^\rho |r - R|} dr.$$

Now by applying Lemma 19.2 we prove relation (19.3).

Proof of Theorem 19.1 Denoting $\tau = 1/u$ and $z = 1/\zeta$ we can pass to a finite curve l located as in Lemma 19.3, and consider the functions

$$f\left(\frac{1}{\zeta}\right) = \exp\left\{ \int_l \frac{\varphi^*(u)du}{|u|^\rho (u - \zeta)} \right\}, \quad \varphi^*(u) = \varphi\left(\frac{1}{u}\right) \in H_l(\mu).$$

Clearly $\varphi^*(0) = \lambda$, where λ is the velocity coefficient. Assuming $|\zeta|$ to be sufficiently small, take $v \in l$ such that $|v| = |\zeta|$, and, for some ϵ, $0 < \epsilon < \mu$, consider the representation

$$\log f\left(\frac{1}{\zeta}\right) = \lambda e^{i\rho \arg v} \int_l \frac{du}{u^\rho (u - \zeta)} +$$

$$+ \int_l \frac{\varphi^*(u) - \lambda) u^{-\epsilon} e^{i\rho \arg u} du}{u^{\rho - \epsilon}(u - \zeta)} + \lambda \int_l \frac{e^{i\rho \arg u} - e^{i\rho \arg v}}{u^\rho (u - \zeta)} du =$$

$$= Q_1(\zeta) + Q_2(\zeta) + Q_3(\zeta).$$

According to [25, p.22,26], we have

$$(\varphi^*(u) - \lambda) u^{-\epsilon} e^{i\rho \arg u} \in H(\mu - \epsilon).$$

Therefore, in view of [25, p.75],

$$Q_2(\zeta) = O(|\zeta|^{-\rho+\epsilon}), \quad \zeta \to 0. \tag{19.5}$$

By Lemma 19.3

$$Q_3(\zeta) = o(|\zeta|^{-\rho}), \quad \zeta \to 0, \tag{19.6}$$

and finally

$$Q_1(\zeta) = -\frac{\pi\lambda}{\sin\rho\pi} \cdot \frac{e^{i\rho\pi}}{\zeta^\rho} + O(1), \quad \zeta \to 0. \tag{19.7}$$

Here the continuous branch of the function ζ^ρ in the domain

$$\Delta_a = \{|z| < a\}\backslash l$$

is chosen such that $\arg\zeta = -\pi$ for $\zeta < 0$. Now relations (19.5)–(19.7) yield

$$\log f\left(\frac{1}{\zeta}\right) = -\frac{\pi\lambda}{\sin\rho\pi} \cdot \frac{e^{-i\rho\pi}}{\zeta^\rho} + o\left(\frac{1}{|\zeta|^\rho}\right), \quad \zeta \to 0, \tag{19.8}$$

where the branch of $\arg\zeta$ is the same as before. Put $\alpha_1 = \arg\zeta$ and denote by α another branch of the same function $\arg\zeta$ satisfying inequality $-2\pi < \arg\zeta \le 0$. Let $c = -(\pi\lambda/\sin\pi\rho)$, $\zeta = Re^{i\alpha}$. Relation (19.8) yields

$$R^\rho \log|f(1/(Re^{i\alpha_1}))| = c\cos\rho(\alpha_1 + \pi) + o(1).$$

Using smoothness of the curve l one can easily see that

$$\lim_{\zeta\to 0}(\cos\rho(\alpha_1+\pi) - \cos\rho(\alpha+\pi)) = 0.$$

Therefore

$$R^\rho \log|f(1/(Re^{i\alpha}))| = c\cos\rho(\alpha + \pi) + o(1), \quad -2\pi \le \alpha \le 0,$$

and, returning to the initial notations, we have

$$r^{-\rho}\log|f(re^{i\theta})| = c\cos\rho(\theta - \pi) + o(1), \quad 0 \le \theta \le 2\pi.$$

Theorem 19.2 *Let conditions(18.3)–(18.5) be fulfilled and $X(z)$ be the function defined by (19.1). Then $X(z)$ is bounded in D and has the order of decreasing ρ and the indicator*

$$h_X(\theta) = \lim_{r\to\infty} r^{-\rho}\log|X(re^{i\theta})| = -\frac{\pi\lambda}{\sin\pi\rho}\cos\rho(\theta - \pi),$$

$$0 \le \theta \le 2\pi. \tag{19.9}$$

Here the convergence is uniform with respect to θ.

Proof. Represent $X(z)$ in the form

$$X(z) = \exp\left\{\frac{z}{2\pi i}\int_L \frac{\log|G(\tau)|}{\tau(\tau - z)}d\tau\right\}\times$$

$$\times \exp\left\{z\int_L \frac{\varphi(\tau)|\tau|^\rho}{\tau(\tau - z)}d\tau\right\} = X_1(z)X_2(z). \tag{19.10}$$

After changing the variables $\tau = 1/u$, $z = 1/\zeta$ we pass to a finite curve l ending at the point $u = 0$. In the neighborhood of zero we have

$$\log\left|X_1\left(\frac{1}{\zeta}\right)\right| = \frac{1}{2\pi}\Im\int_l \frac{\log|G(1/u)|}{u - \zeta}du =$$

$$= \frac{\log|G(\infty)|}{2\pi}\arg\zeta + o(1) = O(1), \tag{19.11}$$

and hence, $\frac{1}{C} < |X_1(z)| < C$ for $|z| > r_0$. Now relation (19.9) follows from (19.10) and Theorem 19.1. Since $\lambda > 0$ and $0 < \rho < 1/2$, we have

$$\max_{0\le\theta\le2\pi} h_X(\theta) = -\pi\lambda\cot\rho\pi < 0.$$

Therefore

$$\max_{0\le\theta\le2\pi} |X(re^{i\theta})| < \exp\left(-\frac{\pi}{2}\lambda r^\rho\cot\rho\pi\right), \quad r > r_0.$$

This implies the statement of the Theorem and in particular, property 5) of the function $X(z)$.

Definition 19.1 A function $X(z)$ satisfying conditions 1)–5) is called the *canonical function of Riemann boundary problem* (18.6).

In the next section we shall prove that such a function is unique.

20 Solution of the homogeneous problem in the class B_L. Description of solutions of order ρ

Definition 20.1 Denote by $\tilde{B}_L$ the class of functions analytic in D and bounded in any disk $|z| \leq R$ and by B_L the class functions analytic and bounded in D .

First consider problem (18.6) in the class $\tilde{B}_L$. Let $\Phi \in \tilde{B}_L$ be a solution. Then the function $F(z) = \Phi(z)/X(z)$ is analytic in D and satisfies the condition $F^+(t) = F^-(t)$ on L. It follows from (19.2) that near the point $z = t_0$ this function can be estimated as $F(z) = O(|z - t_0|^{-\alpha})$, $0 \leq \alpha < 1$, and, hence, $F(z)$ is an entire function. On the other hand, for any entire function $F(z)$ the relation

$$\Phi(z) = F(z)X(z) \tag{20.1}$$

defines a solution of problem (18.6) belonging to the class $\tilde{B}_L$. Thus we have proved the following theorem.

Theorem 20.1 *General solution of problem (18.6) in the class $\tilde{B}_L$ can be expressed by the relation (20.1) where $X(z)$ is the canonical function and $F(z)$ is an arbitrary entire function.*

Let us clarify what additional conditions on the function $F(z)$ provide the inclusion $\Phi(z) \in B_L$, where $\Phi(z)$ is obtained from (20.1). Using the notation of Sect. 18 it means that $\Phi(z)$ is a solution of the homogeneous Riemann boundary problem (18.6).

Theorem 20.2 *If $\Phi(z) \in B_L$, then the order of growth ρ_F of the corresponding function $F(z)$ does not exceed ρ.*

Proof. Let the contrary be true. Then there exists a sequence $z_n \to \infty$ such that

$$\log |F(z_n)| > |z_n|^{(\rho+\rho_F)/2}.$$

Then, by (19.9) we have

$$|F(z_n)| = |X(z_n)F(z_n)| \to \infty,$$

which is impossible.

Theorem 20.3 *For a solution $\Phi(z) \in \tilde{B}_L$ to belong to the class B_L it is necessary and sufficient that the order ρ_F of the corresponding entire function does not exceed ρ and also*

$$\log |F(t)| \leq -\Re\Big\{t \int_L \frac{\varphi(\tau)|\tau|^\rho}{\tau(\tau - t)} d\tau\Big\} + O(1) \tag{20.2}$$

for all $t \in L$, $t \to \infty$.

Proof. Necessity. Let $\Phi(z) \in B_L$. Then

$$\log|F(t)| = \log|\Phi^{\pm}(t)| - \log|X^{\pm}(t)| \leq O(1) - \log|X^{\pm}(t)|. \tag{20.3}$$

Relation (19.10) taking into account the Sohotski formulas yields

$$\log|X^{\pm}(t)| = \pm\frac{1}{2}\log|G(t)| + \Re\Big\{t\int_L \frac{\varphi(\tau)|\tau|^{\rho}}{\tau(\tau-t)}d\tau\Big\} + \Gamma(t), \tag{20.4}$$

where

$$\Gamma(t) = \Im\Big\{\frac{t}{2\pi}\int_L \frac{\log|G(\tau)|}{\tau(\tau-t)}d\tau\Big\}.$$

Since $\log|G(t)| \in H_L$, then $|\log|G(t)|| < C$ and, by (19.11), $|\Gamma(t)| < C$. Now (20.3) and (20.4) yield (20.2). That $\rho_F \leq \rho$ we have proved in the previous theorem.

Sufficiency. Let $\rho_F \leq \rho$ and also relation (20.2) holds. Then (20.3) and (20.4) imply $|\Phi^{\pm}(t)| < C$. For some ρ_1, $\rho < \rho_1 < 1/2$, we have

$$\max_{|z|=r}|\Phi(z)| = \max_{|z|=r}|F(z)X(z)| \leq \exp(r^{\rho_1})$$

for all sufficiently large r. What remains is to apply of the following version of the Phragmën-Lindelöf principle:

Let $\Phi(z) \in \tilde{B}_L$ and also

$$|\Phi^{\pm}(t)| < C, \quad t \in \tilde{L},$$

$$\max_{|z|=r}|\Phi(z)| \leq \exp(r^{\rho_1}), \quad r > r_0, \quad 0 < \rho_1 < 1/2,$$

then $\Phi(z) \in B_L$.

Theorem 20.4 *Let conditions (18.3)-(18.5) be fulfilled. Then the homogeneous Riemann boundary problem (18.6) has infinitely many solutions and the general form of solution is as follows*

$$\Phi(z) = F(z)X(z) =$$

$$= Qz^m \prod_{n=1}^{n_0}\Big(1 - \frac{z}{z_n}\Big)\cdot\exp\Big\{\frac{z}{2\pi i}\int_L \frac{\log G(\tau)}{\tau(\tau-z)}d\tau\Big\},$$

$$Q = \text{const}, \quad n_0 \leq \infty, \quad 0 < |z_1| \leq |z_2| \leq \dots, \quad z_n \to \infty, \tag{20.5}$$

where $F(z)$ is an entire function satisfying (20.2) and having the order $\rho_F \leq \rho$.

This theorem follows directly from Theorem 20.3 if taking into account that due to the inequality $\rho_F < 1$ the function $F(z)$ admits the representation

$$F(z) = Qz^m \prod_{n=1}^{n_0} \left(1 - \frac{z}{z_n}\right), \quad 0 \le m < \infty, \quad 0 \le n_0 \le \infty$$

(for $n_0 = 0$ we assume $F(z) = Qz^m$).

Corollary 20.1 *The homogeneous Riemann boundary problem (18.6) has infinitely many linearly independent solutions.*

Indeed, there are infinitely many linearly independent functions of order less than ρ. Every such function $F(z)$ generates a solution because Theorem 19.1 yields

$$\Re\left\{t \int_L \frac{\varphi(\tau)|\tau|^\rho}{\tau(\tau - t)} d\tau\right\} = -\pi\lambda|t|^\rho \cot \rho\pi + o(|t|^\rho), \quad t \to \infty, \tag{20.6}$$

and,hence, $F(z)$ satisfies condition (20.2).

Corollary 20.2 *The canonical function is unique.*

Actually, if a function $\Phi(z)$ admits representation (20.5) and also possesses properties 1)–5) (see Sect. 19) then $F(z) \equiv 1$ and $\Phi \equiv X(z)$.

Now let us describe the solutions for wich ρ is the order of decreasing. Denote by γ_F the exponent of convergence of the zeros of the function $F(z)$.

Theorem 20.5 *The general form of a solution of the homogeneous problem (18.6) in the class of functions having the order of decreasing ρ and negative indicator is given by relation (20.5) where $F(z)$ is an entire function such that, either*

$$1) \ \rho_F < \rho \quad (\gamma_F < \rho), \tag{20.7}$$

or

$$2) \ \rho_F = \rho, \quad (\gamma_F = \rho), \quad \text{and} \quad h_F(0) < \pi\lambda \cot \rho\pi. \tag{20.8}$$

The indicator of the solution $\Phi(z)$ admits the representation

$$h_\Phi(\theta) = \begin{cases} -\dfrac{\pi\lambda}{\sin \pi\rho} \cos \rho(\theta - \pi) & \rho_F < \rho, \\ \\ h_F(\theta) - \dfrac{\pi\lambda}{\sin \pi\rho} \cos \rho(\theta - \pi) & \rho_F = \rho. \end{cases} \tag{20.9}$$

To prove the theorem we need the following lemma, which can be easily obtained from general properties of trigonometrically convex functions (see [24, p.74]).

Lemma 20.1 *If a function $h(\theta)$ is trigonometrically ρ-convex on the segment $[\theta_1\theta_2]$, and $|\theta_1 - \theta_2| < \pi/\rho$ and $h(\theta_k) < 0$, $k = 1, 2$, then $h(\theta) < 0$ for all $\theta \in [\theta_1, \theta_2]$.*

Proof of Theorem 20.5 Let either (20.7) or (20.8) be fulfilled. By Theorems 19.2 and 20.1 we have

$$h_\Phi(\theta) = -\frac{\pi\lambda}{\sin\pi\rho}\cos\rho(\theta - \pi) + h_F(\theta). \tag{20.10}$$

Here $h_F(\theta) < \infty$. If $\rho_F = \rho$, this follows from (20.8), and in the case $\rho_F < \rho$ we have $h_F(\theta) \equiv 0$. The function $h_F(\theta)$ is trigonometrically ρ-convex on $[0, 2\pi]$. According to (20.10) the function $h_\Phi(\theta)$ is trigonometrically ρ-convex as well and $h_\Phi(0) = h_\Phi(2\pi) < 0$. Since $\rho < 1/2$, Lemma 20.1 provides $h_\Phi(\theta) < 0$, $\theta \in [0, 2\pi]$,and thus $\Phi(z)$ has the order of decreasing ρ and the negative indicator in D.

Conversely, let now $\Phi(z)$ have the order of decreasing ρ and a negative indicator. By Theorem 20.2 we have $\rho_F \leq \rho$. If $\rho_F = \rho$, then, setting $\theta = 0$ in (20.2) and taking into account that $h_\Phi(\theta) < 0$, we obtain $h_F(0) < \pi\lambda\cot\rho\pi$. If $\rho_F < \rho$, then $h_F(\theta) \equiv 0$. Relation (20.9) now follows from (20.10).

Remark 1. The class of solutions given by Theorem 20.5 is smaller than that of Theorem 20.3. At the same time, its structure is simpler because condition (20.8) is easier to verify than condition (20.2).

Remark 2. The solutions of problem (18.6) that have the order of decreasing ρ will be considered in Chapter 5.

21 Formulation of the non-homogeneous problem and an approach to its solution

The non-homogeneous problems have the form

$$\Phi(t) = G(t)\Phi^-(t) + g(t), \quad t \in \tilde{L}. \tag{21.1}$$

Here we consider the case $L = \{t : 1 \le t \le \infty\}$. We also suppose that the functions $G(t)$ and $g(t)$ satisfy the following conditions

$$1) \qquad \arg G(t) = 2\pi\varphi(t)t^\rho, \quad 0 < \rho < l/2, \tag{21.2}$$

$$\text{where } \varphi(t) \in H(\mu_0), \quad \rho < \mu_0 \le 1, \tag{21.3}$$

$$\varphi(\infty) = \lambda > 0, \quad -1 < \varphi(1) \le 0, \tag{21.4}$$

$$2) \qquad \log|G(t)| \in H, \quad g(t) \in H, \tag{21.5}$$

$$3) \qquad g(\infty) = 0, \tag{21.6}$$

$$4) \qquad \text{either } G(1) = 1 \text{ and } g(1) = 0, \text{ or } G(1) \ne 1. \tag{21.7}$$

The condition $\rho < \mu_0$ in relation (21.3) is essential; nevertheless it will be weakened in Ch. 5, Sect. 29. Denote by D the whole complex plane without the ray $1 \le t \le \infty$ and by B the class of functions bounded and analytic in D. (This is a special case of the class B_L). We study problem (21.1) in the class B. Clearly, it suffices to construct a single special solution $\Phi_0(z)$ and then the general solution has the form $\Phi(z) = \Phi_0(z) + \Psi(z)$, where $\Psi(z)$ is the general solution of the corresponding homogeneous problem $\Psi^+(t) = G(t)\Psi^-(t)$, $1 < t < \infty$.

Let $\Psi_0(z) \in B$ be a special solution of problem (18.6), which does not vanish on $\tilde{L}$. Following Gakhov [2, p.111] represent the boundary condition (21.1) in the form

$$\frac{\Phi^+(t)}{\Psi_0^+(t)} - \frac{\Phi^-(t)}{\Psi_0^-(t)} = \frac{g(t)}{\Psi_0^+(t)}.$$

Then a special solution is supposed to be of the form

$$\Phi_0(z) = \frac{\Psi_0(z)}{2\pi i}\int_1^\infty \frac{g(x)dx}{\Psi_0^+(x)(x-z)}, \tag{21.8}$$

and what remains is to choose the function $\Psi_0(z)$ to provide the convergence of the integral in the right-hand side of (21.8) and also the inclusion of the obtained solution into the class B. It should be noted that, generally speaking, one cannot use the canonical function $X(z)$ as the function $\Psi_0(z)$, since (19.9) yields

$$|X^+(t)| < \exp(-at^\rho), \quad 0 < a < \pi\lambda\cot\rho\pi, \quad t > t_0, \tag{21.9}$$

and after setting $\Psi_0(z) = X(z)$ the integral in (21.8) converges only under the additional restriction on the function $g(t)$

$$|g(t)| < \exp(-bt^\rho), \quad b > 0, \quad t > t_0. \tag{21.10}$$

According to Theorem 20.4, one has to construct the function $\Psi_0(z)$ in the form $\Psi_0(z) = X(z)F_0(z)$, where $F_0(z)$ is an entire function which satisfies some special conditions. These conditions have to admit some non-uniqueness of the desired function $F_0(z)$. We shall show that one may control this non-uniqueness in such a way that the obtained function $\Psi_0(z)$ would possess the following properties

$$1)\ \Psi_0(z) \in B, \quad 2)\ 1/|\Psi_0^+(t)| < C, \quad c \le t < \infty, \tag{21.11}$$

$$3)\ g(t)/\Psi_0^+(t) \in H_{[c,\infty]}, \quad \lim_{t\to\infty} g(t)/\Psi_0^+(t) = 0 \tag{21.12}$$

(here and till the end of the chapter $c > 1$ is an arbitrary fixed number).

By Theorem 20.4 the order ρ_{F_0} of the function $F_0(z)$ does not exceed ρ. But, actually, $\rho_{F_0} = \rho$ since otherwise, by virtue of Theorem 20.5, the function $\Psi_0^+(t)$ would satisfy an inequality of the form (21.9) and, again, we need to introduce condition (21.10). Thus we have $\rho_{F_0} = \rho$.

By Theorem 19.2 the indicator of the function $\Psi_0(z)$ admits the representation

$$h_{\Psi_0}(\theta) = -\frac{\pi\lambda}{\sin \pi\rho} \cos \rho(\theta - \pi) + h_{F_0}(\theta), \quad 0 \le \theta \le 2\pi.$$

Since the function $\Psi_0(z)$ satisfies (21.11), we have $h_{\Psi_0}(0) = 0$ and, hence,

$$h_{F_0}(0) = \lambda\pi \cot \rho\pi. \tag{21.13}$$

Further (21.11) yields $h_{\Psi_0}(\theta) \le 0$ and, hence,

$$h_{F_0}(\theta) \le \frac{\pi\lambda}{\sin \pi\rho} \cos \rho(\theta - \pi) = -h_X(\theta), \quad 0 < \theta < 2\pi. \tag{21.14}$$

The geometrical meaning of these inequalities is that the curve $h = h_{F_0}(\theta)$ lies below the curve $h = -h_X(\theta)$ and touches it at the points $\theta = 2\pi n$, $n = 0, \pm 1, \pm 2, \ldots$.

There are infinitely many functions satisfying (21.14) and (21.13). The simplest ones are those having zeros located on one ray. Let us chose the ray $\arg z = \pi$ (by virtue of (21.11) the function $F_0(z)$ does not vanish on the ray $\arg z = 0$) and construct the desired function in the form

$$F_0(z) = \prod_{n=1}^{\infty} \left(1 + \frac{z}{r_n}\right), \quad 0 < r_1 < r_2 < \ldots,$$

for some still unspecified values r_n, $n = 1, 2, \ldots$.

Let $n(r)$ be the counting function of zeros of the function $F_0(z)$. We demand the existence of the limit

$$\lim_{r\to\infty} r^{-\rho} n(r) = \Delta, \quad 0 < \Delta < \infty.$$

The value of this limit is called the *density* of zeros, and has to be specially determined.

Taking into account the identity

$$z\int_0^\infty \frac{dx}{x^{1-\rho}(x-z)} = -\frac{\pi z^\rho}{\sin\pi\rho} e^{-i\rho\pi}, \quad 0 < \arg z < 2\pi,$$

we obtain

$$\log F_0(t) = \int_0^\infty \log\Big(1+\frac{t}{x}\Big) dn(x) =$$

$$= \frac{\pi\Delta}{\sin\pi\rho} t^\rho + t\int_0^\infty \frac{n(x) - \Delta x^\rho}{x(x+t)} dx. \tag{21.15}$$

for $t > 0$.

We need the following lemma.

Lemma 21.1 *The function $X^+(t)$ admits the representation*

$$\log X^+(t) = i(\eta\log t + \lambda\pi t^\rho) - \lambda\pi\cot\pi\rho\cdot t^\rho + q(t), \tag{21.16}$$

where $q(t) \in H_{[c,\infty)}(\sigma)$, $\sigma = \min\{\mu_0 - \rho, \mu\}$ and $\mu < 1$ is the exponent in the Hölder condition in (21.5) and η is a real constant.

Proof. Let us change the variables $\tau = 1/u$, $t = 1/v$ and then denote $\varphi_*(u) = \varphi(1/u)$, $\psi(u) = \log|G(1/u)|$. Applying the Sohotski formulas, we have

$$\log X^+\Big(\frac{1}{\nu}\Big) = \frac{1}{2\pi i}\int_1^0 \frac{\psi(u)}{u-\nu} d\nu + \int_1^0 \frac{\varphi_*(u)-\lambda}{u^\rho(u-\nu)} du +$$

$$+\lambda\int_1^0 \frac{du}{u^\rho(u-\nu)} + \frac{\psi(\nu)}{2} + \frac{\pi i}{\nu^\rho}(\varphi_*(\nu)-\lambda) + \frac{\pi i\lambda}{\nu^\rho}$$

for $0 < v < 1$. Using relations (21.3)–(21.5), we obtain $\psi(u) \in H_{[0,1]}(\mu)$, $\varphi_*(u) \in H_{[0,1]}(\mu_0)$ and $\varphi_*(0) = \lambda$. Taking into account the properties of a singular integral in a neighborhood of point of discontinuity of the density (see [25, p.75]); we find

$$\frac{1}{2\pi i}\int_1^0 \frac{\psi(u)}{u-\nu} du = \frac{\psi(0)}{2\pi i}\log\nu + \psi_0(\nu), \qquad \psi_0(\nu) \in H(\mu).$$

By (21.3) we have $\mu_0 > \rho$, therefore (see [25, p.23])

$$(\varphi_*(u) - \lambda)u^{-\rho} \in H(\mu_0 - \rho), \quad \lim_{u \to 0} (\varphi_*(u) - \lambda)u^{-\rho} = 0.$$

Hence (see [25, p.58])

$$\int_1^0 \frac{\varphi_*(u) - \lambda}{u^\rho (u - \nu)} du \in H(\mu_0 - \rho),$$

and

$$\int_1^0 \frac{du}{u^\rho(u - \nu)} = -\frac{\pi}{\nu^\rho} \cot \rho\pi + \Omega(\nu), \quad \Omega(\nu) \in H(1).$$

Finally we have

$$\log X^+\Big(\frac{1}{\nu}\Big) = i\Big(-\frac{\psi(0)}{2\pi} \log \nu + \frac{\pi\lambda}{\nu^\rho}\Big) - \frac{\pi\lambda}{\nu^\rho} \cot \rho\pi + Q(\nu),$$

where $Q(v) \in H(\sigma)$. To complete the proof one has to return to the initial variables τ and t.

Comparing relations (21.15) and (21.16), define $\Delta = \lambda \cos \rho\pi$. Then we have

$$\log \psi_0^+(t) = i(\eta \log t + \lambda\pi t^\rho) + q(t) + t \int_0^\infty \frac{n(x) - \Delta x^\rho}{x(x+t)} dx. \tag{21.17}$$

Now one has to choose $n(x)$ "close enough" to Δx^ρ such that

$$I(t) = t \int_0^\infty \frac{n(x) - \Delta x^\rho}{x(x+t)} dx \in H_{[c,\infty)}(\rho).$$

Setting $u = \Delta x^\rho$ and $s = \Delta^{1/\rho} t$ and denoting $\tilde{n}(u) = n(\Delta^{-1/\rho} u^{1/\rho})$, we have

$$I(t) = \frac{s}{\rho} \int_0^\infty \frac{\tilde{n}(u) - u}{u(u^{1/\rho} + s)} du = \frac{1}{\rho} I_0(s). \tag{21.18}$$

Let us define the step-function $\tilde{n}(u)$, setting $\tilde{n}(u) = [u + 1/2]$, which corresponds to the system of equalities

$$n(x) = \Big[\Delta x^\rho + \frac{1}{2}\Big], \quad r_n = \Big(\frac{2n-1}{2\lambda \cos \rho\pi}\Big)^{\frac{1}{\rho}}, \quad n = 1, 2, \ldots.$$

This completes the construction of the functions $F_0(z)$ and $\Psi_0(z)$.

22 Solution of the non-homogeneous problem.

Theorem 22.1 *Let*

$$F_0(z) = \prod_{n=1}^{\infty}\Big(1+\frac{z}{r_n}\Big), \quad r_n = \Big(\frac{2n-1}{2\lambda\cos\rho\pi}\Big)^{1/\rho}.$$

Then both the solutions of the homogeneous problem

$$\Psi_0(z) = X(z)F_0(z),$$

and the solution of the non-homogeneous problem

$$\Phi_0(z) = \frac{\Psi_0(z)}{2\pi i}\int_1^{\infty}\frac{g(x)dx}{\Psi_0^+(x)(x-z)}, \tag{22.1}$$

belong to the class B.

We need several auxiliary statements.

Lemma 22.1 *If continuous derivative a function $f(t)$ has continuous derivative on the segment $[c,\infty)$ which satisfies $|f'(t)| < Ct^{-1-\rho}$, for some ρ, $0<\rho\le 1$, then $f(t)\in H_{[c,\infty)}(\rho)$.*

The assertion of the lemma follows from the inequality

$$|f(t_1)-f(t_2)| = |\int_{t_1}^{t_2} f'(t)dt| \le C\Big|\frac{1}{t_1^{\rho}}-\frac{1}{t_2^{\rho}}\Big| \le C\Big|\frac{1}{t_1}-\frac{1}{t_2}\Big|^{\rho}.$$

Lemma 22.2 *Let $f(x)$ and $\varphi(x)$ be real-valued functions on the segment $[c,\infty)$. Suppose that the function $f(t)$ is Riemann integrable on every finite segment $[c,t]$ and*

$$\Big|\int_c^t f(x)\varphi(x)dx\Big| < C,$$

and also the function $\varphi(x)$ is continuous, and has at most n extremal points on $[c,\infty)$ and $\varphi(\infty)=0$. Then

$$\Big|\int_c^t f(x)\varphi(x)dx\Big| < C(2n+2)\max_{c\le x<\infty}|\varphi(x)|$$

for any $t\in[c,\infty)$.

Proof. Take $b > c$ sufficiently large and find $n+2$ points $a = x_0 < x_1 < \ldots < x_{n+1} = b$ such that the function $\varphi(x)$ is monotonic on each interval $[x_k, x_{k+1}]$. The second mean value theorem gives

$$|\int_c^b f(x)\varphi(x)dx| = |\sum_{k=0}^{n}\{\varphi(x_k)\int_{x_k}^{\xi_k} f(x)dx +$$

$$+\varphi(x_{k+1})\int_{\xi_k}^{x_{k+1}} f(x)dx\}| \leq (2n+2)C \max_{c\leq x<\infty} |\varphi(x)|.$$

In the case $b = \infty$ the convergence of the integral follows from the fact that $\varphi(x)$ approaches zero monotonically as $x \to +\infty$, $x \in [x_{n+1}, \infty)$.

Lemma 22.3 *If* $n(x) = [\Delta x^\rho + 1/2]$, *then*

$$I(t) = t\int_0^\infty (n(x) - \Delta x^\rho)\frac{dx}{x(x+t)} \in H_{[c,\infty)}(\rho).$$

Proof. Let $s = t\Delta^{1/\rho}$ and $I_0(s)$ be connected with $I(t)$ by relation (2.18). It suffices to prove that $I_0(s) \in H(\rho)$. We have

$$I_0'(s) = \int_0^\infty \{\tilde{n}(u) - u\}\frac{u^{1/\rho-1}}{(u^{1/\rho}+s)^2}du, \quad \tilde{n}(u) = \left[u + \frac{1}{2}\right].$$

For a fixed $\rho > 0$ denote

$$f(u) = \tilde{n}(u) - u, \quad \varphi(u) = u^{1/\rho \ 1}(u^{1/\rho} + s)^{-2}.$$

The function $\varphi(u)$ has the only extremal point in $[0, \infty)$. This is the maximal point. We have

$$\max_{0\leq u<\infty} \varphi(u) = \frac{1}{4}s^{-1-\rho}(1-\rho)^{1-\rho}(1+\rho)^{1+\rho} < \frac{1}{2}s^{-1-\rho},$$

$$\varphi(\infty) = 0.$$

Since

$$|\int_0^t f(u)du| \leq \frac{1}{8}$$

for any $t > 0$, Lemma 22.2 yields $|I_0'(s)| \leq \frac{5}{16}s^{-1-\rho}$ and, by Lemma 22.1, $I_0(s) \in H_{[c,\infty)}(\rho)$.

Lemma 22.4 *Let* $f(t) \in H_{[c,\infty)}(\mu)$, $f(\infty) = 0$ *and* σ *be a real number. Then*

$$\psi_1(t) = f(t)e^{i\sigma \log t} \in H_{[c,\infty)}(\mu),$$

$$\psi_2(t) = f(t)e^{i\sigma t^\rho} \in H_{[c,\infty)}\left(\frac{\mu}{\rho+1}\right).$$

Proof. We restrict ourselves to considering the function $\psi_2(t)$. The function $\psi_1(t)$ may be studied in the same way and even a little simpler. To have a finite segment $[0, 1/c]$ let us change the variables by setting $t = 1/u$ and denote $g(u) = \psi_2(1/u)$, $h(u) = f(1/u)$. For $0 < u_1 < u_2 < 1/c$ we have

$$|g(u_2) - g(u_1)| \leq |h(u_2) - h(u_1)| + |h(u_1)||\exp(-i\sigma u_2^{-\rho})-$$

$$-\exp(-i\sigma u_1^{-\rho})| \leq C(u_2 - u_1)^{\mu} + C_1 u_1^{\mu} \min\{2, \sigma\rho(u_2 - u_1)u_1^{-1-\rho}\}.$$

If $2 \leq \sigma\rho(u_2 - u_1)u_1^{-1-\rho}$, then $u_1^{\mu} \leq (\sigma\rho/2)^{\mu/(\rho+1)}(u_2 - u_1)^{\mu/(\rho+1)}$ and $\min\{2, \sigma\rho(u_2 - u_1)u_1^{-1-\rho}\} = 2$, that which implies $g(u) \in H_{[0,1/c]}(\mu/(\rho+1))$. If $2 \geq \sigma\rho(u_2 - u_1)u_1^{-1-\rho}$, then

$$u_1^{\mu-\rho-1} \leq (\sigma\rho/2)^{(\mu-\rho-1)/(\rho+1)}(u_2 - u_1)^{(\mu-\rho-1)/(\rho+1)}$$

and

$$\min\{2, \sigma\rho(u_2 - u_1)u_1^{-1-\rho}\} = 6\rho(u_2 - u_1)u_1^{-1-\rho},$$

so we again obtain the desired conclusion.

Proof of Theorem 22.1. The class $H_{[c,\infty)}$ consists of bounded functions. In particular, the function $g(t)$ from relation (22.17) is bounded. Therefore, taking the real part of (22.17) and taking into account Lemma 22.3, we obtain $|\Psi_0^+(t)| \leq C \leq \infty$, $c \leq t < \infty$. By (18.5) we have $0 < C_1 \leq |G(t)| \leq C_2 < \infty$, $c \leq t < \infty$ and relation (18.6) yields $|\Psi_0^-(t)| \leq C$, $c \leq t < \infty$. Since $\Psi_0(z) = X(z)F_0(z)$ where $X(z) \in B$ and $F_0(z)$ is an entire function, then the functions $\Psi_0^{\pm}(t)$ are bounded on $[1, c]$ and, hence, on $[1, \infty)$. Since $\rho_{F_0} = \rho < 1/2$,then by applying the Phragmen-Lindelöf principle, we obtain $\Psi_0(z) \in B$.

Now using (21.17) and Lemma 22.3 we see that

$$\frac{1}{\Psi_0^+(t)} = Q(t)\exp\{-i(\eta \log t + \lambda\pi t^{\rho})\}, \quad c \leq t \leq \infty,$$

where $Q(t) \in H_{[c,\infty)}$. Since $g(t) \in H$ and $g(\infty) = 0$, Lemma 22.4 gives

$$g(t)/\Psi_0^+(t) \in H_{[c,\infty)}, \quad \lim_{t\to\infty} g(t)/\Psi_0^+(t) = 0. \tag{22.2}$$

Rewrite equality (22.1) in the form

$$\Phi_0(z) = \frac{\Psi_0(z)}{2\pi i}\left[\int_1^c + \int_c^{\infty}\right]\frac{g(x)dx}{\Psi_0^+(x)(x - z)}.$$

It follows from (22.2) that the second integral is bounded in any domain $|z-c|>\epsilon$. The first one may be estimated as

$$\int_1^c \frac{g(x)dx}{\Psi_0^+(x)(x-z)} = O\Big(\frac{1}{|z|}\Big), \quad z\to\infty,$$

and bounded in any domain of the form $\{z:\ |z-1|>\epsilon\}\cap\{z:\ |z-c|>\epsilon\}$. Since $\Psi_0(z)\in B$ and c may be chosen arbitrarily close to 1, the function $\Phi_0(z)$ is bounded in $|z-1|>\epsilon$.

It remains to study the behavior of $\Phi_0(z)$ near the point $t=1$. [19] Since $F_0(z)$ does not vanish in a neighborhood of the point $z=1$, we may represent the function $\log\Psi_0(z)$ in the form

$$\log\Psi_0(z) = -\frac{\log G(1)}{2\pi i}\log(z-1)+$$

$$+\frac{1}{2\pi i}\int_1^c \frac{\log G(x)-\log G(1)}{x-z}dx + f_0(z), \qquad (22.3)$$

for all z from a sufficiently small domain $D\cap\{z:\ |z-1|<c\}$, where $f_0(z)$ is analytic in $|z-1|<c$.

Denote

$$-\log G(1)/(2\pi i) = \alpha+\beta i,$$

where by virtue of (21.4) we have $0\le\alpha<1$ due to (21.4). Using the properties of boundary values of the Cauchy-type integrals (see[25,Sect. 22]) we have

$$\log\Psi_0^+(t) = (\alpha+\beta i)\log(t-1)+P(t), \quad 1\le t\le 1+\delta,$$

for sufficiently small $\delta>0$, where $P(t)\in H_{[1,1+\delta]}$. This yields the representation

$$g(t)/\Psi_0^+(t) = \psi(t)(t-1)^{-\alpha-i\beta}, \quad 1\le t\le 1+\delta,$$

where $\psi(t)\in H$, and also, by (21.7), we have if $\alpha+i\beta=0$, then $\psi(1)=0$. Therefore (see [25, p.75]) we have

$$\int_1^\infty \frac{g(x)dx}{\Psi_0^+(x)(x-z)} = \int_1^{1+\delta} + \int_{1+\delta}^\infty = \frac{Q}{(z-1)^{\alpha+i\beta}} + \varphi_0(z), \qquad (22.4)$$

for sufficiently small $|z-1|$. Here $Q=0$ if $\alpha+i\beta=0$ and the function $\varphi_0(z)$ is either bounded if $\alpha=0$, or admits the estimate $|\varphi(z)|=O(|z-1|^{-\alpha_0})$,

[19] There is no verticity at the point $t=1$. Therefore the function $\varphi_0(z)$ may be studied by traditional methods (see [25, Sect. 80]).

$0 < \alpha_0 < \alpha$, $z \to 1$ if $\alpha > 0$. Now (22.3) and (22.4) yield the boundedness of the function $\Phi_0(z)$ in a neighborhood of the point $z = 1$. This completes the proof of Theorem 22.1.

Let us now give the general expression for solutions of the non-homogeneous problem in the class B.

Theorem 22.2 *Let conditions(21.2)–(21.7) be fulfilled . Then the non-homogeneous Riemann boundary problem (21.1) has infinitely many solutions in the class B. The general formula of the solution is as follows:*

$$\Phi(z) = X(z)\Big\{\frac{F_0(z)}{2\pi i}\int_1^\infty \frac{g(x)dx}{X^+(x)F_0(x)(x-z)} + F(x)\Big\}, \qquad (22.5)$$

where

$$F_0(z) = \prod_{n=1}^{\infty}\Big(1+\frac{z}{r_n}\Big), \quad r_n = \Big(\frac{2n-1}{2\lambda\cos\rho\pi}\Big)^{1/\rho},$$

and $F(z)$ is an entire function, whose order does not exceed ρ, and such that

$$\log|F(t)| \le (\pi\lambda\cot\pi\rho)t^\rho + O(1), \quad t \to +\infty. \qquad (22.6)$$

Proof. The assertion of the theorem with relation (20.2) instead of (22.6) directly follows from Theorems 20.4 and 22.1. But under condition (21.3) relations (20.2) and (22.6) are equivalent. Actually the arguments that are used when proving Lemma 21.1 imply, that under, assumption (21.3), we have

$$t\int_1^\infty \frac{\varphi(x)dx}{x^{1-\rho}(x-t)} = -(\lambda\pi\cot\pi\rho)t^\rho + p(t), \quad p(t) \in H_{[c,\infty)}.$$

Remark. In contrast to the homogeneous problem, the non-homogeneous problem (21.1), generally speaking, does not have solutions of positive order of decreasing. This follows from the inequalities

$$|g(t)| \le |G(t)||\Phi^-(t)| + |\Phi^+(t)|, \quad |G(t)| < C$$

that imply that the functions $|\Phi^\pm(t)|$ cannot simultaneously decay faster than $g(t)$.

RIEMANN BOUNDARY PROBLEM WITH INFINITE INDEX IN THE CASE OF VERTICITY OF INFINITE ORDER

23 Statement of the homogeneous problem

Let D be the complex plane cut along the ray $L = \{t:\ 1 \leq t < \infty\}$.

Consider the boundary problem in D

$$\Phi^+(t) = G(t)\Phi^-(t), \quad 1 < t < \infty, \tag{23.1}$$

under the following assumptions

1) $\arg G(t) = 2\pi\varphi(t)t^\rho,$ (23.2)

where

$$\rho > 0, \quad \varphi(t) \in H_{[1,\infty)}(\mu_0),$$

$$\varphi(\infty) = \lambda > 0, \quad -1 < \varphi(1) \leq 0, \tag{23.3}$$

2) $\log|G(t)| \in H_{[1,\infty)}$. (23.4)

In contrast to Chapter 4 we do not demand $0 < \rho < 1/2$, but we suppose that the exponent μ_0 in (23.3) satisfies

$$\max\left\{0, \frac{2p-1}{2p+1}\right\} < \mu_0 \leq 1. \tag{23.5}$$

In the case $\rho \leq 1/2$ the latter condition is equivalent to $0 < \mu_0 \leq 1$, i.e. conditions (23.2)–(23.5) coincide with (18.3)–(18.5) for $L = \{t:\ 1 \leq t < \infty\}$ and condition (23.5) does not impose additional restriction.

In the previous chapter, in the case $0 < \rho < 1/2$, we have described all the solutions of problem (23.1) in the class B (Theorem 20.4) and also all the solutions in the class of functions that decrease with the order ρ (Theorem 20.5). By Theorem 1.1 the order of any function from the class B does not exceed $1/2$. Therefore there exist no solutions of order of decreasing $\rho > 1/2$.A more precise Theorem 4.15 provides the absence of solutions of order of decreasing $\rho = 1/2$ as well. So one can look for solutions of order of decreasing $\sigma < \min(\rho, 1/2)$. In this chapter we describe such solutions. More exactly, we shall describe all solutions of problem (23.1) not in the class B only, but also in the subclasses

$$B_\sigma = B \cap A_{\sigma,0,2\pi}, \quad \bar{B}_\sigma = B \cap \bar{A}_{\sigma,0,2\pi},$$

$$B^*_\sigma = B \cap A^*_{\sigma,0,2\pi}, \quad \bar{B}^*_\sigma = B \cap \bar{A}^*_{\sigma,0,2\pi}, \tag{23.6}$$

where the indexed letters A denote the corresponding classes of functions of c.r.g., introduced in Sect. 4 of Ch. 2. Naturally, according to Theorem 1.1, these classes differ from zero only for $0 \leq \sigma \leq 1/2$. It should also be noted that, by Theorem 4.15, we have $B^*_{1/2} = \bar{B}^*_{1/2} = B$. When looking for solutions from the classes (23.6), we need an additional restriction on the exponent μ_0, namely

$$\max\left\{0, \frac{\rho - \sigma}{1 + \rho - \sigma}\right\} < \mu_0 \leq 1. \tag{23.7}$$

For $\sigma = 1/2$ this condition is equivalent to (23.5). Conditions (23.5) and (23.7) are essential.In Sect. 28 we show that if these conditions fail then problem (23.1), is generally speaking, unsolvable in the appropriate class.

24 Canonical function

Denote

$$X(z) = \exp\Big\{\frac{z^{q+1}}{2\pi i}\int_1^\infty \frac{\log G(t)}{t^{q+1}(t-z)}dt\Big\}, \quad q = [\rho]. \tag{24.1}$$

By virtue of (23.2)–(23.4) we have $\log G(t) = O(t^\rho)$, $t \to \infty$. Therefore the integral in the right-hand side converges. The function $X(z)$ is called *the canonical function of the Riemann boundary problem* (23.1). In the case $\rho < 1/2$, the function $X(z)$ coincides with that of (19.1) and the notion of the canonical function coincides with that of Sect. 19.

Let us establish the following properties of the function $X(z)$.

1) $X(z)$ is analytic in D and continuous up to the open ray $\tilde{L} = \{t : 1 < t < \infty\}$, and also does not vanish in $D \cup \tilde{L}$.

2) $X(z)$ satisfies boundary condition (23.1) i.e.,

$$X^+(t) = G(t)X^-(t), \quad t \in \tilde{L},$$

3) $X(z)$ is normalized by the condition

$$X(z) = 1 + O(|z|^{q+1}), \quad z \to 0.$$

4) The estimate

$$C_1|z-1|^\alpha \le |X(z)| \le C_2, \quad 0 \le \alpha < 1$$

holds in a neighborhood of the starting point $z = 1$ of the curve L.

5) The order of the growth of $X(z)$ in D does not exceed ρ.

We see that properties 1)–4) coincide with the corresponding properties of the canonical function constructed in Sect. 19. As for property 5) from Sect. 19, it fails for $\rho \ge 1/2$. Below we show that, for $\rho > 1/2$, the function $X(z)$ is unbounded and does grow with the order ρ. In the case $\rho = \frac{1}{2}$, the function $X(z)$, generally speaking, may belong to B, but cannot decrease with the order $1/2$ in the angle $[0, 2\pi]$.

One may verify properties 1), 2), 4) in the same way as in Sect. 19. Property 3) is evident. Property 5) is a consequence of the following theorem.

Theorem 24.1 *The canonical function $X(z)$ has the order ρ and satisfies the following conditions*

$$\lim_{r\to\infty} r^{-\rho}\log|X(re^{i\theta})| = -\frac{\pi\lambda}{\sin\pi\rho}\cos\rho(\theta-\pi), \tag{24.2}$$

for non-integer ρ, and

$$\lim_{r\to\infty} r^{-\rho}(\log r)^{-1}\log|X(re^{i\theta})| = -\lambda\cos\rho\theta, \tag{24.3}$$

for integer ρ. In the both cases the convergence is uniform with respect to θ, $0 \leq \theta \leq 2\pi$.

Proof. First, let ρ be a non-integer. We have

$$\log X(z) = z^{q+1} \int_1^\infty \frac{\varphi_*(t)dt}{t^{1+q-\rho}(t-z)},$$

where

$$\varphi_*(t) = \varphi(t) + \frac{1}{2\pi i} t^{-\rho} \log |G(t)|.$$

By virtue of conditions (23.2)–(23.4), we obtain $\varphi_*(t) \in H_{[1,\infty)}$ and $\varphi_*(\infty) = \lambda > 0$. Well-known properties of the Cauchy integral with a singularity with power growth (see [25, p.75]) provide

$$\log X(z) = -\frac{\pi\lambda}{\sin \pi\rho} e^{-i\rho\pi} z^\rho + O(|z|^{\rho-\epsilon}), \quad z \to \infty,$$

for some ϵ and $0 \leq \arg z < 2\pi$. Therefore relation (24.2) is fulfilled uniformly with respect to $\theta \in [0, 2\pi]$.

Now let ρ be an integer. Relation (24.1) takes the form

$$\log X(z) = z^{\rho+1} \int_1^\infty \frac{\varphi_*(t)dt}{t(t-z)}.$$

Therefore by the properties of the Cauchy integral with a power singularity (see [25, p.75]) we get

$$\log X(z) = -\lambda z^\rho \log z + O(|z|^\rho),$$

$$z \to \infty, \quad 0 \leq \arg z \leq 2\pi.$$

and, clearly, relation (24.3) is fulfilled uniformly with respect to $\theta \in [0, 2\pi]$.

Corollary 24.1 *For $\rho > 1/2$ the canonical function $X(z)$ is unbounded and the order of its growth equals ρ.*

To prove this statement one has to mention that, for $\rho > 1/2$, there exists a $\theta \in (0, 2\pi)$ such that either the right-hand side of (24.2)is positive, when ρ is not an integer, or the right-hand side of (24.3) is positive for integer ρ.

Now let us prove that properties 1)–5) *uniquely determine the canonical function $X(z)$*. In other words, let a function $X_0(z)$ possesses these properties. Then the function $F(z) = \frac{X_0(z)}{X(z)}$ is analytic in D. Besides, this function is continuous and non-vanishing in $D \cup L$ except, possibly, at the points $z = 1$ and $z = \infty$. By property 2) we have $F^+(t) = F^-(t)$, $1 < t < \infty$, so the function $F(z)$ is analytic in $0 < |z-1| < \infty$. Using property 4) one can easily see that the point $z = 1$ is a removable singularity, so $F(z)$ is a non-vanishing entire function. Theorem 24.1 gives a lower bound for the function $X(z)$, namely $\log|X(z)| > -|z|^{\rho+\epsilon}$, $|z| > r_\epsilon$, and, since the function $X_0(z)$ possesses property 5), we conclude that $F(z)$ has order no greater than ρ and, hence, $F(z) = \exp\{a_0 z^q + \ldots + a_q\}$. Using the fact that the both functions $X(z)$ and $X_0(z)$ possess property 3) we obtain $a_0 = \ldots = a_q = 0$, i.e. $X(z) = X_0(z)$.

The function $X(z)$ solves boundary problem (23.1) in the class $\tilde{B}$ but, due to Corollary to Theorem 24.1, does not belong to the class B for $\rho > 1/2$. The general expression for the solution of problem (23.1) in the class $\tilde{B}$ is given by the following theorem, which may be proved in the same way as Theorem 20.1.

Theorem 24.2 *Any solution of problem (23.1) in the class $\tilde{B}$ has the form $\Phi(z) = X(z)F(z)$, where $F(z)$ is an arbitrary entire function and $X(z)$ is the canonical function defined in (24.1).*

Now let us look for conditions on the function $F(z)$, that ensure that the product $F(z)X(z)$ belongs to the class B. The following theorem gives a simple necessary condition.

Theorem 24.3 *If the function $\Phi(z) = X(z)F(z)$ belongs to the class B, then the order ρ_F of the function $F(z)$ does not exceed ρ and for $\rho > 1/2$, $\rho_F = \rho$.*

One may prove the inequality $\rho_F \le \rho$ in the same way as in Theorem 20.2 with relations (24.2)and (24.3) instead of (19.9). Now, let $\rho_F < \rho$ and $\rho > 1/2$. Take ϵ such that $\rho_F < \rho - \epsilon$ and $\rho - \epsilon > 1/2$. Using Theorem 24.1, we obtain

$$\limsup_{r\to\infty} r^{-\rho+\epsilon} \log|\Phi(re^{i\theta})| = \limsup_{r\to\infty} r^{-\rho+\epsilon} \log|F(re^{i\theta})| +$$

$$+ \lim_{r\to\infty} r^{-\rho+\epsilon} \log|X(re^{i\theta})| \not\equiv 0.$$

So, $\rho_\Phi \ge \rho - \epsilon > 1/2$, which contradicts Theorem 1.1 since $\Phi(z) \in B$.

25 Asymptotic properties of zero sets of solutions of the homogeneous problem from the classes B and B^*_σ

Since the canonical function does not vanish in the complex plane (except possibly at the point $z = 1$, in which it may have a zero of an order less than 1), then the zero set of any solution of problem (23.1) from the class B coincides with that of the corresponding entire function $F(z)$. Let us consider the properties of this set.

Theorem 25.1 *Let $\Phi(z) \in B$ be a solution of problem (23.1) and $r_n e^{i\theta_n}$, $0 < \theta_n < 2\pi$, be its zeros. Then*

$$\sum_{r_n \neq 0} \frac{1}{\sqrt{r_n}} \sin \frac{\theta_n}{2} < \infty. \tag{25.1}$$

Proof. The function $f(\zeta)\Phi(\sqrt{\zeta})$ is analytic and bounded in the half-plane $\Im z > 0$. According to Theorem 3.1 its zeros satisfy the first condition in (3.4). In terms of zeros $r_n \exp(i\theta_n)$ of the function $\Phi(z)$ this condition may be written as

$$\sum_n \frac{\sqrt{r_n} \sin \frac{\theta_n}{2}}{r_n + 1} < \infty.$$

Since the functions $F(z)$ and, respectively, $\Phi(z)$ have at most finite number of zeros in the unit disk, the latter inequality is equivalent to (25.1).

It should be noted, that the left-hand side of (25.1) would not be altered if one includes boundary zeros of the function $\Phi(z)$, i.e. zeros located on the ray $L = \{t : \ 1 \leq t < \infty\}$. For every such zero we have either $\theta_n = 0$ or $\theta_n = 2\pi$. Thus relation (25.1) is fulfilled independently of the density of the set of the boundary zeros. But this density cannot be arbitrary, since Theorem 24.3 yields $\rho_F \leq \rho$ and, hence, the counting function $n_\Phi(r)$ of the zeros of the function $\Phi(z)$ satisfies

$$n_\Phi(r) = n_F(r) \leq \log M_F(er) + O(1) = O(r^{\rho+\epsilon})$$

as $r \to \infty$. So Theorem 25.1 does not give complete description of distribution of zeros of solutions $\Phi(z) \in B$.

Let $\Phi(z)$ be a solution of problem (23.1) from the class B^*_σ, $0 < \sigma \leq 1/2$. Recall that $B^*_{1/2} = B$ (see Sect. 23). Below we study the properties of the counting function $n_\Phi(r)$ of its zeros. In what follows we assume for convenience that $\log G(t) = 0$ for $0 \leq t \leq 1$. We need several lemmas.

Lemma 25.1 *Let $\Phi(z) \in \tilde{B}$ be a solution of problem (23.1). Then the following representation holds in each domain D_b of the form $D_b = D \cap \{|z| < b\}$:*

$$\Phi(z) = F_b(z)\exp(I_b(z)), \quad I_b(z) = \frac{z}{2\pi i}\int_0^b \frac{\log G(t)}{t(t-z)}dt, \tag{25.2}$$

here $F_b(z)$ is analytic in the disk $|z| < b$.

We omit the proof of the lemma, since it is practically the same as those of Theorems 20.1 and 24.2.

Lemma 25.2 *The function $I_b(z)$, defined by (25.2), satisfies the equality*

$$\Re \int_0^{2\pi} I_b(re^{i\theta})d\theta = -\int_0^r \arg G(t)\frac{dt}{t} \tag{25.3}$$

for any r, b such that $1 < r < b$.

Proof. Denote by T the left-hand side of (25.3). Taking $0 < \epsilon < r-1$, we have ($z = re^{i\theta}$):

$$T = \Im \int_{|z|=r} \frac{dz}{2\pi i}\int_0^{r-\epsilon} \frac{\log G(t)}{t(t-z)}dt +$$

$$+ \Im \int_{|z|=r} \frac{dz}{2\pi i}\int_{r-\epsilon}^{r+\epsilon} \frac{t^{-1}\log G(t) - r^{-1}\log G(r)}{t-z}dt +$$

$$+\Im \frac{\log G(r)}{2\pi i r}\int_{|z|=r} dz \int_{r-\epsilon}^{r+\epsilon}\frac{dt}{t-z} + \Im \int_{|z|=r}\frac{dz}{2\pi i}\int_{r+\epsilon}^b \frac{\log G(t)}{t(t-z)}dt =$$

$$= T_1 + T_2 + T_3 + T_4.$$

Changing the order of integration, we obtain

$$T_1 = -\int_0^{r-\epsilon} \arg G(t)\frac{dt}{t}, \quad T_4 = 0.$$

For $t \in [r-\epsilon, r+\epsilon]$, we have

$$|t^{-1}\log G(t) - r^{-1}\log G(r)| \le$$

$$\le C_1|t-r|^{\nu}/r, \quad 0 < \nu \le 1,$$

hence

$$|T_2| \le C_2\int_{r-\epsilon}^{r+\epsilon}\frac{dt}{|t-r|^{1-\nu}} = O(\epsilon^{\nu}), \quad \epsilon \to 0.$$

and, finally,

$$|T_3| \le \frac{|\log G(r)|}{2\pi} \int_0^{2\pi} d\theta \int_{r-\epsilon}^{r+\epsilon} \frac{dt}{|t - re^{i\theta}|} =$$

$$= |\log G(r)| \int_{r-\epsilon}^{r+\epsilon} \eta(r/t) t^{-1} dt,$$

where

$$\eta(x) = \frac{1}{2\pi} \int_0^{2\pi} |1 - xe^{i\theta}|^{-1} d\theta, \quad 0 \le x < \infty.$$

The function $\eta(x)$ is summable on any finite interval (see [19, p.289]). Therefore $T_3 \to 0$ as $\epsilon \to 0$ and we have

$$T = -\int_0^{r-\epsilon} \arg G(t) \frac{dt}{t} + o(1), \quad \epsilon \to \infty,$$

which yields the assertion of the lemma.

The following important theorem may be considered as a generalization of the well-known Jensen theorem.

Theorem 25.2 *Let $\Phi(z) \in \tilde{B}$ be a solution of problem (23.1). Then, for any $r > 0$, we have*

$$\frac{1}{2\pi} \int_0^{2\pi} \log |\Phi(re^{i\theta})| d\theta =$$

$$= \int_0^r \left(\tilde{n}_\Phi(t) - \frac{1}{2\pi} \arg G(t) \right) \frac{dt}{t} + c_\Phi(r),$$

$$c_\Phi(r) = \log\left(\frac{1}{k!} r^k |\Phi^{(k)}(0)| \right), \tag{25.4}$$

where $\tilde{n}_\Phi(r)$ is the number of zeros of the function $\Phi(z)$ in the annulus $0 < |z| \le r$ taken with regard to their multiplicities and k is the multiplicity of zero of the function $\Phi(z)$ at the point $z = 0$.

Proof. For an arbitrary $b > r$, applying Lemma 25.1, we obtain

$$\frac{1}{2\pi} \int_0^{2\pi} \log |\Phi(re^{i\theta})| d\theta =$$

$$= \frac{1}{2\pi} \int_0^{2\pi} \log |F_b(re^{i\theta})| d\theta + \frac{1}{2\pi} \int_0^{2\pi} \Re I_b(re^{i\theta}) d\theta.$$

The function $F_b(z)$ is analytic in the disk $|z| < b$ and also

$$\tilde{n}_\Phi(r) = n_F(r) - n_F(0), \quad I_b(0) = 0.$$

Therefore the Jensen theorem yields

$$\frac{1}{2\pi}\int_0^{2\pi}\log|F_b(re^{i\theta})|d\theta=\int_0^r\tilde{n}_\Phi(t)\frac{dt}{t}+c_\Phi(r).$$

To obtain relation (25.4) it remains to use Lemma 25.2.

If $\Phi(z)$ belongs to the class $\bar{B}^*_\sigma$, then Theorem 25.2 provides existence of the limit

$$\lim{}^*_{r\to\infty}r^{-\sigma}\int_0^r\Big(\tilde{n}_\Phi(t)-\frac{1}{2\pi}\arg G(t)\Big)\frac{dt}{t}=$$

$$=\frac{1}{2\pi}\int_0^{2\pi}h_\Phi(\alpha)d\alpha \tag{25.5}$$

(The definition of $\lim^*$ see in Sect. 4.) Below we shall prove that this limit exists for solutions from the class B^*_σ as well, whereas the usual limit may not exist even for solutions $\Phi(z)\in\bar{B}^*_\sigma$. (The corresponding example is rather cumbersome and we do not give it here). One may improve the situation by introducing an additional averaging.

Theorem 25.3 *Let $\Phi(z)\in B^*_\sigma$, $0<\sigma\le 1/2$, be a solution of problem (23.1). Then there exists the finite non-negative limit*

$$\lim_{r\to\infty}r^{-\sigma}\int_0^r\frac{dt}{t}\int_0^t\Big(\frac{1}{2\pi}\arg G(x)-\tilde{n}_\Phi(x)\Big)\frac{dx}{x}=$$

$$=-\frac{1}{2\pi\sigma}\int_0^{2\pi}h_\Phi(\alpha)d\alpha, \tag{25.6}$$

where $h_\Phi(\alpha)$ is the indicator with respect to the formal order σ.

Proof. According to Theorem 25.2 we have

$$r^{-\sigma}\int_1^r\frac{dt}{t}\int_0^t\Big(\frac{1}{2\pi}\arg G(x)-\tilde{n}_\Phi(x)\Big)\frac{dx}{x}=$$

$$=-r^{-\sigma}\int_1^r\frac{dt}{t}\cdot\frac{1}{2\pi}\int_0^{2\pi}\log|\Phi(re^{i\theta})|d\theta+r^{-\sigma}\int_1^r c_\Phi(t)\frac{dt}{t}=$$

$$=-\frac{1}{2\pi}\int_0^{2\pi}\Big\{r^{-\sigma}\int_1^r\log|\Phi(te^{i\theta})|\frac{dt}{t}\Big\}d\theta+O(r^{-\sigma}\log^2 r). \tag{25.7}$$

By Corollary 1 to Theorem 6.1 (see relation (6.15)) for $1\le r<\infty$, $0\le\theta\le 2\pi$, the absolute value of the expression in the braces is bounded by a constant independent of r and θ. By Corollary 2 to the same theorem this expression approaches $\sigma^{-1}h_\Phi(\theta)$ as $r \to \infty$, $0 < \theta < 2\pi$. Applying the Lebesgue theorem on bounded convergence, we obtain relation (25.6). It remains to note that $h_\Phi(\alpha) \leq 0$, since $\Phi(z) \in B^*_\sigma \subset B$.

Since $B^*_{1/2} = B$, Theorem 25.3 directly implies

Corollary 25.1 *If $\Phi(z)$ is a solution of the homogeneous Riemann boundary problem (i.e. a solution of (23.1) in the class B), then there exists the finite limit*

$$\lim_{r\to\infty} \frac{1}{\sqrt{r}} \int_0^r \frac{dt}{t} \int_0^t \left(\frac{1}{2\pi} \arg G(x) - \tilde{n}_\Phi(x)\right) \frac{dx}{x} \geq 0. \tag{25.8}$$

Another important corollary of Theorem 25.3 is the following.

Corollary 25.2 *If $\Phi(z) \in B^*_\sigma$, $0 < \sigma \leq 1/2$ is a solution of problem (23.1), then these exists*

$$\lim{}^*_{r\to\infty} r^{-\sigma} \int_0^r \left(\tilde{n}_\Phi(t) - \frac{1}{2\pi} \arg G(t)\right) \frac{dt}{t} =$$

$$= \limsup_{r\to\infty} r^{-\sigma} \int_0^r \left(\tilde{n}_\Phi(t) - \frac{1}{2\pi} \arg G(t)\right) \frac{dt}{t} = \frac{1}{2\pi} \int_0^{2\pi} h_\Phi(\alpha) d\alpha. \tag{25.9}$$

Proof. Denoting

$$F(t) = \frac{1}{2\pi} \int_0^{2\pi} \log |\Phi(te^{i\theta})| d\theta,$$

and combining relation (25.6), (25.7) and Corollary 1 to Theorem 6.1 we see that $F(t)$ satisfies the hypothesis of Theorem 4.11 with $\rho = \sigma$ and

$$a = \frac{1}{2\pi} \int_0^{2\pi} h_\Phi(\alpha) d\alpha.$$

Applying this theorem we obtain the desired assertion.

Remark. If $\Phi(z)$ is a solution of problem (23.1) in the class B_σ then, for any $\epsilon > 0$,

$$\lim_{r\to\infty} r^{-\sigma+\epsilon} \int_0^r \left(\tilde{n}_\Phi(t) - \frac{1}{2\pi} \arg G(t)\right) \frac{dt}{t} = -\infty. \tag{25.10}$$

Actually, since the function $\Phi(z)$ is bounded in D and has the order σ, then σ is the order of decreasing and, at least for one value of θ, $0 < \theta < 2\pi$, we have

$$\lim_{r\to\infty} r^{-\sigma+\epsilon} \log |\Phi(re^{i\theta})| = -\infty.$$

By the properties of the indicator this relation holds for all θ, $0 < \theta < 2\pi$. Therefore (see Theorem 4.1), for any $N > 0$ and sufficiently large r, we have

$$\max_{\pi/2 \leq \theta \leq 3\pi/2} \log |\Phi(re^{i\theta})| < -Nr^{\sigma-\epsilon},$$

$$\frac{1}{2\pi} \int_0^{2\pi} \log |\Phi(re^{i\theta})| d\theta \leq -Nr^{\sigma-\epsilon} + O(1), \quad r \to \infty.$$

Now relation (25.12) follows from (25.4).

Corollary 25.3 *If $\Phi(z) \in B_\sigma$ is a solution of problem (23.1) and $\sigma < \rho$, then there exists the limit*

$$\lim_{t\to\infty} t^{-\rho} n_\Phi(t) = \lambda > 0,$$

where λ is the verticity coefficient (see (23.3)).

Proof. Let the contrary be true. Then one can find either a number $\epsilon > 0$ and a sequence $\{r_n\}$, $r_n \to \infty$, such that

$$n_\Phi(r_n) > (\lambda + \epsilon) r_n^\rho, \tag{25.11}$$

or a number $\epsilon > 0$ and a sequence $\{r_n'\}$, $r_n' \to \infty$, such that

$$n_\Phi(r_n') < (\lambda - \epsilon)(r')_n^\rho. \tag{25.12}$$

Let relation (25.11) hold. Take $\beta > 0$ sufficiently small such that

$$\lambda + \epsilon > \left(\lambda + \frac{2}{3}\epsilon\right)(1 + 3\beta)^\rho. \tag{25.13}$$

By Corollary 2 and the definition of $\lim^*$ for every n there exist points $r_{n,1}$ and $r_{n,2}$ such that $r_n \leq r_{n,1} \leq r_n(1+\beta)$, $r_n(1+2\beta) \leq r_{n,2} < r_n(1+3\beta)$ and

$$\lim_{n\to\infty} r_{n,k}^{-\sigma} \int_0^{r_{n,k}} \left(\tilde{n}_\Phi(t) - \varphi(t)t^\rho\right) \frac{dt}{t} = c \neq \infty, \quad k = 1, 2. \tag{25.14}$$

(Here we have used equality (23.2)).

Now, using monotonicity of the function $n_\Phi(r)$ and also relations (25.11) and (25.13), we have

$$n_\Phi(r) > (\lambda + \epsilon) r_n^\rho > (\lambda + \epsilon)(1 + 3\beta)^{-\rho} t^\rho > \left(\lambda + \frac{2}{3}\epsilon\right) t^\rho \tag{25.15}$$

for $r_{n,1} \leq t \leq r_{n,2}$.

Assume r_1 be sufficiently large so that $\varphi(t) < \lambda + \epsilon/2$ for $t \geq r_1$. Then, applying (25.14) and (25.15), we find

$$\int_0^{r_{n,2}} \big(\tilde{n}_\Phi(t) - \varphi(t)t^\rho\big)\frac{dt}{t} >$$

$$> \int_0^{r_{n,1}} \big(\tilde{n}_\Phi(t) - \varphi(t)t^\rho\big)\frac{dt}{t} + \frac{\epsilon}{3}\int_{r_{n,1}}^{r_{n,2}} t^{\rho-1}dt >$$

$$> \big(c + o(1)\big)r_{n,1}^\sigma + \frac{\epsilon}{3}r^\sigma\big((1+2\beta)^\rho - (1+\beta)^\rho\big),$$

whence, since $\sigma < \rho$, we obtain

$$\lim_{n\to\infty} r_{n,2}^{-\sigma}\int_0^{r_{n,2}} \big(\tilde{n}_\Phi(t) - \varphi(t)t^\rho\big)\frac{dt}{t} = +\infty,$$

in contrast to (25.14).

Relation (25.12) is considered in a similar way.

Corollary 25.4 *If $\rho > 1/2$, of the homogeneous problem (23.1) does not possess bounded solutions with a finite number of zeros.*

Theorem 25.4 *Let $\Phi(z) \in B$ be a solution of the homogeneous problem (23.1). Then the integral*

$$I_a = \int_0^\infty \Big\{\tilde{n}_\Phi(t) - \frac{1}{2\pi}\arg G(x)\Big\}x^{-\frac{3}{2}-p}dx \qquad (25.16)$$

converges (possibly, non-absolutely) for any $p > 0$.

The proof of the theorem uses the following lemma.

Lemma 25.3 *Let the hypothesis of Theorem 25.4 be fulfilled. Then*

$$\int_0^{2\pi} |\log|\Phi(re^{i\theta})||d\theta = O(\sqrt{r}\log r), \quad r\to\infty. \qquad (25.17)$$

First deduce the assertion of Theorem 25.4, using relation (25.17). Denote

$$s(r) = \int_0^r \Big\{\tilde{n}_\Phi(t) - \frac{1}{2\pi}\arg G(x)\Big\}\frac{dx}{x}.$$

By virtue of Theorem 25.2 relation (25.17) yields $s(r) = O(\sqrt{r}\log r)$. Since

$$\int_0^r \Big\{\tilde{n}_\Phi(t) - \frac{1}{2\pi}\arg G(x)\Big\}\frac{dx}{x^{p+3/2}} =$$

$$= \frac{s(r)}{r^{p+1/2}} + \left(\frac{1}{2} + p\right) \int_0^r \frac{s(x)}{x^{p+3/2}} dx,$$

we obtain the assertion of the theorem.

It should be noted that the integral in (25.16), generally speaking, does not converge absolutely. The corresponding example is rather cumbersome and we will omit it.

By relation (23.4) we have $|\log|G(t)|| < C$. Therefore Theorem 25.4 yields the following corollary.

Corollary 25.5 *If $\Phi(z) \in B$ is a solution of the homogeneous problem, then the integral*

$$A_p = \int_0^\infty \left\{\frac{1}{2\pi i} \log G(x) - \tilde{n}_\Phi(x)\right\} x^{-p-3/2} dx \tag{25.18}$$

converges (maybe, non-absolutely) for any $p > 0$.

Proof of Lemma 25.3. We restrict ourselves to estimating the integral which is obtained from that of (25.17) if integrating over the interval $(0, \pi)$. The integral over $[\pi, 2\pi]$ can be considered similarly.

First let us prove that

$$\lim_{r\to\infty} \left\{ \int_0^{r^{-\rho-1}} |\log|\Phi(re^{i\theta})||d\theta + \right.$$

$$\left. + \int_{\pi - r^{-\rho-1}}^{\pi} |\log|\Phi(re^{i\theta})||d\theta \right\} = 0. \tag{25.19}$$

By virtue of Theorem 24.3 we have $\log|\Phi(z)| = \log|X(z)| + \log|F(z)|$ and also $\rho_F \le \rho$.

Theorem 24.1 provides

$$\max_{0\le\theta\le 2\pi} |\log|X(re^{i\theta})|| = o(r^{\rho+\epsilon}), \quad r \to \infty, \quad 0 < \epsilon < 1. \tag{25.20}$$

According to the well-known theorem on meromorphic functions (see [19, p.58]) we have

$$\int_E |\log|F(re^{i\theta})||d\theta \le A\left(\delta \log \frac{2\pi}{\delta}\right) \log M_F(2r), \tag{25.21}$$

where E is an arbitrary measurable subset of $[0, 2\pi]$, $\delta = \operatorname{mes} E$ and A is an absolute constant. Taking $E = [0, r^{-\rho-1}] \cup [\pi - r^{-\rho-1}, \pi]$ and taking into account that $\log M_F(2r) = O(r^{\rho+\epsilon})$ as $r \to \infty$, we obtain relation (25.19) from relations (25.20) and (25.21).

Now let us prove that

$$\int_{r^{-\rho-1}}^{\pi-r^{-\rho-1}} |\log|\Phi(re^{i\theta})||d\theta = O(\sqrt{r}\log r), \quad r\to\infty. \tag{25.22}$$

The function $\Phi(z)$ has the order $\le 1/2$. Therefore, by Theorem 9.1, the following representation is valid in the half-plane $\Im z > 0$.

$$\log|\Phi(z)| = \omega_1(z) + \omega_2(z), \tag{25.23}$$

where

$$\omega_1(z) = \frac{r\sin\theta}{\pi}\int_{-\infty}^{\infty}\frac{\log|\Phi^+(t)|dt}{t^2-2rt\cos\theta+r^2},$$

$$\omega_2(z) = \sum_{\Im z_n>0}\log\left|\frac{z-z_n}{z-\bar z_n}\right|.$$

(Here $\Phi^+(t) = \Phi(t)$ for $t<0$ and $z = re^{i\theta}$ and $\{z_n\}$ are the zeros of the function $\Phi(z)$.)

Denote

$$\beta(r) = \int_1^r \log|\Phi^+(t)|\frac{dt}{t}.$$

By Corollary 1 to Theorem 6.1 (see (6.15)) we have $\beta(r) = O(\sqrt{r})$. Therefore

$$\left|\frac{r\sin\theta}{\pi}\int_{2r}^{\infty}\frac{\log|\Phi^+(t)|dt}{t^2-2rt\cos\theta+r^2}\right| \le \frac{r}{\pi}\int_{2r}^{\infty}\frac{t}{(t-r^2)}d\beta(t) \le$$

$$\le \frac{4r}{\pi}\int_{2r}^{\infty}\frac{d\beta(t)}{t} = -\frac{2\beta(2r)}{\pi} + \frac{4r}{\pi}\int_{2r}^{\infty}\frac{\beta(t)}{t^2}dt \le$$

$$\le C_1\sqrt{r}, \quad 0<\theta<\pi,$$

and,similarly,

$$\left|\frac{r\sin\theta}{\pi}\int_{-\infty}^{-2r}\frac{\log|\Phi^+(t)|dt}{t^2-2rt\cos\theta+r^2}\right| \le C_2\sqrt{r}, \quad 0<\theta<\pi.$$

Thus

$$\int_{r^{-\rho-1}}^{\pi-r^{-\rho-1}} |\omega_1(re^{i\theta})|d\theta \le$$

$$\le C_3 + \sqrt{r}\int_{r^{-\rho-1}}^{\pi-r^{-\rho-1}}\frac{r\sin\theta d\theta}{\pi}\int_{-2r}^{2r}\frac{|\log|\Phi^+(t)||dt}{t^2-2rt\cos\theta+r^2} =$$

$$= C_3\sqrt{r} + \frac{1}{2\pi}\int_{-2r}^{2r}\frac{|\log|\Phi^+(t)||dt}{t}\log\frac{t^2+2rt\cos r^{-\rho-1}+r^2}{t^2-2rt\cos r^{-\rho-1}+r^2}dt.$$

One can easily see that, for $|t| < 2r$,

$$r^2 \sin^2 r^{-\rho-1} \le t^2 \pm 2rt \cos r^{-\rho-1} + r^2 \le (t+r)^2 \le 9r^2.$$

Hence

$$0 \le \log \frac{t^2 + 2rt\cos r^{-\rho-1} + r^2}{t^2 - 2rt\cos r^{-\rho-1} + r^2} \le C_4 \log r.$$

Using once more Corollary 1 to Theorem 6.1 we obtain

$$\int_{r^{-\rho-1}}^{\pi - r^{-\rho-1}} |\omega_1(re^{i\theta})| d\theta = O(\sqrt{r}\log r). \tag{25.24}$$

When estimating $\omega_2(z)$, denote $z_n = r_n \exp(i\theta_n)$. We have

$$\int_{r^{-\rho-1}}^{\pi - r^{-\rho-1}} |\omega_2(re^{i\theta})| d\theta = \frac{1}{2}\int_{r^{-\rho-1}}^{\pi - r^{-\rho-1}} \Big\{ \sum_{\Im z_n > 0} \tau(re^{i\theta}, z_n)\Big\} d\theta,$$

where

$$\tau(z, z_n) = \log \frac{r_n^2 + 2rr_n\cos(\theta + \theta_n) + r^2}{r_n^2 - 2rr_n \cos(\theta - \theta_n) + r^2}.$$

First, let $r_n \ge 2r$. Using inequalities $|z - \bar{z}_n| \ge r/2$ and $\log(1+x) < x$ for $x > 0$, we obtain

$$\tau(z, z_n) \le \frac{4rr_n \sin\theta_n}{|z - \bar{z}_n|^2} \le 16r\frac{\sin\theta_n}{r_n} < 32\sqrt{r}\frac{\sin(\theta_n/2)}{\sqrt{r_n}}.$$

Since, by Theorem 25.1, the series

$$\sum_n \frac{1}{\sqrt{r_n}} \sin\frac{\theta_n}{2}$$

converges, we find

$$\int_{r^{-\rho-1}}^{\pi - r^{-\rho-1}} \Big\{ \sum_{r_n \ge 2r} \tau(z, z_n)\Big\} d\theta = O(\sqrt{r}).$$

Now, let $r_n \le 2r$. Calculating the extremum we have

$$0 \le \tau(z, z_n) \le \log\frac{1 - \cos(\theta + \theta_n)}{1 - \cos(\theta - \theta_n)}.$$

Thus

$$\int_{r^{-\rho-1}}^{\pi - r^{-\rho-1}} |\omega_2(re^{i\theta})| d\theta \le$$

$$\leq O(\sqrt{r}) + \sum_{r_n \leq 2r} \int_{r^{-\rho-1}}^{\pi - r^{-\rho-1}} \log \frac{\sin \frac{1}{2}(\theta + \theta_n)}{\left|\sin \frac{1}{2}(\theta - \theta_n)\right|} d\theta. \tag{25.25}$$

Denote by J_n the n–th integral in the right-hand side of (25.25). To find a bound for J_n assume first that

$$\sin \theta_n > \frac{1}{2} \sin r^{-\rho-1}.$$

We have

$$J_n \leq \int_0^{\pi} \log \frac{\sin \frac{1}{2}(\theta + \theta_n)}{\left|\sin \frac{1}{2}(\theta - \theta_n)\right|} d(\theta - \theta_n) =$$

$$= 2 \sin \theta_n \int_0^{\pi} \frac{\theta - \theta_n}{\cos \theta_n - \cos \theta} d\theta.$$

Since $|\theta - \theta_n| < \pi$ and $\alpha / \sin \alpha \leq \pi/2$ for $0 < \alpha \leq \pi/2$, we get

$$\left|\frac{\theta - \theta_n}{\cos \theta - \cos \theta_n}\right| = \left|\frac{\theta - \theta_n}{2 \sin \frac{1}{2}(\theta - \theta_n)}\right| \cdot \frac{1}{\sin \frac{1}{2}(\theta + \theta_n)} \leq$$

$$\leq \frac{\pi}{2 \sin \frac{1}{2}(\theta + \theta_n)},$$

whence, taking into account that $\sin \theta_n > \sin r^{-\rho-1}$, we obtain

$$J_n < 2\pi \sin \theta_n \cdot \left\{\left|\log \tan \frac{1}{4}(\pi + \theta_n)\right| + \left|\log \tan \frac{1}{4}\theta_n\right|\right\} <$$

$$< C_5 \sin \theta_n \cdot \log r.$$

Now, let

$$\sin \theta_n < \frac{1}{2} \sin r^{-\rho-1}.$$

Then

$$J_n = \int_{r^{-\rho-1}}^{\pi/2} \log \frac{\sin \theta + \sin \theta_n}{\sin \theta - \sin \theta_n} d\theta =$$

$$= \int_{\sin r^{-\rho-1}}^{1} \log \frac{x + \sin \theta_n}{x - \sin \theta_n} \cdot \frac{dx}{\sqrt{1 - x^2}} =$$

$$= 2 \int_{\sin r^{-\rho-1}}^{1} \sum_{k=1}^{\infty} \frac{1}{2k-1} \left(\frac{\sin \theta_n}{x}\right)^{2k-1} \frac{dx}{\sqrt{1 - x^2}} <$$

$$< \frac{8}{3} \sin \theta_n \int_{\sin r^{-\rho-1}}^{1} \frac{dx}{x\sqrt{1 - x^2}} < C_6 \sin \theta_n \cdot \log r.$$

Therefore relation (25.25) with accounting of Theorem 25.1 implies [20]

$$\int_{r^{-\rho-1}}^{r^{\pi}-r^{\rho-1}} |\omega_2(re^{i\theta})|d\theta \leq O(\sqrt{r}) + C_7 \log r \cdot \sum_{r_n \leq 2r} \sin\theta_n \leq$$

$$\leq O(\sqrt{r}) + 2C_7\sqrt{r}\log r \cdot \sum_{r_n \leq 2r} \frac{1}{\sqrt{r_n}} \sin\frac{1}{2}\theta_n = O(\sqrt{r}\log r). \qquad (25.26)$$

Combining (25.23), (25.24), (25.26) we obtain relation (25.22). One may complete the proof of the lemma by applying relation (25.19).

Lemma 25.4 *Let $\Phi(z) \in B$ be a solution of homogeneous boundary problem (23.1). Then the integral*

$$\omega(z) = \int_0^\infty \left\{\frac{1}{2\pi}\log G(x) - \tilde{n}_\Phi(x)\right\}\frac{dx}{x(x-z)}$$

converges uniformly (though, possibly, non-absolutely) on any compact $D_0 \subset D$.

Proof. According to Corollary to Theorem 25.4 (in the special case $p = 1/2$) the function

$$h(x) = \int_0^x \left\{\frac{1}{2\pi}\log G(t) - \tilde{n}_\Phi(t)\right\}t^{-2}dt$$

is bounded and approaches a finite limit as $x \to \infty$. Since

$$\int_{x_1}^{x_2} \left\{\frac{1}{2\pi i}\log G(x) - \tilde{n}_\Phi(x)\right\}\frac{dx}{x(x-z)} =$$

$$= \frac{xh(x)}{x-z}\bigg|_{x_1}^{x_2} + \int_{x_1}^{x_2} \frac{zh(x)dx}{(x-z)^2},$$

and the right-hand side approaches zero as $x_1, x_2 \to \infty$ uniformly with respect to $z \in D_0$, we get the statement of the lemma.

Theorem 25.5 *Let $\Phi(z) \in B_\sigma^*$, $0 < \sigma \leq 1/2$, be a solution of the homogeneous problem (23.1). Then the function*

$$\Omega(z) = \exp\left\{z\int_0^\infty \frac{\frac{1}{2\pi i}\log G(x) - \tilde{n}_\Phi(x)}{x(x-z)}dx\right\} \qquad (25.27)$$

[20] Complete description of possible growth of the value $\int_0^\pi |\omega_2(re^{i\theta})|d\theta$ is given in A.A. Goldberg "A remark on the Blaschke products in the half-plane" (Ukrainian Math. Journal, 1959, Vol.11, No.2, pp.210–213, Russian). — Editor's Note.

satisfies the relation

$$\lim_{r\to\infty} r^{-\sigma}\log|\Omega(re^{i\theta})| = -\frac{\pi\gamma}{\sin\pi\sigma}\cos\sigma(\theta-\pi), \quad 0<\theta<2\pi, \tag{25.28}$$

where

$$\gamma = \lim_{r\to\infty}\frac{\sigma^2}{r^\sigma}\int_0^r\frac{dt}{t}\int_0^t\left\{\frac{1}{2\pi}\arg G(x)-\tilde{n}_\Phi(x)\right\}\frac{dx}{x}\geq 0. \tag{25.29}$$

Here, for any $0<\epsilon<\pi$, *the convergence in (25.28) is uniform with respect to* θ *from the segment* $[\epsilon, 2\pi-\epsilon]$.

Proof. Denote

$$s(t)=\int_0^t\left\{\tilde{n}_\Phi(x)-\frac{1}{2\pi}\arg G(x)\right\}\frac{dx}{x}, \quad p(t)=-\int_0^t s(x)\frac{dx}{x}.$$

Clearly $s(t)=p(t)=0$ for sufficiently small t. Theorem 25.3 ensures the existence of the limit in relation (25.29). It follows from this relation that $p(t)=(\gamma\sigma^{-2}+p_0(t))t^\sigma$ where $\lim_{t\to\infty}p_0(t)=0$. The function $s(t)$ is real and continuous. Therefore, using the relation $p(t)=O(t^\sigma)$ as $t\to\infty$, one can find a sequence $t_k\to\infty$ such that $s(t_k)=O(t_k^\sigma)$ and we get

$$\int_0^\infty\frac{\frac{1}{2\pi}\arg G(x)-\tilde{n}_\Phi(x)}{x(x-z)}dx = -\lim_{k\to\infty}\int_0^{t_k}\frac{ds(x)}{x-z}=$$

$$-\lim_{k\to\infty}\left\{\frac{s(t_k)}{t_k-z}+\int_0^{t_k}\frac{s(x)dx}{(x-z)^2}\right\}=\lim_{k\to\infty}\int_0^{t_k}\frac{xdp(x)}{(x-z)^2}=$$

$$=\lim_{k\to\infty}\left\{\frac{t_kp(t_k)}{(t_k-z)^2}+\int_0^{t_k}\frac{p(x)(x+z)}{(x-z)^3}dx\right\}=$$

$$=\int_0^\infty\frac{p(x)(x+z)}{(x-z)^3}dx.$$

This yields the representation

$$\log\Omega(z)=z\gamma\sigma^{-2}\int_0^\infty\frac{x^\sigma(x+z)}{(x-z)^3}dx+z\int_0^\infty\frac{x^\sigma p(x)(x+z)}{(x-z)^3}dx+$$

$$+\frac{z}{2\pi i}\int_0^\infty\frac{\log|G(x)|}{x(x-z)}dx=\omega_1(z)+\omega_2(z)+\omega_3(z). \tag{25.30}$$

Using the residue theory, we obtain

$$\omega_1(z)=-\frac{\pi\gamma}{\sin\pi\sigma}r^\sigma e^{i\sigma(\theta-\pi)}, \quad z=re^{i\theta}, \quad 0<\theta<2\pi. \tag{25.31}$$

Let us show that, for any fixed ϵ,

$$\lim_{r\to\infty} r^{-\sigma}\omega_2(re^{i\theta}) = 0, \tag{25.32}$$

uniformly with respect to θ, $\epsilon < \theta < 2\pi - \epsilon$. Actually, if $\epsilon \leq \theta \leq 2\pi - \epsilon$ and $x > 0$, then $|x - re^{i\theta}| > r\sin\epsilon$ and $|x - e^{i\theta}| \geq |x - e^{i\epsilon}|$. Setting $N = \max_{0<t<\infty} |p_0(t)|$, we have

$$|\omega_2(re^{i\theta})| \leq Nr\int_0^{\eta r} \frac{x^\sigma(x+r)}{|x-re^{i\theta}|^3}dx+$$

$$+r\int_{\eta r}^{\infty} \frac{x^\sigma(x+r)}{|x-re^{i\theta}|^3}|p_0(x)|dx \leq 2Nr^{-1}(\sin\epsilon)^{-3}\int_0^{\eta r} x^\sigma dx+$$

$$+r^\sigma \sup_{\epsilon r\leq x<\infty} |p_0(x)|\int_\eta^\infty \frac{u^\sigma(u+1)}{|u-e^{i\epsilon}|^3}du \tag{25.33}$$

for any $\eta \in (0,1)$. Therefore

$$\limsup_{r\to\infty} r^{-\sigma}|\omega_2(re^{i\theta})| \leq 2N(\sin\epsilon)^{-3}(\sigma+1)^{-1}\eta^{\sigma+1},$$

which implies (25.32) because η may be taken arbitrary small. One can easily check the similar relation for the function $\omega_3(z)$ by using the relations $\log|G(x)| = 0$ for $0 \leq x \leq 1$ and $|\log|G(x)|| < C$ for $1 \leq x < \infty$.

Remark. Let us single out the properties of the function

$$\psi(x) = \frac{1}{2\pi i}\log G(x) - \tilde{n}_\Phi(x)$$

that were used when proving the theorem:

1) Boundedness of $\Im\psi(x)$ for $0 \leq x < \infty$.
2) Existence of the limit

$$\gamma = \lim_{r\to\infty} \frac{\sigma^2}{r^\sigma}\int_0^r \frac{dt}{t}\int_0^t \frac{\Re\psi(x)}{x}dx.$$

3) Fulfillment of the equality $\psi(x) = 0$ for sufficiently small $x > 0$.
4) Convergence of the integral $\int_0^\infty \Re\psi(x)x^{-2}dx$ (which provides convergence of the integral $\int_0^\infty \psi(x)(x-z)^{-1}x^{-1}dx$.)

Since $B^*_{1/2} = B$, Theorem 25.5 implies the following statement.

Corollary 25.6 *If $\Phi(z) \in B$ is a solution of the homogeneous boundary problem (23.1), then the function $\Omega(z)$, defined in (25.27), satisfies the relation*

$$\lim_{r\to\infty} r^{-1/2}\log|\Omega re^{i\theta})| = -\pi\gamma\sin\frac{1}{2}\theta, \tag{25.34}$$

where

$$\gamma = \lim_{r\to\infty} \frac{1}{4\sqrt{r}} \int_0^r \frac{dt}{t} \times$$

$$\times \int_0^t \left\{ \frac{1}{2\pi} \arg G(x) - \tilde{n}_\Phi(x) \right\} \frac{dx}{x} \geq 0,$$

and, for any $\epsilon, 0 < \epsilon < pi$ the convergence in (25.34) is uniform with respect to θ for $\epsilon < \theta < 2\pi - \epsilon$.

26 General form of solutions of the homogeneous problem in the class B

According to Theorem 24.3 any solution $\Phi(z)$ from the class B admits the representation $\Phi(z) = F(z)X(z)$ where $X(z)$ is the canonical function and $F(z)$ is an entire function having order $\rho_F \leq \rho$; here $\rho_F = \rho$, if $\rho > 1/2$. To describe all the solutions one has to completely describe the structure of admissible functions $F(z)$. But a more convenient way is to use another representation of solutions, which is obtained in the following theorem.

Theorem 26.1 *Every solution $\Phi(z) \in B$ of the homogeneous problem (23.1) admits the representation*

$$\Phi(z) = Qz^m \exp\Big\{\frac{z}{2\pi i}\int_0^\infty \frac{\log G(z) - 2\pi i \tilde{n}_\Phi(x)}{x(x-z)}\Big\}\times$$

$$\times \prod_{n=1}^{n_0} \frac{1 - z/z_n}{1 - z/|z_n|}, \tag{26.1}$$

where Q = const, $m \geq 0$ is an integer, $n_0 \leq \infty$ and $z_n \neq 0$ are the zeros of the function $\Phi(z)$ and $\tilde{n}_\Phi(r)$ is the number of zeros in the annulus $0 < |z| \leq r$. In the case $n_0 = \infty$ the infinite product converges absolutely and uniformly on any compact in D.

Before proving the theorem let us mention its consequence.

Corollary 26.1 *Let two solutions $\Phi_1(z), \Phi_2 \in B$ of the homogeneous problem (23.1) have the same zeros (with account of multiplicities). Then $\Phi_1(z) = Q\Phi_2(z)$, where Q is a constant.*

To prove the theorem we need the following lemma.

Lemma 26.1 *Let the sequence of points $z_n = r_r \exp(i\theta_n)$, where $0 < \theta_n < 2\pi$ and $r_n \to \infty$, satisfy condition (25.1). Then the infinite product*

$$b(z) = \prod_n \frac{1 - z/z_n}{1 - z/r_n} \tag{26.2}$$

converges absolutely and uniformly on any compact in D and also has c.r.g. with zero indicator with respect to the formal order $1/2$ in every angle $[\epsilon, 2\pi - \epsilon]$, $0 < \epsilon < \pi$.

Proof. The convergence of the product is a consequence of a simple estimate, which is valid for $r_n > 2|z|$:

$$|\log(1 - z/z_n) - \log(1 - z/r_n)| =$$
$$= \Big|\sum_{k=1}^{\infty} \frac{1}{k}(\exp(-ik\theta_n) - 1)\Big(\frac{z}{r_n}\Big)^k\Big| \leq$$
$$\leq \sum_{k=1}^{\infty} \frac{1}{k} 2\Big|\sin\Big(\frac{1}{2}k\theta_n\Big)\Big| \cdot \Big|\frac{z}{r_n}\Big|^k \leq$$
$$\leq 2\sin\Big(\frac{1}{2}\theta_n\Big)\sum_{k=1}^{\infty}\Big|\frac{z}{r_n}\Big|^k \leq 4\sin\Big(\frac{1}{2}\theta_n\Big)\Big|\frac{z}{r_n}\Big|.$$

To complete the proof let us split the product into two parts $b(z) = b_1(z)b_1(z)$, where the product $b_1(z)$ is taken over the set

$$N_1 = \{n:\ 0 < \theta_n < \epsilon/2\} \cup \{n:\ 2\pi - \epsilon/2 < \theta_n < 2\pi\}$$

and $b_2(z)$ is taken over the set

$$N_2 = \{z:\ \epsilon/2 \leq \theta_n < 2\pi - \epsilon/2\}.$$

First, we estimate the product $b_1(z)$ from below. We have

$$\log\frac{1}{|b_1(z)|} = \sum_{n \in N_1} \log\Big|1 + z\frac{1 - e^{i\theta}}{z_n - z}\Big| \leq$$
$$\leq 2|z| \sum_{n \in N_1} \frac{\sin(\theta_n/2)}{|z_n - z|}.$$

If $\epsilon < \theta < 2\pi - \epsilon$ and either $0 < \theta_n < \epsilon/2$ or $2\pi - \epsilon/2 < \theta_n < 2\pi$, then

$$\frac{2\sqrt{rr_n}}{|z_n - re^{i\theta}|} \leq \Big\{\frac{(r_n + r)^2}{(r_n + r)^2 - 4rr_n\cos^2\frac{1}{2}(\theta - \theta_n)}\Big\}^{1/2} \leq$$
$$\leq \frac{1}{\Big\{1 - \cos^2\frac{1}{2}(\theta - \theta_n)\Big\}^{1/2}} \leq \frac{1}{\sin(\epsilon/4)}. \qquad (26.3)$$

Therefore, for arbitrary fixed $R > 0$, we have

$$\log\frac{1}{|b_1(z)|} \leq 2r \sum_{n \in N_1,\ r_n \leq R} \frac{\sin(\theta_n/2)}{|z_n - z|} +$$

$$+\frac{\sqrt{r}}{\sin(\epsilon/4)}\sum_{n\in N_1,\ r_n>R}\frac{\sin(\theta_n/2)}{\sqrt{r_n}}.$$

Hence the relation

$$\limsup_{r\to\infty}\frac{1}{\sqrt{r}}\log\frac{1}{|b_1(re^{i\theta})|}\le\frac{1}{\sin(\epsilon/4)}\sum_{n\in N_1,\ r_n>R}\frac{\sin(\theta_n/2)}{\sqrt{r_n}}$$

is fulfilled uniformly with respect to θ, $\epsilon<\theta<2\pi-\epsilon$. Since R is arbitrary and series (25.1) converges, we obtain

$$\limsup_{r\to\infty}\frac{1}{\sqrt{r}}\log\frac{1}{|b_1(re^{i\theta})|}\le 0$$

uniformly with respect to $\theta\in[\epsilon,2\pi-\epsilon]$.

To estimate the product $b_1(z)$ from above we write

$$\log|b_1(z)|=\sum_{n\in N_1}\log\left|1+z\frac{1-\exp(-i\theta_n)}{r_n-z}\right|\le$$

$$\le 2|z|\sum_{n\in N_1}\frac{\sin(\theta_n/2)}{|r_n-z|}.$$

Arguing as above but using the simple estimate

$$\frac{2\sqrt{rr_n}}{|r_n-r^{i\theta}|}\le\frac{1}{\sin(\epsilon/2)},\qquad \epsilon\le\theta\le 2\pi-\epsilon,$$

instead of (26.3), we obtain

$$\limsup_{r\to\infty}r^{-1/2}\log|b_1(re^{i\theta})|\le 0$$

uniformly with respect to $\theta\in[\epsilon,2\pi-\epsilon]$.

Thus we have proved that the limit

$$\lim_{r\to\infty}r^{-1/2}\log|b_1(re^{i\theta})|=0 \tag{26.4}$$

exists uniformly with respect to $\theta\in[\epsilon,2\pi-\epsilon]$.

Consider now the function $b_2(z)$. It follows From (25.1) it follows that

$$\sum_{n\in N_2}r_n^{-1/2}\le(\sin(\epsilon/4))^{-1},\qquad\sum_{n\in N_2}r_n^{-1/2}\sin(\theta_n/2)<\infty. \tag{26.5}$$

Therefore one may represent $b_2(z)$ as $b_2(z)=b_3(z)/b_4(z)$, where

$$b_3(z) = \prod_{n\in N_2} \Big(1 - \frac{z}{z_n}\Big),$$

$$b_4(z) = \prod_{n\in N_2} \Big(1 - \frac{z}{r_n}\Big).$$

Both $b_3(z)$ and $b_4(z)$ are entire functions with the same counting function of zeros $n_0(r)$. By (26.5) this function admits the estimate $n_0(r) = o(\sqrt{r})$. By virtue of Theorems 4.2 and 4.3 the both functions $b_3(z)$ and $b_4(z)$ are functions of c.r.g.with zero indicator with respect to the order 1/2. Hence, uniformly with respect to $\theta \in [0, 2\pi]$, we have

$$\limsup_{r\to\infty} r^{-1/2} \log |b_3(re^{i\theta})| =$$

$$= \lim{}^*_{r\to\infty} r^{-1/2} \log |b_3(re^{i\theta})| = 0. \tag{26.6}$$

The same equality holds for the function $b_4(z)$ as well. Moreover, since $b_4(z)$ has only positive zeros, we have (see [24, p.151])

$$\lim_{r\to\infty} r^{-1/2} \log |b_4(re^{i\theta})| = 0 \tag{26.7}$$

uniformly with respect to $\theta \in [\epsilon, 2\pi - \epsilon]$.

Combining relations (26.4),(26.6) and (26.7), we see that, uniformly with respect to $\theta \in [\epsilon, 2\pi - \epsilon]$,

$$\limsup_{r\to\infty} r^{-1/2} \log |b(re^{i\theta})| =$$

$$= \lim{}^*_{r\to\infty} r^{-1/2} \log |b(re^{i\theta})| = 0,$$

which proves the lemma.

Definition 26.1 The product of form (26.2) is called the *Blachke-type product.*

Proof of the theorem. As was already mentioned, we have the representation $\Phi(z) = X(z)F(z)$, where $F(z)$ in an entire function of order $\leq \rho$. The Hadamard theorem gives

$$F(z) = z^m e^{P(z)} \prod_{n=1}^{n_0} E\Big(\frac{z}{z_n}, q\Big), \tag{26.8}$$

where $m \geq 0$ is an integer, $n_0 \leq \infty$, $q = [\rho]$ and $E(u, q)$ is the canonical Weierstrass multiplier and $P(z) = a_q z^q + \cdots + a_0$ is a polynomial of degree not exceeding q. Consider the auxiliary entire function

$$F_0(z) = \prod_{n=1}^{n_0} E\Big(\frac{z}{z_n}, q\Big),$$

where $r_n = |z_n|$. The logarithm of this function may be represented as the Stiltjes integral with respect to the measure $d\tilde{n}_\Phi(x)$. After integration by parts (this is a well-known method for estimating entire functions with positive zeros, see, for example, [19, p.92]), we obtain

$$F_0(z) = \exp\left\{-z^{q+1}\int_0^\infty \frac{\tilde{n}_\Phi(x)dx}{x^{q+1}(x-z)}\right\}. \tag{26.9}$$

Now rewrite the representation of $\Phi(z)$ in the form

$$\Phi(z) = \{X(z)F_0(z)\}\{F(z)/F_0(z)\}$$

and consider each of expressions in braces separately. By virtue of Lemma 26.1 the second expression admits the representation

$$F(z)/F_0(z) = z^m e^{a_0+Q(z)} b(z), \tag{26.10}$$

where $b(z)$ is some Brachke-type product (26.2) and

$$Q(z) = c_q z^q + \cdots + c_1 z$$

is a polynomial, whose coefficients c_k have the form

$$c_k = a_k + \frac{1}{k}\sum_{n=1}^{n_0}(z_n^{-k} - r_n^{-k}), \quad k = 1,\ldots,q. \tag{26.11}$$

It should be noted that, in the case $n_0 = \infty$, the series in (26.11) converges absolutely, since

$$|z_n^{-k} - r_n^{-k}| = 2r_n^{-k}\left|\sin\left(\frac{1}{2}k\theta_n\right)\right| \le 2kr_n^{-k}\sin\left(\frac{1}{2}\theta_n\right),$$

and we may apply relation (25.1).

As for the product $X(z)F_0(z)$, relations (24.1) and (26.9) imply

$$X(z)F_0(z) = \exp\left\{z^{q+1}\int_0^\infty \frac{\frac{1}{2\pi i}\log G(t) - \tilde{n}_\Phi(t)}{t^{q+1}(t-z)}dt\right\}.$$

Let us use the identity

$$\frac{z^{q+1}}{t^{q+1}(t-z)} = \frac{z}{t(t-z)} - zt^{-2} - \ldots - z^q t^{-q-1}.$$

This identity together with Lemma 25.4 and the fact, that the integrals in (25.18) with $p = k - 1/2$, $k = 1,\ldots,q$ converge, yield

$$X(z)F_0(z) = e^{R(z)}\Omega(z), \tag{26.12}$$

where $\Omega(z)$ is defined by (25.27) and

$$R(z) = d_q z^q + \cdots + d_1 z$$

is the polynomial with the coefficients

$$d_k = -\int_0^\infty \Big\{\frac{1}{2\pi i}\log G(x) - \tilde{n}_\Phi(x)\Big\} x^{-k-1}dx, \quad k = 1, \ldots, q.$$

Now from (26.10) and (26.12) we obtain

$$\Phi(z) = z^m e^{a_0 + S(z)}\Omega(z)b(z),$$

where

$$S(z) = s_q z^q + \cdots + s_1 z$$

is the polynomial with the coefficients $s_k = c_k + d_k$, $k = 1, \ldots, q$. To get representation (26.1) it remains to prove that $S(z) \equiv 0$. To this end, let us mention that, in view of Theorem 25.5 and Lemma 26.1, we have

$$\lim_{r\to\infty} r^{-1}\log|\Omega(re^{i\theta})| \equiv 0,$$

$$\lim{}^*_{r\to\infty} r^{-1}\log|b(re^{i\theta})| \equiv 0, \quad 0 < \theta < 2\pi,$$

respectively. On the other hand, since $\Phi(z) \in B = B^*_{1/2}$, we have

$$\lim{}^*_{r\to\infty} r^{-1}\log|\Phi(re^{i\theta})| \equiv 0, \quad 0 < \theta < 2\pi.$$

Therefore

$$\lim{}^*_{r\to\infty} r^{-1}\Re S(re^{i\theta}) \equiv 0, \quad 0 < \theta < 2\pi,$$

which implies $S(z) \equiv 0$.

Remark 1. We have also proved that the coefficients of the polynomial $P(z)$ in canonical representation (26.8) of the function $F(z)$ have the form

$$a_k = \int_0^\infty \Big\{\frac{1}{2\pi i}\log G(x) - \tilde{n}_\Phi(x)\Big\} x^{-k-1}dx + \\ + \frac{1}{k}\sum_{n=1}^{n_0}(1 - e^{-ik\theta_n})r_n^{-k}. \tag{26.13}$$

Remark 2. As follows from the proof of Theorem 26.1, the function $\Phi(z)$ has one more representation

$$\Phi(z) = Qz^m e^{T(z)}\exp\Big\{\frac{z^{q+1}}{2\pi i}\int_0^\infty \frac{\log G(x)dx}{x^{q+1}(x-z)}\Big\}\times$$

$$\times \prod_{n=1}^{n_0} \Big(1 - \frac{z}{z_n}\Big) \exp\Big\{\frac{z}{r_n} + \frac{z^2}{2r_n^2} + \cdots + \frac{z^q}{qr_n^q}\Big\}, \tag{26.14}$$

where the polynomial $T(z)$ has the form

$$T(z) = \sum_{k=1}^{q} \frac{z^k}{2\pi i} \int_0^\infty \{\log G(x) - 2\pi i \tilde{n}_\Phi(x)\} x^{-k-1} dx.$$

Relation (26.14) is rather more complicated than (26.1) but has the advantage that all its entries belong to the class $\tilde{B}$, i.e. are bounded in any disk $|z| \le R$, in contrast to the Blachke-type products in (26.1) that have poles on the positive ray.

Remark 3 In the case $0 < \rho < 1$ relation (26.14) takes the form

$$\Phi(z) = Qz^m \exp\Big\{\frac{z}{2\pi i} \int_0^\infty \frac{\log G(x) dx}{x(x-z)}\Big\} \prod_{n=1}^{n_0} \Big(1 - \frac{z}{z_n}\Big).$$

This expression coincides with formula (20.5), which was proved under the assumption $\rho < 1/2$. If $\rho > 1/2$, then Corollary 4 of Theorem 25.3 gives $n_0 = \infty$. If $\rho < 1/2$, one cannot exclude the case $n_0 < \infty$.

Theorem 26.2 *The general solution of the homogeneous problem (23.1) in the class B is given by expression (26.1) in which Q is a constant, $m \ge 0$ is an integer and $\{z_n\}$, $z_n \to \infty$, $z_n \ne 0$ is a sequence of points such that*

1) The series

$$\sum_n r_n^{-1/2} \sin \frac{\theta_n}{2}$$

converges, here $z_n = r_n e^{i\theta_n}$.

2) Let $\tilde{n}_\Phi(r)$ be the counting function of the sequence $\{z_n\}$. Then there exists a finite non-negative limit

$$\lim_{r\to\infty} \frac{1}{4\sqrt{r}} \int_0^r \frac{dt}{t} \int_0^t \Big\{\frac{1}{2\pi} \arg G(x) - \tilde{n}_\Phi(x)\Big\} \frac{dx}{x} = \gamma \ge 0.$$

3) The integral

$$\int_0^\infty \{\arg G(x) - 2\pi \tilde{n}_\Phi(x)\} x^{-2} dx \tag{26.15}$$

converges.

4) For $t>1$, $t \neq r_n$ the following relation holds

$$m \log t + \sum_n \log\left|\frac{t-z_n}{t-r_n}\right| + \frac{t}{2\pi}\int_0^\infty \frac{\arg G(x) - 2\pi\tilde{n}_\Phi(x)}{x(x-t)}dx < C \quad (26.16)$$

The indicator of the solution $\Phi(z)$ with respect to the formal order $1/2$ has the form

$$h_\Phi(\theta) = \limsup_{r\to\infty} r^{-1/2}\log|\Phi(re^{i\theta})| =$$

$$= -\pi\gamma \sin\frac{\theta}{2}, \quad 0 \le \theta \le 2\pi. \quad (26.17)$$

Proof. Let $\Phi(z) \in B$ be a solution of problem (23.1). Then, combining Theorem 25.1, Corollary 1 to Theorem 25.3, Theorem 25.4 and Theorem 26.1, we see that $\Phi(z)$ admits representation (26.1) and also conditions 1)–3) are fulfilled. Applying the Sohotski formula to (26.1) we obtain

$$\log|\Phi^+(t)| = \log|Q| + m\log t + \sum_n \log\left|\frac{t-z_n}{t-r_n}\right| +$$

$$+\frac{1}{2}\log|G(t)| + \frac{t}{2\pi}\int_0^\infty \frac{\arg G(x) - 2\pi\tilde{n}_\Phi(x)}{x(x-t)}dx.$$

Since $\Phi(z) \in B$, then $\log|\Phi^+(t)| < C$ for $1 < t < \infty$ and, taking into account the inequality $|\log|G(t)|| < C$, for $1 < t < \infty$, which follows from (23.4), we obtain relation (26.16).

Now let conditions 1)–4) be fulfilled. Condition 1) together with Lemma 26.1 provide convergence of the infinite product in (26.1). Using condition 3) and the arguments from the proof of Lemma 25.4, we see that the integral in (26.1) converges uniformly on every compact in D. So we may consider the function $\Phi(z)$ defined by relation (26.1). Let us prove that this function is a solution of problem (23.1) belonging to the class B.

Condition 2) together with relations (23.2) and (23.3) imply the estimate $\tilde{n}_\Phi(x) = O(x^\mu)$, $x \to +\infty$, where $\mu = \max(\rho, 1/2)$. Therefore we may convert the transformations of the function $\Phi(z)$, that were made when proving Theorem 26.1. We obtain the representation $\Phi(z) = X(z)F(z)$, where $X(z)$ is the canonical function and $F(z)$ is the entire function defined by equality (26.8) in which the coefficients a_k of the polynomial $P(z)$ are defined by (26.13). This representation together with Theorem 24.2 give $\Phi(z) \in \tilde{B}$.

Let us prove that $\Phi(z) \in B$. Define the function $\Omega(z)$ as in (25.27). Using the remark to Theorem 25.5 with $\sigma = 1/2$, we can see that equality (25.28) is valid and approaching the limit in this equality is uniform on every segment $[\epsilon, 2\pi - \epsilon]$. Applying Lemma 26.1 to the Blachke-type product, we see that the order of the function $\Phi(z)$ does not exceed $1/2$ in any angle $[\epsilon, 2\pi - \epsilon]$, and in addition if the order equals $1/2$, it is the order of decreasing, since, by virtue of (25.28) with $\sigma = 1/2$ and Lemma 26.1, we have

$$\limsup_{r \to \infty} r^{-1/2} \log |\Phi(re^{i\theta})| = -\pi\gamma \sin\frac{\theta}{2}, \quad 0 < \theta < 2\pi.$$

To estimate the function $\Phi(z)$ in the angle $[0, 2\pi]$, let us again consider the representation $\Phi(z) = X(z)F(z)$. Since the both functions $F(z)$ and $X(z)$ have a finite order (we have used Theorem 24.1), then $\Phi(z)$ also has a finite order in $[0, 2\pi]$. It follows from condition 4) that the boundary values $\Phi^{\pm}(t)$ are bounded on the ray $0 \le t < \infty$. Taking ϵ sufficiently small and applying the Phragmen -Lindelöf principle to the function $\Phi(z)$ in the angles $[0, \epsilon]$ and $[2\pi - \epsilon, 2\pi]$, we see that the growth of the function $\Phi(z)$ in $[0, 2\pi]$ does not exceed the minimal type with respect to the order $1/2$. Now once more using the boundedness of $\Phi^{\pm}(t)$, $0 \le t < \infty$, and applying the Phragmën -Lindelöf principle in the whole angle $[0, 2\pi]$, we claim that $\Phi(z) \in B$.

Equality (26.17) was already obtained in the course of our reasoning.

It is still unknown whether there exists at least one sequence $\{z_n\}$ satisfying conditions 1)–4) of Theorem 26.2. Therefore the question about existence of a solution $\Phi(z) \in B$ is still open. We postpone this question till Sect. 28 in which the affirmative answer will be given for even a more general situation, i.e. for the classes B_{σ}^{*}, $0 < \sigma \le 1/2$.

27 General form of solutions of the homogeneous problem in the class B_σ^*.

Since $B_\sigma^* \subset B$, every solution from the class B_σ^* admits representation (26.1) and its zeros satisfy conditions 1)–4) of Theorem 26.2. If a solution of the homogeneous problem belong to the class B_σ^*, $\sigma < 1/2$, then its zeros satisfy additional conditions and we begin with determining these conditions. Some important knowledge about modulus of zeros is given by Theorem 25.3. It will be convenient to reformulate this theorem, by using the following definition.

Definition 27.1 We say that *the zero set of a solution $\Phi(z)$ of problem (23.1) has the modular σ-density which is equal to* $-\gamma$ if there exists a finite limit

$$\lim_{r\to\infty} \frac{\sigma^2}{r^\sigma} \int_0^r \frac{dt}{t} \int_0^t \left\{ \tilde{n}_\Phi(x) - \frac{1}{2\pi} \arg G(x) \right\} \frac{dx}{x} = -\gamma. \tag{27.1}$$

Theorem 25.3 is equivalent to the following statement.

Lemma 27.1 *The zero set of any solution $\Phi(z) \in B_\sigma^*$ of problem (23.1) has non-positive modular σ-density.*

Now consider the distribution of zeros of a solution $\Phi(z) \in B_\sigma^*$ over their arguments. We use notation and results of Sect. 7 of Ch. 2.

Definition 27.2 We say that *the set of zeros $z_n = r_n e^{i\theta_n}$ of a solution $\Phi(z)$ of problem (23.1) has the modified argument σ-density* if the following conditions are fulfilled:

1) For all $\eta_1, \eta_2 \in (0, 2\pi)\backslash N$, $\eta_1 < \eta_2$, where N is an at most countable set, there exists the angular density

$$\Delta(\eta_1, \eta_2) = \lim_{r\to\infty} r^{-\sigma} n(r, \eta_1, \eta_2). \tag{27.2}$$

2) The following estimate is valid

$$\sum_{z_n \in (r,0,2\pi)} \sin(\theta_n/2) = O(r^\sigma), \quad r \to \infty. \tag{27.3}$$

3) There exists the finite limit

$$\lambda_* = \lim_{r\to\infty} r^{-\sigma} \sum_{z_n \in (r,0,2\pi)} \sin\theta_n. \tag{27.4}$$

It should be mentioned that $\sin\theta_n$ may have various signs.

Lemma 27.2 *The zero set of any solution $\Phi(z) \in B_\sigma^*$ of problem (23.1) has the modified argument σ-density.*

Proof. Condition 1) is a direct consequence of Theorem 4.12 because the function $\Phi(z)$ possesses c.r.g. in $(0, 2\pi)$. Since both $\Phi(z)$ and $\Phi(-z)$ possess c.r.g. in the half-plane $\Im z > 0$, Theorem 7.1 (see relation (7.6)) gives

$$\sum_{z_n \in (r,0,\pi)} \sin\theta_n = O(r^\sigma),$$

$$\sum_{z_n \in (r,\pi,2\pi)} |\sin\theta_n| = O(r^\sigma), \quad r \to \infty.$$

Besides, by condition 1) we have $n(r, \frac{\pi}{2}, \frac{3\pi}{2}) = O(r^\sigma)$. Combining these three asymptotic estimates, we obtain (27.3), i.e. condition 2) is also fulfilled.

Let us verify condition 3). The function $\Phi(z)$ possesses c.r.g. in $\Im z > 0$ and due to Theorem 7.1 its zero set has the argument boundary density. The representation $\Phi(z) = X(z)F(z)$, where $X(z)$ is the canonical function and $F(z)$ is an entire function, implies that $\Phi(z)$ does not have a singular boundary function. This follows from the fact that $X(z)$ has not such a function (because $\log G(t)$ satisfies the Hölder condition). Therefore the existence of the argument-boundary density $\lambda(0, \pi)$ (see relation (7.5)) of the zero set of the function $\Phi(z)$ means the existence of the finite limit

$$\lim_{r\to\infty} r^{-\sigma}\Big\{ \sum_{z_n \in (r,0,\pi)} \sin\theta_n - \frac{1}{2\pi}\int_1^r \log|\Phi^+(t)\Phi(-t)|\frac{dt}{t}\Big\}.$$

Similarly, applying Theorem 7.1 to the function $\Phi(-z)$ we prove the existence of the finite limit

$$\lim_{r\to\infty} r^{-\sigma}\Big\{ \sum_{z_n \in (r,\pi,2\pi)} \sin(\theta_n - \pi) - \frac{1}{2\pi}\int_1^r \log|\Phi^-(t)\Phi(-t)|\frac{dt}{t}\Big\}.$$

Hence, there exists the finite limit

$$\lim_{r\to\infty} r^{-\sigma}\Big\{ \sum_{z_n \in (r,0,2\pi)} \sin\theta_n - \frac{1}{2\pi}\int_1^r \log|\Phi^+(t)/\Phi^-(t)|\frac{dt}{t}\Big\}.$$

Since $\Phi^+(t)/\Phi^-(t) = G(t)$ and, by virtue of (23.4), the relation $|\log|G(t)|| = O(1)$, $t \to \infty$, holds, the finite limit in (27.4) exists as well. This completes the proof.

Remark 1 Using Lemma 6.4 (see(6.38)), one can easily see that

$$\lambda_* = \frac{\sigma^2 - 1}{2\pi\sigma} \int_0^{2\pi} h_\Phi(\theta) \sin\theta d\theta.$$

Remark 2 Applying Lemma 7.1 (and changing the angle $(0, \pi)$ to $(0, 2\pi)$), one can easily see that, for all $\eta_1, \eta_2 \in (0, 2\pi) \backslash N$, N is the same as in condition 1), there exist finite limits

$$\lim_{r \to \infty} r^{-\sigma} \sum_{z_n \in (r, \eta_1, \eta_2)} \sin(\theta_n/2) = \int_{\eta_1}^{\eta_2} \sin(\psi/2) d\Delta(\psi),$$

$$\lim_{r \to \infty} r^{-\sigma} \sum_{z_n \in (r, \eta_1, \eta_2)} \sin\theta_n = \int_{\eta_1}^{\eta_2} \sin\psi d\Delta(\psi). \tag{27.5}$$

Remark 3 It follows from the previous remark and condition 2), that

$$\int_{+0}^{2\pi - 0} \sin(\psi/2) d\Delta(\psi) < \infty.$$

Theorem 27.1 *Let a set $\{z_n\}$ have the modified argument σ-density with $\sigma < 1/2$. Then the Blachke-type product $b(z)$ defined by (26.2) converges absolutely and uniformly on every compact in D and also possesses c.r.g. with respect to the formal order σ in every angle $[\epsilon, 2\pi - \epsilon]$. Its indicator admits the representation*

$$h_b(\theta) = \frac{\pi}{\sin\pi\sigma} \Big\{ \int_{+0}^{2\pi - 0} \big(\cos\sigma(|\theta - \psi| - \pi) - \cos\sigma(\theta - \pi)\big) d\Delta(\psi) +$$

$$+ \Big(\lambda_* - \int_{+0}^{2\pi - 0} \sin\psi d\Delta(\psi)\Big) \sigma \sin\sigma(\theta - \pi) \Big\}. \tag{27.6}$$

Proof. As can be easily seen, condition (27.3) in the case $\sigma < 1/2$ implies (25.1) and, hence, convergence of the product follows from Lemma 26.1.

Take a number $\delta \in (0, \epsilon/2)$ such that $\delta, 2\pi - \delta \notin N$, where N is the exceptional set from condition 1) in definition 27.2. Split the product $b(z)$ into two parts $b(z) = b_1(z) b_2(z)$. Here the product $b_1(z)$ is taken over the set

$$N_1 = \{n : \ 0 < \theta_n < \delta\} \cup \{n : \ 2\pi - \delta < \theta_n < 2\pi\}$$

and $b_2(z)$ is taken over the set

$$N_2 = \{n : \ \delta < \theta_n < 2\pi - \delta\}.$$

First, consider the product $b_2(z)$. It may be written as $b_2(z) = b_3(z)/b_4(z)$ where

$$b_3(z) = \prod_{n\in N_2}\left(1-\frac{z}{z_n}\right),\quad b_4(z) = \prod_{n\in N_2}\left(1-\frac{z}{r_n}\right).$$

By virtue of the condition 1) in definition 27.2 the zero sets of the functions $b_3(z)$ and $b_4(z)$ possess the angular σ-density. Theorems 4.2 and 4.3 imply that these functions have c.r.g. with respect to the formal order σ and their indicators admit representations

$$h_{b_3}(\theta) = \frac{\pi}{\sin\pi\sigma}\int_\delta^{2\pi-\delta}\cos(|\theta-\psi|-\pi)d\Delta(\psi),$$

$$h_{b_4}(\theta) = \frac{\pi\cos\sigma(\theta-\pi)}{\sin\pi\sigma}\int_\delta^{2\pi-\delta}d\Delta(\psi).$$

All zeros of the function $b_4(z)$ lie on the positive ray. Therefore (see [24, p.15]) the function $r^{-\sigma}\log|b_4(re^{i\theta})|$ approaches $h_{b_4}(\theta)$ as uniformly for $\theta\in[\epsilon,2\pi-\epsilon]$ as $r\to\infty$. This implies that the function $b_2(z)$ possesses c.r.g. with respect to the formal order σ in the angle $[\epsilon,2\pi-\epsilon]$ with the indicator

$$h_{b_2}(\theta) = \frac{\pi}{\sin\pi\sigma}\int_\delta^{2\pi-\delta}\big(\cos(|\theta-\psi|-\pi)- \\ -\cos\sigma(\theta-\pi)\big)d\Delta(\psi). \tag{27.7}$$

Now let us estimate the product $b_1(z)$. We need the following asymptotic representation of its factor (which is similar to that of Lemma 9.3):

$$\log\left\{\left(1-\frac{z}{z_n}\right)\Big/\left(1-\frac{z}{r_n}\right)\right\} = \frac{iz\sin\theta_n}{r_n-z}\{1+\beta(z,z_n)\} \tag{27.8}$$

for

$$\theta_n\in(0,\delta)\cup(2\pi-\delta,2\pi),\quad \epsilon\le\theta\le2\pi-\epsilon,\quad \delta<\epsilon/2,$$

where

$$|\beta(z,z_n)|\le6\frac{\sin(\theta_n/2)}{\sin(\epsilon/2)}. \tag{27.9}$$

We will obtain this representation in the case $\theta_n\in(0,\delta)$, since the case $\theta_n\in(2\pi-\delta,2\pi)$ is similar. We have

$$\log\left\{\left(1-\frac{z}{z_n}\right)\Big/\left(1-\frac{z}{r_n}\right)\right\} =$$

$$=\int_0^{\theta_n}\frac{\partial}{\partial\alpha}\log\left(1-\frac{ze^{-i\alpha}}{r_n}\right)d\alpha = iz\int_0^{\theta_n}\frac{e^{-i\alpha}d\alpha}{r_n-ze^{-i\alpha}},$$

whence

$$\beta(z,z_n)=\frac{r_n-z}{\sin\theta_n}\int_0^{\theta_n}\frac{e^{-i\alpha}d\alpha}{r_n-ze^{-i\alpha}}-1=$$

$$=\frac{r_n-z}{\sin\theta_n}\int_0^{\theta_n}\Big(\frac{e^{-i\alpha}}{r_n-ze^{-i\alpha}}-\frac{\cos\alpha}{r_n-z}\Big)d\alpha=$$

$$=\frac{1}{\sin\theta_n}\int_0^{\theta_n}\Big(-i\sin\alpha+\frac{ze^{-i\alpha}(e^{-i\alpha}-1)}{r_n-ze^{-i\alpha}}\Big)d\alpha.$$

Since for $\epsilon<\theta<2\pi-\epsilon$ and $0<\alpha<\epsilon/2$ we have

$$|r_n-ze^{-i\alpha}|\geq r\sin(\epsilon/2),$$

we obtain

$$|\beta(z,z_n)|\leq\frac{1}{\sin\theta_n}\int_0^{\theta_n}\Big(\sin\alpha+\frac{2\sin(\alpha/2)}{\sin(\epsilon/2)}\Big)d\alpha\leq$$

$$\leq\frac{6\sin^2(\theta_n/2)}{\sin\theta_n\cdot\sin(\epsilon/2)},$$

which implies (27.9).

Now, using (27.8) we have, for $\epsilon\leq\theta\leq 2\pi-\epsilon$,

$$\log b_1(z)=\lim_{R\to\infty}\sum_{n\in N_1,\ r_n\leq R}\log\Big\{\Big(1-\frac{z}{z_n}\Big)\Big/\Big(1-\frac{z}{r_n}\Big)\Big\}=$$

$$=\lim_{R\to\infty}\Big\{\sum_{n\in N_1,\ r_n\leq R}\frac{iz\sin\theta_n}{r_n-z}+\sum_{n\in N_1,\ r_n\leq R}\frac{iz\sin\theta_n\cdot\beta(z,z_n)}{r_n-z}\Big\}=$$

$$=\lim_{R\to\infty}\Big\{\Sigma_R^{(1)}+\Sigma_R^{(2)}\Big\}. \tag{27.10}$$

We shall prove that both the sums $\Sigma_R^{(1)}$, $\Sigma_R^{(2)}$ approach the corresponding limits and calculate the values of these limits.

Denote

$$\eta_1(R)=\sum_{n\in N_1,\ r_n\leq R}\sin\theta_n,$$

$$\eta_2(R,z)=\sum_{n\in N_1,\ r_n\leq R}\beta(z,z_n)\sin\theta_n.$$

Since

$$\eta_1(R)=\sum_{r_n\leq R}\sin\theta_n-\sum_{n\in N_2,\ r_n\leq R}\sin\theta_n,$$

by virtue of condition 3) in Definition 27.2 and Remark 2 to Lemma 27.2, there exists the limit

$$c_\delta = \lim_{R\to\infty} R^{-\sigma}\eta_1(R) = \lambda_* - \int_\delta^{2\pi-\delta} \sin\psi d\Delta(\psi). \tag{27.11}$$

Using (27.9) we have

$$|\eta_2(R,z)| \le \frac{6}{\sin(\epsilon/2)} \sum_{n\in N_1,\ r_n\le R} |\sin\theta_n| \sin(\theta_n/2) \le$$

$$\le \frac{6\sin\delta}{\sin(\epsilon/2)} \sum_{n\in N_1,\ r_n\le R} \sin(\theta_n/2),$$

whence, by virtue of condition 2) in Definition 27.2, we have

$$|\eta_2(R,z)| \le C\frac{\sin\delta}{\sin(\epsilon/2)} R^\sigma. \tag{27.12}$$

Here C is independent of δ, ϵ and R.

Since

$$\Sigma_R^{(1)} = iz\int_0^R \frac{d\eta_1(t)}{t-z} = \frac{iz\eta_1(R)}{R-z} + iz\int_0^R \frac{\eta_1(t)dt}{(t-z)^2},$$

relation (27.11) yields that there exists the limit

$$\lim_{R\to\infty} \Sigma_R^{(1)} = iz\int_0^\infty \frac{\eta_1(t)dt}{(t-z)^2}.$$

Likewise, relation (27.12) yields that there exists the limit

$$\lim_{R\to\infty} \Sigma_R^{(2)} = iz\int_0^\infty \frac{\eta_2(t,z)dt}{(t-z)^2}.$$

Returning to (27.10), we obtain the representation

$$\log b_1(z) = iz\int_0^\infty \frac{\eta_1(t)dt}{(t-z)^2} + iz\int_0^\infty \frac{\eta_2(t)dt}{(t-z)^2} = \omega_1(z) + \omega_2(z).$$

By virtue of (27.11), we have

$$\eta_1(t) = t^\sigma(c_\delta + \eta_0(t)), \quad \lim_{t\to\infty} \eta_0(t) = 0.$$

Using the residue theory and then obtaining the estimates that are similar to those of the proof in Theorem 25.5 (see (25.30)–(25.33)), we get

$$\omega_1(z) = izc_\delta \int_0^\infty \frac{t^\sigma dt}{(t-z)^2} + iz \int_0^\infty \frac{t^\sigma \eta_0(t)dt}{(t-z)^2} =$$

$$= -\frac{\pi i c_\delta \sigma}{\sin \pi\sigma} e^{i\sigma(\theta-\pi)} r^\sigma + o(r^\sigma), \quad r \to \infty,$$

uniformly with respect to $\theta \in [\epsilon, 2\pi - \epsilon]$. According to (27.12) we obtain, for $\theta \in [\epsilon, 2\pi - \epsilon]$,

$$|\omega_2(z)| \le C \frac{\sin \delta}{\sin(\epsilon/2)} r \int_0^\infty \frac{t^\sigma dt}{|t-z|^2} =$$

$$= C \frac{\sin \delta}{\sin(\epsilon/2)} r^\sigma \int_0^\infty \frac{u^\sigma du}{|u - e^{i\theta}|^2} \le C_\epsilon \delta r^\sigma,$$

where $C_\epsilon > 0$ is independent of δ and r. So, we see that, for sufficiently large r, the following relation holds

$$\max_{\epsilon \le \theta \le 2\pi-\epsilon} \left| \log |b_1(re^{i\theta})| - \frac{\pi c_\delta \sigma}{\sin \pi\sigma} \sin \sigma(\theta - \pi) \cdot r^\sigma \right| < 2C_\epsilon \delta r^\sigma. \tag{27.13}$$

To complete the proof denote by $h(\theta)$ the expression in the right-hand side of (27.6). Let $\eta > 0$ be an arbitrarily small number. According to (27.7) and (27.11), one may choose a small enough δ such that

$$\max_{\epsilon \le \theta \le 2\pi-\epsilon} \left| h(\theta) - h_{b_2}(\theta) - \frac{\pi c_\delta \sigma}{\sin \pi\sigma} \sin \sigma(\theta - \pi) \right| < \eta. \tag{27.14}$$

Since the function $b_2(z)$ possesses c.r.g. in the angle $[\epsilon, 2\pi - \epsilon]$, there exists a set E of zero relative measure such that

$$\max_{\epsilon \le \theta \le 2\pi-\epsilon} \left| \log |b_2(re^{i\theta})| - h_{b_2}(\theta) r^\sigma \right| < \eta r^\sigma$$

for $r \notin E$. Therefore (27.13) and (27.14) imply that, for $r \notin E$,

$$\max_{\epsilon \le \theta \le 2\pi-\epsilon} \left| \log |b(re^{i\theta})| - h(\theta) r^\sigma \right| \le 2(\eta + C_\epsilon \delta) r^\sigma.$$

Since η and δ may be chosen arbitrarily small, by applying Theorem 4.7 we see that the function $b(z)$ possesses c.r.g. in the angle $[\epsilon, 2\pi - \epsilon]$ with respect to the formal order σ and $h_B(\theta) \equiv h(\theta)$.

Theorem 27.2 *The general solution of the homogeneous problem (23.1) in the class* B_σ^*, $0 < \sigma < \min(1/2, \rho)$, *has the form (26.1) where* Q *is a constant,* $m \geq 0$ *is an integer and* $z_n \to \infty$ *is a sequence of non-zero points such that:*

1) The set $\{z_n\}$ *possesses the modified argument* σ*-density* λ_*.

2) The set $\{z_n\}$ *possesses the modular* σ*-density* $-\gamma \leq 0$.

Here and in what follows $\tilde{n}_\Phi(r)$ *is the counting function of the set* $\{z_n\}$.

3) Integral (26.15) converges.

4) Relation (26.16) holds for $t > 1$, $t \neq r_n = |z_n|$.

The indicator of the solution $\Phi(z)$ *with respect to the formal order* σ *admits the representation*

$$h_\Phi(\theta) = \frac{\pi}{\sin \pi\sigma} \Big\{ \int_{+0}^{2\pi-0} \big(\cos\sigma(|\theta - \psi| - \pi) - \cos\sigma(\theta - \pi)\big) d\Delta(\psi) +$$

$$+ \Big(\lambda_* - \int_{+0}^{2\pi-0} \sin\psi d\Delta(\psi)\Big) \sigma \sin\sigma(\theta - \pi) -$$

$$-\gamma \cos\sigma(\theta - \pi)\Big\}, \tag{27.15}$$

where $\Delta(\psi)$ *is the angular density of the set* $\{z_n\}$.

Proof. Let $\Phi(z) \in B_\sigma^*$ be a solution of problem (23.1) and $0 < \sigma < \min(\rho, 1/2)$. Then $\Phi(z) \in B$ and, by virtue of Theorem 26.2, relation (26.1) holds and also conditions 3) and 4) are fulfilled. Conditions 1) and 2) follow from Lemmas 27.2 and 27.1 respectively.

Now let conditions 1)–4) be fulfilled. Conditions 3) and 4) coincide with those of Theorem 26.1 and conditions 1) and 2) are stronger than those of Theorem 26.1. By Theorem 26.1 both the integral and the infinite product in (26.1) are convergent and the function $\Phi(z)$ defined by (26.1) is a solution of problem (23.1) from the class B. That $\Phi(z) \in B_\sigma^*$ and also relation (27.15) follow from Theorems 25.5 and 27.1.

Theorem 27.3 *The description of the general solution of problem (23.1) from the class* B_σ, $0 < \sigma < \min(\rho, 1/2)$, *may be obtained by adding one more condition to the hypothesis of Theorem 27.2:*

5) If $\gamma = 0$ *in condition 2), then*

$$\lim_{r\to\infty} r^{-\sigma+\epsilon} \int_0^r \Big\{ \frac{1}{2\pi} \arg G(x) - \tilde{n}_\Phi(x) \Big\} \frac{dx}{x} = +\infty$$

for any $\epsilon > 0$.

Proof. If $\Phi(z) \in B_\sigma$ is a solution of problem (23.1), then condition 5) follows from relation (25.10).

Now let conditions 1)–4) of Theorem 27.2 be fulfilled as well as condition 5). By Theorem 27.2 we have $\Phi(z) \in B_\sigma^*$ and to prove that $\Phi(z) \in B_\sigma$ one has to check that the order ρ_Φ of the function $\Phi(z)$ is not less than σ. In the case $\gamma > 0$ this follows from (27.15) in view of the relation

$$h_\Phi(\pi) = \frac{\pi}{\sin \pi\sigma}\Big\{\int_{+0}^{2\pi-0} \big(\cos\sigma(|\pi-\psi|-\pi)-1\big)d\Delta(\psi)-\gamma\Big\} \leq$$

$$\leq -\frac{\pi\gamma}{\sin\pi\sigma} < 0.$$

Now let $\gamma = 0$. Assume that $\rho_\Phi < \sigma$. Take $\epsilon > 0$ sufficiently small such that $\rho_\Phi < \sigma - \epsilon$. According to Definition 1.4 there exists a sequence $R_n \to \infty$ such that

$$\lim_{n\to\infty} R_n^{-\sigma+\epsilon} \log|\Phi(R_n e^{i\theta})| = 0$$

uniformly with respect to $\theta \in (0, 2\pi)$. By Theorem 25.2 we have

$$\int_0^{R_n}\Big(\frac{1}{2\pi}\arg G(x) - \tilde{n}_\Phi(x)\Big)\frac{dx}{x} =$$

$$= \frac{1}{2\pi}\int_0^{2\pi} \log|\Phi(R_n e^{i\theta})|d\theta + O(\log R_n).$$

Multiplying the both sides by $R_n^{-\sigma+\epsilon}$ and letting $R_n \to \infty$, we obtain a contradiction with condition 5).

It should be noted that the question on the solvability of problem (23.1) in the classes B_σ^* and B_σ still remains open, because we have to construct examples of sequences $\{z_n\}$ satisfying conditions 1)–4), or moreover, conditions 1)–5). This construction is given in the following section.

28 An example of a solution of the homogeneous problem in the class B_σ. Importance of the restriction on the exponent in the Hölder condition for the function $\varphi(t) = \arg G(t)/(2\pi t^\rho)$

Until now we did not use condition (23.7). This condition turns out to be important when constructing a sequence $\{z_n\}$ satisfying conditions 1)–5) from Theorems 27.2 and 27.3. First let us prove the following lemma.

Lemma 28.1 *Let a real function* $\psi(x)$, $0 \le x < \infty$, *satisfy the Hölder condition* $H_{[0,\infty)}(\mu)$, *where*

$$\frac{\rho - \nu}{1 + \rho - \nu} < \mu < 1, \quad 0 < \nu < \rho < \infty, \tag{28.1}$$

and also $\lim_{x\to\infty} \psi(x) > 0$. *Then*

$$|\max_{1\le x\le t} \{\psi(x)x^\rho\} - \psi(t)t^\rho| = O(t^\nu), \quad t \to \infty. \tag{28.2}$$

Proof. Denote $r(t) = t^{1-\rho+\nu}$ and prove that

$$\max_{1\le x\le t} \{\psi(x)x^\rho\} = \max_{t-r(t)\le x\le t} \{\psi(x)x^\rho\} \tag{28.3}$$

for sufficiently large t. To this end, it suffices to check that, for $r(t) \le h \le t$, we have

$$\Delta(t,h) = \psi(t)t^\rho - \psi(h-t)(t-h)^\rho \ge 0, \quad t > t_0.$$

Let $K = \max_{0\le t<\infty} |\psi(t)|$ and $\lambda = \lim_{t\to\infty} \psi(t)$. Put

$$b = 1 - (4K\lambda^{-1} + 1)^{-1/\rho}.$$

Since

$$\Delta(t,h) = \psi(t)\big(t^\rho - (t-h)^\rho\big) + (t-h)^\rho\big(\psi(t) - \psi(t-h)\big), \tag{28.4}$$

we get

$$\Delta(t,h) \ge \frac{1}{2}\lambda t^\rho\big(1 - (1-b)^\rho\big) - t^\rho(1-b)^\rho 2K = 0$$

for $bt \le h \le t$ and sufficiently large t. Now let $r(t) \le h \le bt$. Then $t^\rho - (t-h)^\rho \ge \rho h t^{\rho-1}(1-b)^\rho$. Besides, from the Hölder condition we have

$$|\psi(t) - \psi(t-h)| \le \frac{Ch^\mu}{t^\mu (t-h)^\mu}. \tag{28.5}$$

Therefore, for sufficiently large t, relation (28.4) yields

$$\Delta(t,h) \geq \frac{1}{2}\lambda\rho h t^{\rho-1}(1-b)^{\rho} - Ch^{\mu}(1-b)^{-\mu}t^{\rho-2\mu} =$$

$$ht^{\rho-1}\Big\{\frac{1}{2}\lambda\rho(1-b)^{\rho} - Ch^{\mu-1}(1-b)^{-\mu}t^{1-2\mu}\Big\}.$$

Since

$$h^{\mu-1}t^{1-2\mu} \leq t^{(1-\rho+\nu)(\mu-1)+1-2\mu}$$

and, by (28.1),

$$(1-\rho+\nu)(\mu-1)+1-2\mu < 0,$$

then $\Delta(t,h) \geq 0$ for sufficiently large t and $r(t) \leq h \leq bt$. This proves (28.3).

Now, to obtain (28.2) it suffices to prove

$$\max_{t-r(t)\leq x\leq t} |\psi(x)x^{\rho} - \psi(t)t^{\rho}| = O(t^{\nu}), \quad t \to \infty. \tag{28.6}$$

Let $0 \leq h \leq r(t) = t^{1-\rho+\nu}$. Then (28.4) and (28.5) yield

$$|\Delta(t,h)| \leq 2K\rho h t^{\rho-1} + \frac{Ch^{\mu}t^{\rho}}{t^{\mu}(t-h)^{\mu}} \leq$$

$$\leq 2K\rho t^{\nu} + 2Ct^{\rho-2\mu+(1-\rho+\nu)\mu}$$

for sufficiently large t. Using (28.1), we see that $|\Delta(t,h)| = O(t^{\nu})$ as $t \to \infty$, $0 < h < r(t)$, which is equivalent to (28.6).

Theorem 28.1 *Let positive numbers σ, $0 < \sigma < \min(\rho, 1/2)$, and $\tau > 0$ be given. If conditions (23.2)–(23.4) and (23.7) are fulfilled, then the sequence of positive numbers $\{z_n\}$ with the counting function*

$$n_{\Phi}(r) = [\max_{1\leq x\leq r}\{\varphi(x)x^{\rho} - \tau x^{\sigma}\}^{+}]$$

satisfies conditions 1)–5) of Theorems 27.2 and 27.3 and, hence, the function

$$\Phi(z) = \exp\Big\{\frac{z}{2\pi i}\int_0^{\infty}\frac{\log G(x) - 2\pi i n_{\Phi}(x)}{x(x-z)}dx\Big\} \tag{28.7}$$

is a solution of problem (23.1) in the class B_{σ}.

Proof. Denote

$$\psi(x) = \{\varphi(x) - \tau x^{\sigma-\rho}\}^{+}.$$

We have $\psi(x) = 0$ for sufficiently small x. In view of (23.3) we have $\varphi(x) \in H(\mu_0)$ and $\varphi(\infty) = \lambda > 0$. Besides $x^{\sigma-\rho} \in H_{[1,\infty)}(\eta)$ where $\eta = \min(1, \rho - \sigma)$. Therefore $\psi(x) \in H(\mu)$ where $\mu = \min(\mu_0, \eta)$ and $\lim_{x\to\infty} \psi(x) > 0$. It follows from (23.7) that $\mu > (\rho-\sigma)/(1+\rho-\sigma)$, hence one can choose a number ν such that

$$0 < \nu < \sigma, \quad \mu > (\rho - \nu)/(1 + \rho - \nu). \tag{28.8}$$

Applying Lemma 28.1 to the function $\psi(x)$ and using (28.2), we obtain

$$n_\Phi(r) = \varphi(r)r^\rho - \tau r^\sigma + \beta(r)r^\nu,$$

$$\beta(r) = O(1), \quad r \to \infty. \tag{28.9}$$

Now let us verify conditions 1)–5) from Theorems 27.2 and 27.3. Condition 1) is fulfilled because $z_n > 0$ and also $\lambda_* = 0$. One can easily obtain conditions 2), 3) and 5) from equality (28.9) with $\gamma = \tau$ in condition 2). It remains to check condition 4). Taking into account relation (26.16) and the equality $z_n = r_n$, one may transform this condition as

$$J(t) = t \int_0^\infty \frac{\varphi(x)x^\rho - n_\Phi(x)}{x(x-t)} dx < C, \quad t > 1, \quad t \neq r_n. \tag{28.10}$$

Using representation (28.9), rewrite $J(t)$ in the form

$$J(t) = \tau t \int_0^\infty \frac{dx}{x^{1-\sigma}(x-t)} - \tau t \int_{t-t^{-\rho}}^{t+t^{-\rho}} \frac{x^\sigma - t^\sigma}{x(x-t)} dx -$$

$$-\tau t^{\sigma+1} \int_{t-t^{-\rho}}^{t+t^{-\rho}} \frac{dx}{x(x-t)} - t \int_0^{t+t^{-\rho}} \frac{\beta(x)dx}{x^{1-\nu}(x-t)} -$$

$$-t \int_{t+t^{-\rho}}^{\infty} \frac{\beta(x)dx}{x^{1-\nu}(x-t)} + t \int_{t-t^{-\rho}}^{t+t^{-\rho}} \frac{\varphi(x) - \varphi(t)}{x^{1-\rho}(x-t)} dx +$$

$$+t\varphi(t) \int_{t-t^{-\rho}}^{t+t^{-\rho}} \frac{x^\rho - t^\rho}{x(x-t)} dx + t^{\rho+1}\varphi(t) \int_{t-t^{-\rho}}^{t+t^{-\rho}} \frac{dx}{x(x-t)} +$$

$$+ \int_{t-t^{-\rho}}^{t+t^{-\rho}} n_\Phi(t) \frac{dx}{x} - \int_{t-t^{-\rho}}^{t+t^{-\rho}} \frac{n_\Phi(x)dx}{x-t} = \sum_{j=1}^{10} J_j.$$

The integral J_1 may be calculated explicitly:

$$J_1 = -\tau(\pi \cot \sigma\pi) t^\sigma < 0.$$

Since

$$|x^\sigma - t^\sigma| \le \sigma(t/2)^{\sigma-1}|x-t|, \quad t-t^{-\rho} \le x \le t+t^{-\rho}$$

for sufficiently large t, we get

$$|J_2| \le 2^{1-\sigma}\tau t^\sigma \int_{t-t^{-\rho}}^{t+t^{-\rho}} \frac{dx}{x} < C_1 t^{\sigma-\rho-1} < C_1.$$

Futher we have

$$|J_3| = \tau t^\sigma \log\frac{t+t^{-\rho}}{t-t^{-\rho}} < C_2 t^{\sigma-\rho-1} < C_2.$$

Taking into account that $\beta(r) = O(1)$, we obtain

$$|J_4| \le C_3 t\Big\{\int_0^{t/2} \frac{dx}{x^{1-\nu}(t-x)} + \frac{2^{1-\nu}}{t^{1-\nu}}\int_{t/2}^{t-t^{-\rho}} \frac{dx}{t-x}\Big\} < C_4 t^\nu \log t,$$

$$|J_5| \le C_5 t\Big\{\frac{1}{t^{1-\nu}}\int_{t+t^{-\rho}}^{2t} \frac{dx}{x-t} + \int_{2t}^{\infty} \frac{dx}{x^{1-\nu}(x-t)}\Big\} < C_6 t^\nu \log t.$$

Now, using that $\varphi(t) \in H_{[1,\infty)}(\mu)$, we have

$$|J_6| \le C_7 t^{1-\mu} \int_{t-t^{-\rho}}^{t+t^{-\rho}} \frac{|x-t|^{\mu-1}}{x^{1-\rho+\mu}} dx \le$$

$$\le C_8 t^{\rho-2\mu} \int_{t-t^{-\rho}}^{t+t^{-\rho}} |x-t|^{\mu-1} dx \le C_9 t^{\rho-2\mu-\rho\mu}.$$

Since (28.8) implies $\rho - 2\mu - \rho\mu < \nu$, we obtain

$$|J_6| < C_9 t^\nu.$$

One may estimate the integrals J_7 and J_8 in the same way as the integrals J_2 and J_3.

$$|J_7| \le C_{10} t^\rho \int_{t-t^{-\rho}}^{t+t^{-\rho}} \frac{dx}{x} < C_{11},$$

$$|J_8| \le C_{12} t^{\rho+1} \log\frac{t+t^{-\rho}}{t-t^{-\rho}} \le C_{13}.$$

In view of (28.9) we have $n_\Phi(r) = O(r^\rho)$ as $r \to \infty$. Therefore

$$|J_9| \le C_{14}/t.$$

At last, since the function $n_\Phi(x)$ is non-decreasing,

$$J_{10} = -\int_0^{t^{-\rho}} \big(n_\Phi(t+x) - n_\Phi(t-x)\big)\frac{dx}{x} < 0.$$

Combining the obtained estimates of integrals $J_1,\ldots,J_{10}$, we get

$$J(t) < -(\pi \cot \sigma\pi)t^\sigma + C_{15}t^\nu \log t < 0,$$

which proves relation (28.10) and at the same time completes the proof of the theorem.

Remark 1. One can prove that the constructed solution of problem (23.1) belongs to the class $\bar{B}_\sigma$.

Now let us prove that conditions (23.5) and (23.7) are important. If they are not fulfilled, then problem (23.1) may appear to be unsolvable in the classes B and B^* respectively.

Theorem 28.2 *Define the function $G(t)$ by the relations*

$$|G(t)| \equiv 1, \quad \arg G(t) = 2\pi\varphi(t)t^\rho, \quad 1 \le t < \infty, \tag{28.11}$$

where

$$\varphi(t) = 1 + t^{-\alpha}\sin(t^{\rho-\sigma}) - 2t^{-\rho}, \quad \frac{1}{2} < \rho < \infty, \quad 0 < \sigma \le \frac{1}{2},$$

$$\sigma < \alpha < \rho - \sigma,$$

Then the homogeneous boundary problem (23.1) is unsolvable in the class B^*_σ. In particular (when $\sigma = 1/2$), it has not solutions in the class B.

Proof. According to Lemma 22.4 we have

$$\varphi(t) \in H_{[1,\infty)}(\mu_0), \quad \mu_0 = \frac{\alpha}{1+\rho-\sigma} < \frac{\rho-\sigma}{1+\rho-\sigma}.$$

Nevertheless $\varphi(t) \notin H_{[1,\infty)}(\mu)$ for $\mu > \mu_0$, since, if taking

$$t_n = (\pi n)^{1/(\rho-\sigma)}, \quad \tau_n = \left(\pi n + \frac{1}{2}\pi\right)^{1/(\rho-\sigma)},$$

we obtain

$$|\varphi(t_n) - \varphi(\tau_n)| > C\left|\frac{1}{t_n} - \frac{1}{\tau_n}\right|^{\mu_0}, \quad n = 1, 2, \ldots.$$

Thus, condition (23.7) (which in the case $\sigma = 1/2$ is the same as (23.5)) is not fulfilled.

Consider the function

$$X_0(z) = (1-z)^2 \exp\{2\pi(J_1(z) + J_2(z))\},$$

where

$$J_1(z) = \frac{z^{q+1}}{2\pi} \int_1^\infty \frac{x^{\rho-q-1}}{x-z} dx, \quad q = [\rho],$$

$$J_2(z) = \frac{z}{2\pi} \int_1^\infty \frac{x^{\rho-\alpha-1} \sin(x^{\rho-\sigma})}{x-z} dx$$

(one can easily verify that the integral J_2 is convergent; this may be done with accounting the boundedness of the integral $\int_1^x u^{\rho-\alpha-1} \sin(u^{\rho-\sigma}) du$, $1 \le x < \infty$, and then integrating by parts). The function $X(z)$ is analytic in D, continuous up to the boundary, satisfies condition (23.1) and also has a zero of the order β, $0 < \beta < 1$, at the point $z = 1$. In what follows we assume ρ is not an integer (the reasoning for integer ρ is similar).

One can easily see that

$$J_1(z) = -\frac{e^{-i\rho\pi}}{2\sin\rho\pi} z^\rho - \frac{z^{q+1}}{2\pi} \int_0^1 \frac{x^{\rho-q-1}}{x-z} dx =$$

$$= -\frac{e^{-i\rho\pi}}{2\sin\rho\pi} z^\rho + \sum_{n=1}^{q} c_n z^n + O(1), \quad z \to \infty,$$

where $c_1 \ldots, c_q$ are real constants.

Let

$$D_0^+ = \{z : \ |z| > 1\} \cap \{z : \ 0 < \arg z < \psi\},$$

$$D_0^- = \{z : \ |z| > 1\} \cap \{z : \ -\psi < \arg z < 0\},$$

$$\psi = \min\left\{\frac{\pi}{4(\rho-\sigma)}, \frac{\pi}{4}\right\}.$$

We have

$$J_2(z) = \frac{z}{4\pi i} \int_{\partial D_0^+} \frac{\exp(i\zeta^{\rho-\sigma})}{\zeta^{\alpha+1-\rho}(\zeta - z)} d\zeta -$$

$$-\frac{z}{4\pi i} \int_{\partial D_0^+ \cap \{z: \Im z > 0\}} \frac{\exp(i\zeta^{\rho-\sigma})}{\zeta^{\alpha+1-\rho}(\zeta - z)} d\zeta +$$

$$+\frac{z}{4\pi i} \int_{\partial D_0^-} \frac{\exp(-i\zeta^{\rho-\sigma})}{\zeta^{\alpha+1-\rho}(\zeta - z)} d\zeta -$$

$$-\frac{z}{4\pi i}\int_{\partial D_0^-\cap\{z:\Im z<0\}}\frac{\exp(-i\zeta^{\rho-\sigma})}{\zeta^{\alpha+1-\rho}(\zeta-z)}d\zeta =$$

$$= P_1^+ - P_2^+ + P_1^- - P_2^-$$

(the branch of $\arg\zeta$ is defined by the condition $|\arg z| \le \psi$). The Cauchy theorem yields

$$P_1^+ = \begin{cases} \frac{1}{2}z^{\rho-\alpha}\exp(iz^{\rho-\sigma}), & z\in D_0^+, \\ 0, & z\in D_0^-, \end{cases}$$

$$P_1^- = \begin{cases} 0, & z\in D_0^+, \\ \frac{1}{2}z^{\rho-\alpha}\exp(-iz^{\rho-\sigma}), & z\in D_0^-, \end{cases}$$

A simple estimate shows that $P_2^+ = o(1)$, $P_2^- = o(1)$ as $z\to\infty$ for $|\arg z| \le \psi/2$. Therefore, for $|\theta| < \psi/2$, we have

$$\log|X_0(re^{i\theta})| = -\frac{r^\rho}{2\sin\rho\pi}\cos\rho(\theta-\pi)+$$

$$+\sum_{n=1}^{q} c_n r^n\cos n\theta + \frac{1}{2}r^{\rho-\alpha}\exp\{\mp r^{\rho-\sigma}\sin(\rho-\sigma)\theta\}\times$$

$$\times\cos\{\theta(\rho-\alpha)\pm r^{\rho-\sigma}\cos(\rho-\sigma)\theta\} + O(1). \tag{28.12}$$

Here the upper sign is taken for $\theta\in[0,\psi/2]$ and the lower one for $\theta\in[-\psi/2,0]$ respectively. Besides, since $|G(t)|\equiv 1$, we see that $|X_0(z)|$ is continuous on the ray $1\le t<\infty$.

Denote

$$a_{n1} = \left(2\pi n + \frac{1}{2}\pi\right)^{1/(\rho-\sigma)}, \quad b_{n1} = \left(2\pi n + \frac{2}{3}\pi\right)^{1/(\rho-\sigma)},$$

$$b_{n2} = \left(2\pi n + \frac{4}{3}\pi\right)^{1/(\rho-\sigma)}, \quad a_{n2} = \left(2\pi n + \frac{3}{2}\pi\right)^{1/(\rho-\sigma)},$$

and choose $\epsilon > 0$ sufficiently small such that $\alpha - 2\epsilon > 0$ and $\alpha + 2\epsilon < \rho-\sigma$. Consider the sequence of the annular sectors

$$D_n = \{z:\ a_{n1} < |z| < a_{n2}\}\cap\{z:\ |\arg z| < a_{n1}^{\sigma-\rho+\epsilon}\}.$$

Since $\sigma-\rho+\epsilon < 0$, we have

$$D_n \subset \{z:\ |\arg z| < \frac{1}{2}\psi\}$$

for sufficiently large n. Clearly

$$\max_{\zeta\in D_n} |\zeta - a_{n1}| \leq 2a_{n2}a_{n1}^{\sigma-\rho+\epsilon} - 2(a_{n2} - a_{n1}) \leq$$

$$\leq C_1 a_{n1}^{1+\sigma-\rho+\epsilon}. \tag{28.13}$$

Denote by $f_1(z)$ the sum of the first two summands in (28.12) and by $f_2(z)$ the sum of the remaining ones. By (28.13) we obtain

$$\max_{\zeta\in D_n} |f_1(\zeta) - f_1(a_{n1})| \leq C_2 a_{n1}^{\sigma+\epsilon}. \tag{28.14}$$

For $|\theta| < a_{n1}^{\sigma-\rho+\epsilon}$ we have

$$\theta(\rho-\alpha) \pm a_{nk}^{\rho-\sigma}\cos(\rho-\sigma)\theta = \pm a_{nk}^{\rho-\sigma} + \varphi_{nk}(\theta),$$

$$|\varphi_{nk}(\theta)| \leq C_3 a_{n1}^{\sigma-\rho+2\epsilon}.$$

Therefore, in view of $\cos a_{n,k}^{\rho-\sigma} = 0$, we see that

$$|\cos\{\theta(\rho-\alpha) \pm a_{nk}^{\rho-\sigma}\cos(\rho-\sigma)\theta\}| \leq C_3 a_{n1}^{\sigma-\rho+2\epsilon},$$

and taking into account that $f_2(a_{n1}) = 0$ we obtain

$$\max_{|\theta|\leq a_{n1}^{\sigma-\rho+\epsilon}} |f_2(a_{nk}e^{i\theta}) - f_2(a_{n1})| \leq$$

$$\leq C_4 a_{n1}^{\sigma-\alpha+2\epsilon} < C_4 a_{n1}^{\sigma}. \tag{28.15}$$

Besides,

$$\max_{a_{n1}\leq r\leq a_{n2}} |f_2(r\exp(\pm i a_{n1}^{\sigma-\rho+\epsilon})) - f_2(a_{n1})| \leq C_5. \tag{28.16}$$

Relations (28.12)and (28.14)–(28.16) yield

$$\min_{z\in\partial D_n} \log|X_0(z)| \geq \log|X_0(a_{n1})| - C_6 a_{n1}^{\sigma+\epsilon}. \tag{28.17}$$

On the other hand, for $b_{n1} \leq t \leq b_{n2}$, we have

$$f_2(t) - f_2(a_{n1}) = \frac{1}{2}t^{\rho-\alpha}\cos(t^{\rho-\sigma}) \leq$$

$$\leq -\frac{1}{4}t^{\rho-\alpha} \leq -C_7 a_{n1}^{\sigma+2\epsilon}.$$

Therefore, by virtue of (28.12) we see

$$\max_{b_{n1}\leq t\leq b_{n2}} \log|X_0(t)| < \log|X_0(a_{n1})| - C_8 a_{n1}^{\sigma+2\epsilon}. \tag{28.18}$$

Now assume, that problem (23.1) with the function $G(t)$ defined by (28.11) has a solution $\Phi(z) \in B^*_\sigma$. In the same way as in Sect. 24 one can prove that $\Phi(z) = X_0(z)F(z)$ where $F(z)$ is an entire function. By (28.17)

$$\max_{z\in\partial D_n} \log|F(z)| \leq -\log|X_0(a_{n1})| + C_9 a_{n1}^{\sigma+\epsilon},$$

and the maximum principle yields

$$\max_{b_{n1}\leq t\leq b_{n2}} \log|F(t)| \leq -\log|X_0(a_{n1})| + C_9 a_{n1}^{\sigma+\epsilon}.$$

By relation (28.18) we obtain

$$\log|\Phi(t)| \leq -C_{10} a_{n1}^{\sigma+2\epsilon} < -C_{11} t^{\sigma+2\epsilon}$$

for $b_{n1} \leq t \leq b_{n2}$, whence, taking into account that $|\Phi(t)| \leq C_{12}$ for all t, $1 \leq t < \infty$, we see

$$b_{n2}^{-\sigma} \int_1^{b_{n2}} \log|\Phi(t)| \frac{dt}{t} \leq b_{n2}^{-\sigma} \sum_{j=n_0}^{n} \int_{b_{j1}}^{b_{j2}} (-C_{11} t^{\sigma+2\epsilon}) \frac{dt}{t} +$$

$$+C_{13} b_{n2}^{-\sigma} \log b_{n1} \leq -C_{14} n^{-\sigma/(\rho-\sigma)} \sum_{j=n_0}^{n} j^{(-\rho+2\sigma+2\epsilon)/(\rho-\sigma)} \leq$$

$$\leq -C_{15} n^{2\epsilon/(\rho-\sigma)} \to -\infty.$$

Since $\Phi(z) \in B^*_\sigma \subset A^*_\sigma$, we have got a contradiction to Theorem 6.1.

Remark. By complicating the proof one can obtain an example in which the function $\varphi(t)$ belongs to $H(\mu_0)$ for $\mu_0 = (\rho-\sigma)/(1+\rho-\sigma)$ instead of $\mu_0 < (\rho-\sigma)/(1+\rho-\sigma)$.

29 Statement of the non-homogeneous problem and an approach to its solution

Let D be the complex plane without the ray $L = \{1 \le t < \infty\}$. Consider the boundary problem in D

$$\Phi^+(t) = G(t)\Phi^-(t) + g(t) \tag{29.1)}$$

under the following assumptions:

1) $\arg G(t) = 2\pi\varphi(t)t^\rho, \quad 0 < \rho < \infty,$

where

$$\varphi(t) \in H(\mu), \quad \rho/(\rho+1) < \mu \le 1, \tag{29.2}$$

$$\varphi(\infty) = \lambda > 0, \quad -1 < \varphi(1) \le 0. \tag{29.3}$$

2)

$$\log|G(t)| \in H, \quad g(t) \in H, \tag{29.4}$$

3)

$$g(\infty) = 0, \tag{29.5}$$

4)

$$\text{either } G(1) = 1 \text{ and } g(1) = 0, \text{ or } G(1) \ne 1. \tag{29.6}$$

It should be noted, that here the exponent μ in Hölder's condition satisfies stronger restrictions than that in the homogeneous problem (see (23.5) and (23.7)) though the weaker restrictions than those in Sect. 21.

Lemma 29.1 *Let $0 < \sigma < \sigma_0 < \min(\rho, 1/2)$ and $\Phi(z)$ be a bounded solution of problem (29.1) having the order of decreasing less than σ_0 in any angle $[\epsilon, 2\pi - \epsilon]$. Then the general solution of problem (29.1) in the class B_σ has the form $\Phi(z) = \Phi_0(z) + \Psi(z)$ where $\Psi(z)$ is the general solution in the class B_σ of the corresponding homogeneous problem $\Psi^+(t) = G(t)\Psi^-(t)$, $1 \le t < \infty$. In addition, the indicators of the functions $\Psi(z)$ and $\Phi(z)$ coincide for all $\theta \in (0, 2\pi)$.*

The proof is trivial and we omit it.

Following F.D. Gakhov (compare Sect. 21), we look for a solution $\Phi_0(z)$ of the form

$$\Phi_0(z) = \frac{\Psi_0(z)}{2\pi i}\int_1^\infty \frac{g(x)dx}{\Psi_0^+(x)(x-z)}, \tag{29.7}$$

where $\Psi_0(z)$ is a solution of the homogeneous problem, which has to be chosen such that

1) $\Psi_0(z) \in B$,

2)

$$\max_{\epsilon\leq\theta\leq 2\pi-\epsilon} |\Psi_0(re^{i\theta})| < \exp(-C_\epsilon r^{\sigma_0}),$$

$$\sigma < \sigma_0 < \min(\rho, 1/2), \tag{29.8}$$

for any $\epsilon > 0$

3)

$$g_0(t) = g(t)/\Psi_0^+(t) \in H_{[t_0,\infty)}, \quad g_0(\infty) = 0, \tag{29.9}$$

for any $t_0 > 1$

4) For all t, $0 \leq t-1 \leq \epsilon_0$, where ϵ_0 is sufficiently small, the representation

$$g_0(t) = \frac{\psi_0(t)}{(t-1)^{\alpha+i\beta}}, \quad \alpha + i\beta = -\frac{\log G(1)}{2\pi i}, \quad 0 \leq \alpha < 1, \tag{29.10}$$

holds with $\psi_0(t) \in H$ and in addition, if $\alpha + i\beta = 0$, then $\psi_0(1) = 0$. If $\Psi_0(z)$ is such a solution, then the function $\Phi_0(z)$ defined by (29.7) satisfies the conditions of Lemma 29.1. Actually, it follows from conditions 3) and 4) (see Sect. 22) that the integral in (29.7) is bounded for $|z-1| > \epsilon > 0$. Using condition 4) in the same way as in Sect. 22 we see that it is bounded for $|z-1| < \epsilon$ as well, i.e. $\Phi_0(z) \in B$. That the function $\Phi_0(z)$ has the order of decreasing no less than σ, follows from condition 2).

Let us proceed to the construction of the solution $\Psi_0(z)$. We shall obtain it as a product

$$\Psi_0(z) = \Psi_{1,0}(z)\Psi_{2,0}(z),$$

where the functions $\Psi_{k,0}(z)$, $k = 1, 2$ are bounded solutions of the problems

$$\Psi_{1,0}^+(t) = \exp\{2\pi i(\varphi(t)t^\rho - t^{\sigma_0})\}\Psi_{1,0}^-(t), \tag{29.11}$$

$$\Psi_{2,0}^+(t) = |G(t)| \exp(2\pi i t^{\sigma_0})\Psi_{2,0}^-(t). \tag{29.12}$$

In addition, the product $\Psi_0^+(t) = \Psi_{1.0}^+\Psi_{2,0}^+(t)$ should satisfy conditions 3) and 4) and also $\Psi_{2,0}(z) \in B_{\sigma_0}$ with $h_{\Psi_{2,0}}(\theta) < 0$ for $\theta \in (0, 2\pi)$. According to Theorem 26.1, one has to construct the function $\Psi_{1,0}(z)$ in the form:

$$\Psi_{1,0}(z) = \exp\Big\{z\int_1^\infty \frac{\varphi(x)x^\rho - x^{\sigma_0} - n(x)}{x(x-z)}dx\Big\}\prod_{n=1}^\infty \frac{1 - z/\zeta_n}{1 - z/t_n}, \tag{29.13}$$

where $t_n = |\zeta_n|$ and $n(r)$ is the counting function of the sequence $\{\zeta_n\}$, which is defined as follows. First, define the values $t_n = |\zeta_n|$ by setting

$$n(r) = \Big[\max_{1\leq x\leq r}\Big\{\varphi(x)x^\rho - x^{\sigma_0} + \frac{1}{2}\Big\}^+\Big], \tag{29.14}$$

and then put

$$\zeta_n = t_n \exp(it_n^{-\rho}). \tag{29.15}$$

Below we show that, with such a choice of the points ζ_n, both the integral and the infinite product in (29.13) are convergent.

The function $\Psi_{2,0}(z)$ is constructed in such a form as in Theorem 22.1, i.e.

$$\Psi_{2,0}(z) = X(z)F(z), \tag{29.16}$$

where

$$X(z) = \exp\left\{\frac{z}{2\pi i}\int_1^\infty \frac{\log|G(x)| + 2\pi i x^{\sigma_0}}{x(x-z)}dx\right\}, \tag{29.17}$$

$$F(z) = \prod_{n=1}^{\infty}\left(1+\frac{z}{s_n}\right), \quad s_n = \left(\frac{2n-1}{2\cos\sigma_0\pi}\right)^{1/\sigma_0}. \tag{29.18}$$

Thus, both the functions $\Psi_{1,0}(z)$ and $\Psi_{2,0}(z)$ are constructed. It remains to prove that they satisfy the desired conditions. The proof is rather cumbersome and is carried out in the following sections.

30 Auxiliary statements

First consider the asymptotic behavior of the sequence $t_n = |\zeta_n|$. Without loss of generality we may assume

$$\rho/(\rho+1) < \mu < (2\rho - \sigma_0)/(\rho+1), \tag{30.1}$$

Denote $\mu_0 = \mu - \rho/(\rho+1)$.

Lemma 30.1 *Let conditions (29.2),(29.3) be fulfilled and also*

$$\psi(t) = \varphi(t)t^\rho - t^{\sigma_0}, \quad \sigma_0 < \min(\rho, 1/2).$$

Then the following equation with respect to h is solvable:

$$\psi(t+h) - \psi(t) = 1, \quad t \geq 0, \quad 1 - t < h < \infty$$

and all its solutions satisfy

$$\frac{1}{\lambda\rho} t^{1-\rho}(1 - ct^{-\mu_0(1+\rho)}) < h < \frac{1}{\lambda\rho} t^{1-\rho}(1 + ct^{-\mu_0(1+\rho)}),$$

where c is a positive constant independent of t and h.

Proof. Denote

$$\Delta(t,h) = \psi(t+h) - \psi(t) = \big(\varphi(t+h) - \lambda\big)\big((t+h)^\rho - t^\rho\big) +$$
$$+\lambda\big((t+h)^\rho - t^\rho\big) + t^\rho\big(\varphi(t+h) - \varphi(t)\big) -$$
$$-\big((t+h)^{\sigma_0} - t^{\sigma_0}\big) = s_1 + s_2 + s_3 + s_4.$$

First let us prove that, for sufficiently large t, all solutions h of the equation $\Delta(t,h) = 1$ satisfy $h > t/2$. Indeed, if t is large enough, then

$$s_1 + s_2 > \frac{\lambda}{2}\big((t+h)^\rho - t^\rho\big) >$$
$$> \frac{\lambda}{2}\rho h^\rho \int_{1/2}^1 (x + t/h)^{\rho-1} dx > \lambda\rho h^\rho/(3 \cdot 2^{\rho+1}),$$
$$|s_3| < C_1 t^\rho \Big|\frac{1}{t+h} - \frac{1}{t}\Big|^\mu = C_1 t^\rho \Big(\frac{h}{t+h}\Big)^\mu t^{-\mu} \leq 2^\rho C_1 h^\rho t^{-\mu},$$
$$|s_4| \leq (3h)^{\sigma_0} + (2h)^{\sigma_0}.$$

Therefore $\Delta(t,h) > 1$ for $t > t_0$. Similarly one can prove that, for sufficiently large t, all solutions of the equation $\Delta(t,h) = 1$ satisfy $h < -t/2$.

Now consider the value $\Delta(t,h)$ under the condition

$$2^{\rho+3}t^{1-\rho}/(\lambda\rho) < |h| < t/2.$$

In this case, for sufficiently large t, we have

$$|s_1+s_2| > \frac{2}{3}\lambda\rho t^{\rho-1}\int_0^{|h|}(1\pm y/t)^{\rho-1}dy >$$

$$> \frac{2}{9}\lambda\rho 2^{-\rho+1}t^{\rho-1}|h| > \lambda\rho 2^{-\rho-2}|h|t^{\rho-1},$$

$$|s_3| < C_1 t^\rho\left|\frac{1}{t+h}-\frac{1}{t}\right|^\mu < 2^\mu C_1 t^{\rho-2\mu}|h|^\mu <$$

$$< C_2|h|t^{\rho-1-\mu_0(1+\rho)}$$

and (we recall that $\mu_0 = \mu - \rho/(\rho+1)$)

$$|s_4| < C_3|h|t^{\sigma_0-1},$$

whence

$$|\Delta(t,h)| > |h|t^{\rho-1}(2^{-\rho-2}\lambda\rho - C_2 t^{-\mu_0(1+\rho)} - C_3 t^{\sigma_0-\rho}) > 3/2$$

for sufficiently large t.

At last consider the asymptotic behavior of the quantity $\Delta(t,h)$ as $t\to\infty$ and $|h| < 2^{\rho+3}t^{1-\rho}/(\lambda\rho)$. We have

$$|s_1| < \frac{C_1}{(t+h)^\mu}\cdot 2^{\rho+1}|h|t^{\rho-1} < C_4|h|t^{\rho-1-\mu} < C_5 t^{-\mu},$$

$$s_2 = \lambda\rho h t^{\rho-1} + \lambda\rho(\rho-1)t^{\rho-2}\int_0^h dx\int_0^x(1+y/t)^{\rho-2}dy =$$

$$= \lambda\rho h t^{\rho-1} + O(h^2t^{\rho-2}) = \lambda\rho h t^{\rho-1} + O(t^{-\rho}),$$

$$s_3 = O(t^{-\mu_0(1+\rho)}),\quad s_4 = O(t^{\sigma_0-\rho}).$$

Thus $\Delta(t,h) = \lambda\rho h t^{\rho-1} + O(t^{-\mu(1+\rho)})$ which implies both the existence of a solution of the equation $\Delta(t,h) = 1$ and its asymptotic behavior is $h = t^{1-\rho}/(\lambda\rho) + O(t^{1-\rho-\mu_0(1+\rho)})$ as $t\to\infty$.

Lemma 30.2 *Let $\{t_n\}$ be the sequence of positive numbers, whose counting function $n(t)$ is defined by (29.14). Then, for all $n = 1,2,\ldots$ and $t \in [t_n, t_{n+1}]$ we have*

$$C_1 t^{1-\rho} < t_{n+1} - t_n < C_2 t^{1-\rho}, \tag{30.2}$$

$$C_1 < t_{n+1}^\rho - t_n^\rho < C_2, \tag{30.3}$$

where C_1 and C_2 are positive constants independent of n and t.

Proof. By virtue of Lemma 30.1 we have

$$\frac{1}{\lambda\rho} t_n^{1-\rho}(1 - c t_n^{-\mu_0(1+\rho)}) <$$

$$< t_{n+1} - t_n < \frac{1}{\lambda\rho} t_n^{1-\rho}(1 + c t_n^{-\mu_0(1+\rho)}) \tag{30.4}$$

for sufficiently large n. Hence, for sufficiently large n, we get $t_{n+1} \leq 2t_n \leq 2t$, which implies relation (30.2). Relation (30.3) is a direct consequence of (30.2).

Lemma 30.3 *Let $f(t)$ and $g(t)$ be bounded complex-valued functions in $[t_0, \infty)$, $t_0 \geq 1$ satisfying the following conditions:*

1) If $t \geq t_1 \geq t_0$ and $0 < h < C_1 t^{1-\rho}$, then

$$|f(t+h) - f(t)| \leq C_2\Big(\Big|\frac{1}{t+h} - \frac{1}{t}\Big|^{\mu_0} + h t^{\rho-1}\Big),$$

2) $g(t) \in H_{[t_1,\infty)}(\delta), \quad 0 < \delta \leq \mu_0, \quad g(\infty) = 0.$

Then

$$g_0(t) = g(t)f(t) \in H_{[t_1,\infty)}\Big(\frac{\delta}{\rho+1}\Big), \quad g_0(\infty) = 0.$$

Proof. By the hypothesis we have, for $t > t_1$,

$$|f(t)| \leq C_3, \quad |g(t+h) - g(t)| \leq$$

$$\leq C_4\Big|\frac{1}{t+h} - \frac{1}{t}\Big|^{\delta}, \quad |g(t)| \leq C_5 t^{-\delta}.$$

Let $t \geq t_1 \geq 1$ and $0 < h < C_1 t^{1-\rho}$. Then

$$|g_0(t+h) - g_0(t)| \leq |f(t+h)|\,|g(t+h) - g(t)| +$$

$$+ |g(t)|\,|f(t+h) - f(t)| < C_3 C_4\Big|\frac{1}{t+h} - \frac{1}{t}\Big|^{\delta} +$$

$$+ C_5 C_2 t^{-\delta}\Big(\Big|\frac{1}{t+h} - \frac{1}{t}\Big|^{\mu_0} + h t^{\rho-1}\Big) <$$

$$< C_6(1 + t^{-\delta})\Big|\frac{1}{t+h} - \frac{1}{t}\Big|^{\delta} + C_7 h t^{\rho-1-\delta} \leq$$

$$\leq C_8\Big|\frac{1}{t+h} - \frac{1}{t}\Big|^{\delta/(\rho+1)}\{1 + h^{1-\delta/(\rho+1)} t^{\rho-1-\delta\rho/(\rho+1)} \times$$

$$\times (t+h)^{\delta/(\rho+1)}\} < C_9\Big|\frac{1}{t+h} - \frac{1}{t}\Big|^{\delta/(\rho+1)}.$$

Now let $h > C_1 t^{1-\rho}$, then

$$|g_0(t+h) - g_0(t)| < C_3\big(|g(t+h)| + |g(t)|\big) <$$

$$< C_{10} t^{-\delta} = C_{10}\left|\frac{1}{t+h} - \frac{1}{t}\right|^{\delta/(\rho+1)} \times$$

$$\times \{t^{-\delta\rho/(\rho+1)}(1+t/h)^{\delta/(\rho+1)}\} < C_{11}\left|\frac{1}{t+h} - \frac{1}{t}\right|^{\delta/(\rho+1)}$$

This completes the proof.

Lemma 30.4 *Let $\{t_n\}_{n=1}^{\infty}$ be the sequence of positive numbers with the counting functions $n(r)$ defined by (29.14). Consider the sequence of functions $\{n_m(t)\}_{m=1}^{\infty}$ defined by the relations: $n_1(t) = n(t)$*

$$n_m(t) = \begin{cases} n(t), & t \in [1, t_{m-1}) \cup [t_{m+2}, \infty), \\ m, & t \in [t_{m-1}, t_{m+2}). \end{cases} \tag{30.5}$$

for $m \geq 2$ Then the following representation holds

$$\varphi(t)t^{\rho} - t^{\sigma_0} - n_m(t) = \beta_m(t) + \gamma_m(t), \quad t \geq 1, \tag{30.6}$$

in which the functions $\beta_m(t)$ and $\gamma_m(t)$ satisfy the conditions

1) $|\beta_m(t)| < C_1,$ (30.7)

2) $|\int_1^t \beta_m(x)x^{\rho-1}dx| < C_2,$ (30.8)

3) $|\gamma_m(t+h) - \gamma_m(t)| < C_3\left|\frac{1}{t+h} - \frac{1}{t}\right|^{\mu_0}, \quad h \geq 0,$ (30.9)

4) $\gamma_m(1) = \gamma_m(\infty) = 0.$ (30.10)

The positive constants C_1, C_2, C_3 are independent of t and m.

Proof. First, note that

$$(t_{m+1}^{\rho} - t_m^{\rho})^{-1} = \lambda + O(t_m^{-\mu_0(1+\rho)}), \quad t_m \to \infty. \tag{30.11}$$

Indeed, denoting $h_m = t_{m+1} - t_m$ and using (30.4), we have

$$(t_{m+1}^{\rho} - t_m^{\rho})^{-1} - \lambda = t_m^{-\rho}\{(1 + h_m/t_m)^{\rho} - 1\}^{-1} - \lambda =$$

$$= t_m^{1-\rho}/(\rho h_m) + O(t_m^{-\rho}) - \lambda =$$

$$= \lambda(1 + O(t_m^{-\mu_0(1+\rho)})) - \lambda + O(t_m^{-\rho}) = O(t_m^{-\mu_0(1+\rho)}).$$

Define the function $\beta_m(t)$ by the equality

$$\beta_m(t) = \begin{cases} \dfrac{2t^\rho - t_{n+1}^\rho - t_n^\rho}{2(t_{n+1}^\rho - t_n^\rho)}, & t \in [t_n, t_{n+1}), \\ & n = 0, \ldots, m-2, m+2, m+3, \ldots \\ 3\dfrac{2t^\rho - t_{m+2}^\rho - t_{m-1}^\rho}{2(t_{m+2}^\rho - t_{m-1}^\rho)}, & t \in [t_{m-1}, t_{m+2}), \end{cases} \tag{30.12}$$

where $t_0 = 1$. One can easily see that (30.12) and (29.14) imply

$$\beta_m(t_n) = \varphi(t_n)t_n^\rho - t_n^{\sigma_0} - n_m(t_n) = \\ = \begin{cases} -\dfrac{1}{2}, & n \neq m-1, m, m+1, \\ -\dfrac{3}{2}, & n = m-1. \end{cases} \tag{30.13}$$

Besides, (30.12) yields $|\beta_m(t)| \leq 3/2$, $t \geq 1$, $m = 1, 2, \ldots$, which proves (30.7).

For a fixed m define a sequence $\{\tilde{t}_n\}_{n=1}^\infty$ by setting

$$\tilde{t}_n = \begin{cases} t_n, & n = 0, 1, \ldots, m-1, \\ t_{n+2}, & n - m, m+1, \ldots \end{cases}$$

For any $t \geq 1$ find $\tilde{t}_n$ such that $t_n \leq t < \tilde{t}_{n+1}$. We have

$$\int_1^t \beta_m(x)x^{\rho-1}dx =$$

$$= \sum_{k=0}^{n-1} \frac{1}{\tilde{t}_{k+1}^\rho - \tilde{t}_k^\rho} \int_{\tilde{t}_k}^{\tilde{t}_{k+1}} \left(x^\rho - \frac{1}{2}(\tilde{t}_{k+1}^\rho + \tilde{t}_k^\rho)\right)x^{\rho-1}dx +$$

$$+ \frac{1}{\tilde{t}_{n+1}^\rho - \tilde{t}_n^\rho} \int_{\tilde{t}_n}^t \left(x^\rho - \frac{1}{2}(\tilde{t}_{n+1}^\rho + \tilde{t}_n^\rho)\right)x^{\rho-1}dx =$$

$$= \sum_{k+1}^{n-1} J_k + J_n(t).$$

One can easily see that $J_k = 0$ for $k = 1, 2, \ldots, n-1$. Using (30.3), we find that $|J_n(t)| < C$, which proves relation (30.8).

Define the function $\gamma_m(t)$ by relation (30.6) and prove that it satisfies conditions (30.9) and(30.10). Let, again, $t \geq 1$ and $\tilde{t}_n \leq t < \tilde{t}_{n+1}$. By virtue of (30.2), there exists $C_1 > 0$ independent of m, n and t, and such that, for $0 < h < C_1 t^{1-\rho}$, at least one of the following inequalities holds:

$$\tilde{t}_n \le t < t+h \le \tilde{t}_{n+1}, \quad \tilde{t}_n \le t < \tilde{t}_{n+1} < t+h < \tilde{t}_{n+2}. \tag{30.14}$$

Let the first of these inequalities hold. Then

$$\gamma_m(t+h) - \gamma_m(t) = \Big\{\varphi(t+h) - \frac{\delta_n^m}{\tilde{t}_{n+1}^\rho - \tilde{t}_n^\rho}\Big\}\times$$

$$\times\big((t+h)^\rho - t^\rho\big) + \big(\varphi(t+h) - \varphi(t)\big)t^\rho + \big((t+h)^{\sigma_0} - t^{\sigma_0}\big),$$

where $\delta_n^m = 3$ for $n = m-1$ and $\delta_n^m = 1$ for $n \ne m-1$. Since $\varphi(t) \in H(\mu)$, taking into account (30.11), we have

$$|\gamma_m(t+h) - \gamma_m(t)| <$$

$$< \{C_2 t^{-\mu} + C_2 t^{-\mu_0(1+\rho)}\} h t^{\rho-1} + C_3 h^\mu t^{\rho-2\mu} + C_4 h t^{\sigma_0 - 1}.$$

It follows from (30.1) that $\mu > \mu_0(1+\rho)$ and $\sigma_0 - 1 < \rho - 1 - \mu_0(1+\rho)$, therefore inequality $0 < h < C_1 t^{1-\rho}$ implies

$$h t^{\rho - 1 - \mu_0(1+\rho)} < C_5 h^\mu t^{\rho - 2\mu},$$

whence

$$|\gamma_m(t+h) - \gamma_m(t)| < C_6 h^\mu t^{\rho-2\mu} =$$

$$= C_6 \Big|\frac{1}{t+h} - \frac{1}{t}\Big|^{\mu_0} h^{\mu-\mu_0}(t+h)^{\mu_0} t^{\rho - 2\mu + \mu_0} \le$$

$$\le C_7 \Big|\frac{1}{t+h} - \frac{1}{t}\Big|^{\mu_0}.$$

Let $0 < h < C_1 t^{1-\rho}$ and the second of inequalities (30.14) hold. In this case we have

$$|\gamma_m(t+h) - \gamma_m(t)| \le$$

$$\le |\gamma_m(t+h) - \gamma_m(\tilde{t}_{n+1})| + |\gamma_m(\tilde{t}_{n+1}) - \gamma_m(t)| \le$$

$$\le C_7\Big(\Big|\frac{1}{t+h} - \frac{1}{\tilde{t}_{n+1}}\Big|^{\mu_0} + \Big|\frac{1}{\tilde{t}_{n+1}} - \frac{1}{t}\Big|^{\mu_0}\Big) \le 2C_7 \Big|\frac{1}{t+h} - \frac{1}{t}\Big|^{\mu_0}.$$

Thus we have proved (30.9) for $0 < h < C_1 t^{1-\rho}$.

Let us prove (30.9) for $h > C_1 t^{1-\rho}$. Suppose $\tilde{t}_n \le t < \tilde{t}_{n+1}$. From the definition of the function $\gamma_m(t)$ and (30.12) we see that $\gamma_m(t_n) = 0$. Therefore we have

$$|\gamma_m(t)| = |\gamma_m(t) - \gamma_m(\tilde{t}_n)| \le 2C_7 \Big|\frac{1}{\tilde{t}_{n+1} - \frac{1}{\tilde{t}-n}}\Big|^{\mu_0} \le C_8 t^{-\mu_0(1+\rho)}, \tag{30.15}$$

whence

$$|\gamma_m(t+h)-\gamma_m(t)| \le 2C_8 t^{-\mu_0(1+\rho)} =$$

$$\le 2C_8\Big|\frac{1}{t+h}-\frac{1}{t}\Big|^{\mu_0}\Big(\frac{t+h}{t^\rho h}\Big)^{\mu_0} < C_9\Big|\frac{1}{t+h}-\frac{1}{t}\Big|^{\mu_0},$$

which proves the relation (30.9).

The second of conditions (30.10) is a direct consequence of (30.15) and the first one is evident.

Using the obtained properties of the sequence $\{t_n\}$, we can start to study the function $\Psi_{1,0}(z)$ defined in Sect. 29. The most difficult part of this studying is the following statement.

Lemma 30.5 *Let $\{t_n\}$ be the sequence of positive numbers with the counting function $n(r)$ defined by (29.14) and $\{\zeta_n\}$ be the sequence defined by (29.15). Then both the integral and the infinite product in the right-hand side of (29.13) converge absolutely and uniformly on every compact in D and the function $\Psi_{1,0}(z)$ satisfies the conditions:*

1) $\sup_{t\ge c}|\log|\Psi^+_{1,0}(t)||<\infty$ *for any* $c>1$;

2) $|\log\Psi^+_{1,0}(t+h)-\log\Psi^+_{1,0}(t)| < C_2\Big(\Big|\frac{1}{t+h}-\frac{1}{t}\Big|^{\mu_0}+ht^{\rho-1}\Big)$

for sufficiently large $t>t_*$ *and* $0<h<C_1t^{1-\rho}$.

Proof. It follows from equality (30.6) with $m = 1$ that the function $\varphi(x)x^\rho - x^{\sigma_0} - n(x)$ is bounded on the ray $[1,\infty)$. Therefore the integral in (29.13) converges absolutely and uniformly on any compact in D. The convergence of the infinite product follows from Lemma 26.1, since, by (29.15),

$$\sum_{n=1}^{\infty}\frac{1}{\sqrt{t_n}}\sin(t_n^{-\rho}/2) \le \sum_{n=1}^{\infty} t_n^{-\rho-\frac12} =$$

$$= \Big(\rho+\frac12\Big)\int_0^\infty \frac{n(x)}{x^{\rho+3/2}}dx < \infty. \tag{30.16}$$

Using the Sohotski formula, we have

$$\log\Psi^+_{1,0}(t) = t\int_1^\infty \frac{\varphi(x)-x^{\sigma_0}-n(x)}{x(x-t)}dx +$$

$$+\pi i\{\varphi(t)t^\rho - t^{\sigma_0} - n(t)\} + \sum_{n=1}^{\infty}\log\frac{t_n - t\exp(-it_n^{-\rho})}{t_n-t}, \tag{30.17}$$

where

$$\log\frac{t_n - t\exp(-it_n^{-\rho})}{t_n-t}$$

is the boundary value on the ray $\{t:\ 1\le t<\infty\}$ of that branch of the logarithm in the domain $\{t:\ 0<\Im t<|t|\sin|t|^{-\rho},\Re t>0\}$, which vanishes as $t\to\infty$. For every $m\ge 2$ consider the function $\log\Psi_{1,0}(t)$ in the interval $t_m\le t<t_{m+1}$. Using notation (30.5), rewrite (30.17) in the form

$$\log\Psi_{1,0}^{+}(t)=t\int_1^\infty\frac{\varphi(x)x^\rho-x^{\sigma_0}-n_m(x)}{x(x-t)}dx+$$

$$+\pi i\{\varphi(t)t^\rho-t^{\sigma_0}-n_m(t)\}+\sum_{n=1}^{\infty}{}''\log\frac{t_n-t\exp(-it_n^{-\rho})}{t_n-t}+$$

$$+\log\frac{(1-t/\zeta_m)(1-t/\zeta_{m+1})}{(1-t/\zeta_{m-1})(1-t/\zeta_{m+2})}=$$

$$=I_m(t)+Q_m(t)+V_m(t)+U_m(t),\tag{30.18}$$

where the sign $''$ denotes that the summands corresponding to $n=m,m+1$ are omitted and summands corresponding to $n=m-1,m+2$ are doubled.

To prove the lemma it suffices to obtain, for m large enough and $t\in[t_m,t_{m+1})$, the estimates

$$|\log|\Psi_{1,0}^{+}(t)||<C_2,$$

$$|\log\Psi_{1,0}^{+}(t+h)-\log\Psi_{1,0}^{+}(t)|<$$

$$<C_3\Big(\Big|\frac{1}{t+h}-\frac{1}{t}\Big|^{\mu_0}+ht^{\rho-1}\Big),$$

where $0<h<\frac{1}{2}(t_{m+2}-t_{m+1})$. (It should be noted that, in view of Lemma 30.2, the latter condition is equivalent to $0<h<C_1t^{1-\rho}$.)

Using equality (30.6), one may reduce the integral $I_m(t)$ to the form

$$I_m(t)=t\int_1^\infty\frac{\gamma_m(x)}{x(x-t)}dx+t\int_{E_m}\frac{\beta_m(x)x^{\rho-1}}{x^\rho(x-t)}dx+$$

$$+t\int_{t_{m-1}}^{t_{m+2}}\frac{\beta_m(x)}{x(x-t)}dx=I_{m1}(t)+I_{m2}(t)+I_{m3}(t),\tag{30.19}$$

where $E_m=[1,t_{m-1}]\cup[t_{m+2},\infty)$. Since, according to Lemma 30.4,

$$|\gamma_m(t+h)-\gamma_m(t)|<C_4\Big|\frac{1}{t+h}-\frac{1}{t}\Big|^{\mu_0},$$

$$t\ge 1,\quad h>0,\quad\text{and}\quad \gamma_m(1)=\gamma_m(\infty)=0,$$

then, by the appropriate properties of singular integrals (see [2], p.53), we have

$$|I_{m1}(t+h) - I_{m1}(t)| \leq C_5 \Big|\frac{1}{t+h} - \frac{1}{t}\Big|^{\mu_0},$$

$$|I_{m1}(t)| \leq C_6, \quad t \geq 1.$$

When passing to the integral $I_{m2}(t)$, let us note that the function $\nu(x) = x^{-\rho}(x-t)^{-1}$ has only one extremum, located at the point $x_0 = \rho t/(\rho+1)$. Since

$$\min_{x \in E_m} |x - t| \geq C_7 t^{1-\rho},$$

this implies

$$\max_{x \in E_m} |\nu(x)| \leq C_8/t.$$

By virtue of Lemmas 22.2 and 30.4 we have

$$|I_{m2}(t)| \leq C_9 \quad \text{for } t_m \leq t < t_{m+1}.$$

In what follows we also need the estimate

$$|I'_{m2}(t)| = \Big|\int_{E_m} \frac{\beta_m(x) x^{\rho-1}}{x^{\rho-1}(x-t)^2} dx\Big| \leq C_{10} t^{\rho-1},$$

$$t \in \Big[t_m, \frac{1}{2}(t_{m+1} + t_{m+2})\Big).$$

To estimate the integral $I_{m3}(t)$ draw a half-circle s_m of diameter $[t_{m-1}, t_{m+2}]$ lying in the half-plane $\Im z > 0$. According to (30.12), the function $\beta_m(z)$ is analytic within this half-circle. Using the Cauchy theorem and the Sohocki formulas we obtain

$$I_{m3}(t) = \pi i \beta_m(t) - t \int_{s_m} \frac{\beta_m(z)}{z(z-t)} dz.$$

Since $|\beta_m(t)| \leq 3/2$, $t \in [t_{m-1}, t_{m+2})$, by using Lemma 30.2 we obtain

$$|\beta_m(z)| \leq |\beta_m(t)| + \Big|\int_t^z \beta'_m(u) du\Big| \leq$$

$$\leq \frac{3}{2} + 2(t_{m+2} - t_{m-1}) \frac{3}{2(t^\rho_{m+2} - t^\rho_{m-1})} \cdot \rho t^{\rho-1}_{m+2} \leq C_{11}$$

for $z \in s_m$. Therefore $|I_{m3}(t)| \leq C_{12}$ for $t \in [t_m, t_{m+1})$ and also

$$|I'_{m3}(t)| = \Big|\frac{3\pi i \rho t^{\rho-1}}{t^\rho_{m+2} - t^\rho_{m-1}} - \int_{s_m} \frac{\beta_m(z)}{(z-t)^2} dz\Big| \leq$$

$$\leq C_{13}t^{\rho-1}, \quad t \in \Big[t_m, \frac{1}{2}(t_{m+1}+t_{m+2})\Big).$$

Combining the obtained estimates and substituting them into (30.19), we see that, for $t \in [t_m, t_{m+1})$ and $0 < h < \frac{1}{2}(t_{m+2} - t_{m+1})$,

$$|I_m(t)| \leq C_{14}, \tag{30.20}$$

$$|I_m(t+h) - I_m(t)| \leq C_{15}\Big(\Big|\frac{1}{t+h} - \frac{1}{t}\Big|^{\mu_0} + ht^{\rho-1}\Big). \tag{30.21}$$

The estimate of the second summand in (30.18) may be obtained rather simply. Since

$$Q_m(t) = \pi i(\beta_m(t) + \gamma_m(t)),$$

we have by Lemma 30.4 and relation (30.6) that

$$|Q_m(t)| \leq C_{16}, \tag{30.22}$$

$$|Q_m(t+h) - Q_m(t)| \leq C_{17}\Big(\Big|\frac{1}{t+h} - \frac{1}{t}\Big|^{\mu_0} + ht^{\rho-1}\Big) \tag{30.23}$$

for $t \in [t_m, t_{m+1})$ and $0 < h < \frac{1}{2}(t_{m+2} - t_{m+1})$.

When passing to the third summand in (30.18), represent it in the form

$$V_m(t) = \int_{L_m} \log\frac{x - t\exp(-ix^{-\rho})}{x-t}dn(x)+$$

$$+\int_{l_m} \log\frac{x - t\exp(-ix^{-\rho})}{x-t}d[n_m(x) - n_m(t)], \tag{30.24}$$

where

$$L_m = \Big[1, \frac{1}{2}t_m\Big] \cup \Big[\frac{3}{2}t_{m+1}, \infty\Big), \quad l_m = \Big(\frac{1}{2}t_m, \frac{3}{2}t_{m+1}\Big).$$

Assume that m is sufficiently large such that $t_m < 2t_{m-1}$, $t_{m+2} < \frac{3}{2}t_{m+1}$. Since, by virtue of (29.14),we have $n(t) = O(t^\rho)$ as $t \to \infty$, we obtain integrating by parts in (30.24)

$$V_m(t) = t\int_{L_m}\frac{(1-\exp(-ix^{-\rho}))n(x)}{(x\exp(ix^{-\rho}) - t)(x-t)}dx - i\rho t\int_{L_m}\frac{x^{-\rho-1}n(x)dx}{x\exp(ix^{-\rho}) - t}+$$

$$+t\int_{l_m}\frac{(1-\exp(-ix^{-\rho}))[n_m(x) - n_m(t)]}{(x\exp(ix^{-\rho}) - t)(x-t)}dx-$$

$$-i\rho t\int_{l_m}\frac{x^{-\rho-1}[n_m(x) - n_m(t)]}{x\exp(ix^{-\rho}) - t}dx + [n_m(x) - n_m(t)]\times$$

$$\times \log\Big(1+\frac{t(1-\exp(-ix^{-\rho}))}{x-t}\Big)\Big|_{x=\frac{1}{2}t_m}^{\frac{3}{2}t_{m+1}} =$$

$$= V_{m1}(t)+V_{m2}(t)+V_{m3}(t)+V_{m4}(t)+V_{m5}(t). \tag{30.25}$$

Since $|\exp(\pm ix^{-\rho})-1|n(x)=O(1)$, and $|x\exp(ix^{-\rho})-t|\ge|x-t|$, we have for $t\in[t_m,t_{m+1})$

$$|V_{m1}(t)|\le C_{18}t\Big\{\int_0^{t/2}\frac{dx}{(x-t)^2}+\int_{3t/2}^{\infty}\frac{dx}{(x-t)^2}\Big\}\le C_{19},$$

$$|\Re V_{m2}(t)|\le C_{20}t\int_{L_m}\frac{dx}{x^\rho|x\exp(ix^{-\rho})-t|^2}\le C_{21}.$$

Consider the integral

$$\Re V_{m3}(t)=-t\int_{l_m}\frac{(x+t)(1-\cos(x^{-\rho}))[n_m(x)-n_m(t)]}{(x^2-2xt\cos(x^{-\rho})+t^2)(x-t)}dx.$$

Let us verify that if m is sufficiently large, then

$$|n_m(x)-n_m(t)|\le C_{22}|x-t|t^{\rho-1},$$

$$x\in\Big[\frac{1}{4}t,\frac{7}{4}t\Big],\quad t\in[t_m,t_{m+1}). \tag{30.26}$$

Denote $x=t+\Delta t$. Since $n_m(x)-n_m(t)=0$ for $x\subset(t_{m-1},t_{m+2})$. We have to prove relation (30.26) only in the case

$$\frac{3}{4}t>|\Delta t|>|\max(t_m-t_{m-1},t_{m+2}-t_{m+1})>C_{23}t^{1-\rho}.$$

Using (30.1),we have (compare with the similar estimates in the proof of Lemma 30.1):

$$|n_m(t+\Delta t)-n_m(t)|=$$

$$=|\varphi(t+\Delta t)\big((t+\Delta t)^\rho-t^\rho\big)+t^\rho(\varphi(t+\Delta t)-\varphi(t))-$$

$$-\big((t+\Delta t)^{\sigma_0}-t^{\sigma_0}\big)+O(1)|\le$$

$$\le C_{24}t^{\rho-1}|\Delta t|(1+t^{-\mu_0(1+\rho)}),$$

which is equivalent to (30.26). Now, using (30.26),we obtain

$$|\Re V_{m3}(t)|\le C_{25}t^{1-\rho}\int_{t/4}^{7t/4}\frac{dx}{(x-t)^2+4xt\sin^2(x^{-\rho}/2)}\le$$

$$\leq C_{26}t^{-\rho}\int_{-3/4}^{3/4}\frac{du}{u^2+C_{27}t^{-2\rho}} < C_{28}.$$

Similarly, with account of (30.26) we obtain the estimate

$$|\Re V_{m4}(t)| \leq C_{29}t^{\rho}\int_{t/4}^{7t/4}x^{-\rho-1}dx \leq C_{30}.$$

Besides, (30.26) implies $|V_{m5}(t)| \leq C_{31}$. Substituting all these estimates in (30.25),we claim that

$$|\Re V_m(t)| \leq C_{32}, \quad t \in [t_m, t_{m+1}). \tag{30.27}$$

Now let us prove that

$$|V_m'(t)| \leq C_3/t, \quad t \in \Big[t_m, \frac{1}{2}(t_{m+1}+t_{m+2})\Big). \tag{30.28}$$

Relation (30.24) implies

$$V_m'(t) = t\int_{L_m}\frac{x(\exp(ix^{-\rho})-1)dn(x)}{(x\exp(ix^{-\rho})-t)(x-t)}+$$

$$+\int_{l_m}\frac{x(\exp(ix^{-\rho})-1)d[n_m(x)-n_m(t)]}{(x-t)(x\exp(ix^{-\rho})-t)}.$$

Denote these integrals by $v_1(t)$, $v_2(t)$. We have

$$|v_1(t)| \leq \Big(\int_1^{t/2}+\int_{3t/2}^{\infty}\Big)\frac{x^{1-\rho}}{(x-t)^2}dn(x) \leq$$

$$\leq \frac{2^{1+\rho}}{t^{1+\rho}}\Big\{n\Big(\frac{1}{2}t\Big)+3^{1-\rho}n\Big(\frac{3}{2}t\Big)\Big\}+$$

$$+\Big(\int_1^{t/2}+\int_{3t/2}^{\infty}\Big)\frac{(\rho+1)x-(\rho-1)t}{x^{\rho}(x-t)^3}n(x)dx.$$

Since $n(x) = O(x^{\rho})$, this implies $|v_1(t)| \leq C_{34}/t$. By taking into account that $n_m(x) - n_m(t) = 0$ for $x \in (t_{m-1}, t_{m+2})$, we obtain

$$|v_2(t)| \leq \int_{t/4}^{7t/4}\frac{x^{1-\rho}}{(x-t)^2}d[n_m(x)-n_m(t)] =$$

$$= \Big(\frac{n_m(x)-n_m(t)}{x^{\rho-1}(x-t)^2}\Big)\Big|_{x=t/4}^{x=7t/4}+$$

$$+\Big(\int_{t/4}^{t_{m-1}}+\int_{t_{m+2}}^{7t/4}\Big)\frac{(\rho+1)x-(\rho-1)t}{x^{\rho}(x-t)^3}[n_m(x)-n_m(t)]dx,$$

whence, granting relation (30.26), we obtain $|v_2(t)|\le C_{35}/t$. Thus estimate (30.28) is proved. It follows from this estimate that

$$|V_m(t+h)-V_m(t)|\le C_{36}ht^{\rho-1}. \qquad (30.29)$$

for $t\in[t_m,t_{m+1})$ and $0<h<\frac{1}{2}(t_{m+2}-t_{m+1})$.

Finally, consider the last summand $U_m(t)$ in (30.18).

$$\Re U_m(t)|=\log\Big|\frac{\zeta_m-t}{\zeta_{m-1}-t}\Big|+\log\Big|\frac{\zeta_{m+1}-t}{\zeta_{m+2}-t}\Big|+\log\frac{t_{m-1}t_{m+2}}{t_mt_{m+1}}.$$

Since, by Lemma 30.2

$$|t_k-t|<C_{37}t^{1-\rho}\quad ,k=m-1,m,m+1,m+2,$$

and

$$|\zeta_k-t|=\{(t_k-t)^2+4tt_k\sin^2\frac{t_k^{-\rho}}{2}\}^{1/2},$$

we get

$$C_{38}t^{1-\rho}<|\zeta_k-t|<C_{39}t^{1-\rho},$$

whence

$$|\Re U_m(t)|\le C_{40},\quad t\in[t_m,t_{m+1}), \qquad (30.30)$$

and also

$$|U_m'(t)|=\Big|\frac{1}{\zeta_{m-1}-t}+\frac{1}{\zeta_{m+2}-t}-\frac{1}{\zeta_m-t}-\frac{1}{\zeta_{m+1}-t}\Big|\le$$
$$\le C_{41}t^{\rho-1}$$

for $t_m\le t\le\frac{1}{2}(t_{m+1}+t_{m+2})$.

The latter estimate yields

$$|U_m(t+h)-U_m(t)|\le C_{42}ht^{\rho-1}. \qquad (30.31)$$

for

$$t\in[t_m,t_{m+1}),\quad 0<h<\frac{1}{2}(t_{m+2}-t_{m+1})$$

Relations (30.18), (30.20), (30.22),(30.27) and(30.30) yield the first assertion of the lemma, and relations (30.18), (30.21), (30.23), (30.29), (30.31) yield the second one.

31 Solution of the homogeneous problem

Lemma 31.1 *Let $\Psi_0(z) = \Psi_{1,0}(z)\Psi_{2,0}(z)$ where $\Psi_{1,0}(z)$ is defined by equations (29.13)-(29.15) and $\Psi_{2,0}(z)$ is defined by equations (29.16)-(29.18). Then the function $\Psi_0(z)$ belongs to the class B and satisfies conditions (29.8)-(29.10).*

Proof. That $\Psi_{1,0}(z)$ belongs to the class B, follows from Theorem 26.2. Actually,by Lemma 30.5, we have $|\Psi_{1,0}^{+}(t)| \le C$ for $t \ge 1$, which provides condition 4) of this theorem. Conditions 3) and 4) follow from the fact that the function $\varphi(t)t^{\rho} - t^{\sigma_0} - n(t)$ is bounded, which in turn is a consequence of (29.14). (It should be remcalled that the function $\Psi_{1,0}(t)$ is a solution of boundary problem (29.11) with the function $\exp\{2\pi i(\varphi(t)t^{\rho} - t^{\sigma_0})\}$ instead of $G(t)$). The checking of condition 1) was already done when proving Lemma 30.5 (see(30.16)).

Lemma 30.5 and the inequality

$$|e^z - e^{\zeta}| \le |z - \zeta| e^{\max(|\Re z|, |\Re \zeta|)}$$

imply that the function $1/\Psi_{1,0}^{+}(t)$ is bounded on $[c, \infty)$ for any $c > 1$ and also the inequality

$$\left| \frac{1}{\Psi_{1,0}^{+}(t+h)} - \frac{1}{\Psi_{1,0}^{+}(t)} \right| < C_2 \left(\left| \frac{1}{t+h} - \frac{1}{t} \right|^{\mu_0} + h t^{\rho-1} \right) \tag{31.1}$$

holds for sufficiently large $t > t_*$ and $0 < h < C_1 t^{1-\rho}$.

Now consider the function $\Psi_{2,0}(z)$. It has the same structure as the function $\Psi_0(z)$ in Theorem 22.1 with only one difference: instead of the function $\varphi(t)$ in (21.2) we use (see (29.12)) the function t^{σ_0}, which does not satisfy the condition $-1 < \varphi(t) \le 0$. This fact does not affect the behavior of the function $\Psi_{2,0}(z)$ outside a neighborhood of the point $z = 1$. Therefore, by means of the same arguments as were used in the proof of Theorem 22.1, one may show that the function $\Psi_{2,0}$ is bounded for $|z - 1| \ge \epsilon > 0$ and also satisfies the condition

$$\left| \frac{1}{\Psi_{2,0}^{+}(t+h)} - \frac{1}{\Psi_{2,0}^{+}(t)} \right| < C_3 \left| \frac{1}{t+h} - \frac{1}{t} \right|^{\nu} \tag{31.2}$$

on the ray $[1 + \epsilon, \infty)$ for some $\nu \in (0, 1]$.

Thus the function $\Psi_0(z) = \Psi_{1,0}(x)\Psi_{2,0}(z)$ is bounded for $|z - 1| > \epsilon$. It follows from the definition of the functions $\Psi_{1,0}(z)$ and $\Psi_{2,0}(z)$ that

$$\Psi_0(z) = \exp\left\{ \frac{z}{2\pi i} \int_1^{\infty} \frac{\log G(z) - 2\pi i n(x)}{x(x - z)} dx \right\} \times$$

$$\times \prod_{n=1}^{\infty} \frac{1 - z/\zeta_n}{1 - z/t_n}(1 + z/s_n), \tag{31.3}$$

where the values $n(x)$, ζ_n, t_n, and s_n are defined in relations (29.14), (29.15), (29.18) respectively. Since $n(x) = 0$ on some segment $[1, 1+\epsilon_0]$, then, if $\epsilon > 0$ is sufficiently small, we have the representation

$$\log \Psi_0(z) = \frac{1}{2\pi i} \int_1^{1+\epsilon_0} \frac{\log G(x)}{x - z} dx + \omega(z) \tag{31.4}$$

for $|z - 1| \leq \epsilon$. Here $\omega(z)$ is an analytic function in the disk $|z - 1| \leq \epsilon$. Using the same arguments as when proving relation (19.2), one may obtain that $|\Psi_0(z)|$ is bounded in $\{z : \ |z - 1| \leq \epsilon\}$ and, hence, $\Psi_0(z) \in B$. A more complete consideration of relation (29.7) developed when proving Theorem 22.1 shows that the function $g_0(t) = g(t)/\Psi_0^+(t)$ satisfies (29.10).

Using conditions (31.1), (31.2) and the fact that $1/\Psi_{1,0}^+(t)$ is bounded for $t \in [c, \infty)$, one can easily prove that $1/\Psi_0^+(t)$ is bounded for $t \in [c, \infty)$ and also

$$\left| \frac{1}{\Psi_0^+(t+h)} - \frac{1}{\Psi_0^+(t)} \right| \leq C_4 \left(\left| \frac{1}{t+h} - \frac{1}{t} \right|^{\delta_0} + ht^{\rho-1} \right)$$

for sufficiently large $t > t_*$ and $0 < h < C_1 t^{1-\rho}$, here $\delta_0 = \min(\mu_0, \nu)$. Since $g(t) \in H$ and $g(\infty) = 0$ (see (29.4), (29.5)) then,applying Lemma 30.3, one can see that the function $g_0(t) = g(t)/\Psi_0^+(t)$ belongs to $H_{[t_*, \infty)}$ and also $g_0(\infty) = 0$. That $g_0(t)$ belongs to the class $H_{[t_0, t_*]}$, for any $t_0 \in (1, t_*)$, follows from the well-known properties of the Cauchy-type integral. So, we have proved relation (29.9).

It remains to verify (29.8). Let the functions $X(z)$ and $F(z)$ be defined by equalities (29.17) and (29.18). By Theorem 19.2 the function $X(z)$ decreases in the angle $[0, 2\pi]$ with the order σ_0 and with the indicator

$$h_X(\theta) = -\frac{\pi}{\sin \pi\sigma_0} \cos \sigma_0(\theta - \pi), \quad 0 \leq \theta \leq 2\pi.$$

Since

$$n(r) = \left[r^{\sigma_0} \cos \sigma_0 \pi + \frac{1}{2} \right],$$

we see (cf. [24, p.129]) that the function $F(z)$ possesses c.r.g. with respect to the order σ_0 and with the indicator $h_F(\theta) = \pi(\cot \sigma_0 \pi) \cos \sigma_0 \theta$, $-\pi \leq \theta \leq \pi$. Therefore, (we recall that $\sigma_0 < 1/2$) the indicator of the function $\Psi_{2,0}(z) = X(z)F(z)$ is negative for all $\theta \in (0, 2\pi)$. Since $\Psi_{1,0}(z) \in B$ and $\Psi_0(z) = \Psi_{1,0}(z)\Psi_{2,0}(z)$, we obtain relation (29.8), which completes the proof.

Remark 1 One can prove that the function $\Psi_{1,0}(z)$ is of zero order and, moreover, outside the union of the disks

$$|z-\zeta_n| < C|\zeta_n|^{1-\rho}, \quad n=0,1,\dots, \quad \zeta_0 = 1,$$

the inequality

$$|\log|\Psi_{1,0}(z)|| < C_1$$

holds.

Remark 2 Denote

$$\Psi_1(z) = \exp\Big\{\frac{z}{2\pi i}\int_1^\infty \frac{\log G(z) - 2\pi i n_1(x)}{x(x-z)}dx\Big\}\times$$
$$\times \prod_{n=1}^\infty \frac{1-\frac{z}{\tau_n}\exp(-i\tau_n^{-\rho})}{1-\frac{z}{\tau_n}}, \tag{31.5}$$

where the sequence $\{\tau_k\}$, $1<\tau_1<\tau_2\dots$ is defined by its counting function

$$n_1(x) = \max_{1\le u\le x}\Big\{\varphi(u)u^\rho + \frac{1}{2}\Big\}^+.$$

The function $\Psi_1(z)$ is a solution of the homogeneous problem $\Psi_1^+(t) = G(t)\Psi_1^-(t)$, $1\le t<\infty$, and belongs to the class B. This function satisfies conditions (29.9) and (29.10) (they are even simpler to check), but condition (29.8) is not satisfied.

Remark 3 If, in contrast to (29.3), one chooses a branch of the function $\arg G(x) = 2\pi\varphi(x)x^\rho$ in (31.5) in such a way that $0\le\varphi(1)<1$, then the function defined by (31.5) still is a solution of the homogeneous problem satisfying (29.9) and (29.10). This function is bounded outside any neighborhood of the point $z=1$ and admits the representation $\Psi_1(z) = (z-1)^{-\varphi(1)}\psi(z)$, $\psi(z) = O(1)$ as $z\to 1$.

Combining Lemmas 29.1, 31.1 and Theorem 27.3, we obtain the final result.

Theorem 31.1 *Let conditions (29.2)–(29.6) be fulfilled. The general solution of the non-homogeneous boundary problem (29.1) in the class* B_σ, $0<\sigma<\min(\rho,1/2)$ *admits the representation*

$$\Phi(z) = \frac{\Psi_0(z)}{2\pi i}\int_1^\infty \frac{g(x)dx}{\Psi_0^+(x)(x-z)} + \Psi(z), \tag{31.6}$$

where $\Psi_0(z)$ *is defined by (31.3) and* $\Psi(z)$ *is the general solution in the class* B_σ *of the corresponding homogeneous problem (described by Theorem 27.3). In addition,*

$$h_\Phi(\theta) = h_\Psi(\theta), \quad 0<\theta<2\pi.$$

The equality of indicators follows from the fact that $\Psi(z) \in B_\sigma$ and the first summand in the right-hand side of (31.6) has the order of decreasing $\sigma_0 > \sigma$ within any angle $[\epsilon, 2\pi - \epsilon]$. We recall that the expression of the indicator $h_\Psi(\theta)$ by means of the angular, modular, and modified angular σ-density of the zero set is given by the right-hand side of (27.15).

Combining Remark 2 to Lemma 31.1 and Theorem 26.2 we obtain the following result.

Theorem 31.2 *Let conditions (29.2)–(29.6) be fulfilled. The general solution of the non-homogeneous boundary problem (29.1) in the class B has the form*

$$\Phi(z) = \frac{\Psi_1(z)}{2\pi i} \int_1^\infty \frac{g(x)dx}{\Psi_1^+(x)(x-z)} + \Psi(z), \tag{31.7}$$

where $\Psi_1(z)$ is defined by (31.5) and $\Psi(z)$ is the general solution in the class B of the corresponding homogeneous problem (this solution is described by Theorem 26.2).

In contrast to Theorem 31.1 we do not state the identity of the indicators $h_\Phi(\theta) = h_\Psi(\theta)$ (with respect to the order 1/2) which, generally speaking, may be false. Naturally the statement of Theorem 31.2 remains true if substitute $\Psi_0(z)$ in (31.7) instead of $\Psi_1(z)$ from (31.6). Nevertheless the function $\Psi_1(z)$ has a simpler structure.

RIEMANN BOUNDARY PROBLEM WITH A NEGATIVE INDEX

32 An example of a solvable homogeneous problem with a negative index

Let L be a smooth curve, the function $\log G(t)$ satisfy the Hölder condition on L and let

$$\mathrm{Ind}_L G(t) = \frac{1}{2\pi}\Delta_L \arg G(t) < 0.$$

As well known (see [2, pp.110, 155]) the homogeneous boundary problem $\Phi^+(t) = G(t)\Phi^-(t)$, $t \in L$ is unsolvable in the class of bounded functions. It should be noted that the Hölder condition implies that $\mathrm{Ind}_L G$ is finite.

It appears that if not to impose the Hölder condition for $\log G(t)$,is not imposed then the problem may be solvable both in the case $0 > \mathrm{Ind}_L G > -\infty$ and in the case $\mathrm{Ind}_L = -\infty$. In this section we construct the corresponding examples.

First let us prove several lemmas.

Lemma 32.1 *Let Q be a smooth convex*[21] *curve of the length s, $\psi(\tau)$ be a continuous function on Q and z be an arbitrary point of the plane. Then*

$$\left|\int_Q \psi(\tau) d \arg(\tau - z)\right| \leq$$

$$\leq \max_{\tau\in Q} |\psi(\tau)| \cdot \min\{6\pi, s/\mathrm{dist}(z, Q)\}.$$

Proof. Clearly one may divide Q into three parts Q_1, Q_2, Q_3, such that the function $\arg(\tau - z)$ changes monotonically on each of them and also $|\Delta_{Q_n} \arg(\tau - z)| \leq 2\pi$, $n = 1, 2, 3$. Since

$$\left|\int_{Q_n} \psi(\tau) d \arg(\tau - z)\right| \leq$$

$$\leq \max_{\tau\in Q_n} |\psi(\tau)| \cdot |\Delta_{Q_n} \arg(\tau - z)|,$$

$$|\Delta_{Q_n} \arg(\tau - z)| \leq \int_{Q_n} \frac{|d\tau|}{|\tau - z|} \leq \frac{\mathrm{length}\, Q_n}{\mathrm{dist}(z, Q_n)},$$

we have the desired estimate.

---

[21] That is, that either Q is a segment or each straight line intersects Q at most in two points.

Lemma 32.2 *Let D be a domain whose boundary is a smooth open curve L with the end at the point $t = 0$ and let $u(z)$ be a function harmonic in D and satisfying*

$$u^{\pm}(t) < -C_1\big(\log(1/|t|)\big)^{\nu_1}, \quad t \in L, \quad |t| < \epsilon,$$

$$u(z) < C_2\big(\log(1/|z|)\big)^{\nu_2}, \quad z \in D, \quad |z| < \epsilon,$$

here $0 < \epsilon < 1$. Then

$$u(z) < -C_3\big(\log(1/|z|)\big)^{\nu_1}, \quad z \in D, \quad |z| < \epsilon.$$

The lemma is a direct consequence of the Phragmen-Lindelöf principle and we omit the proof.

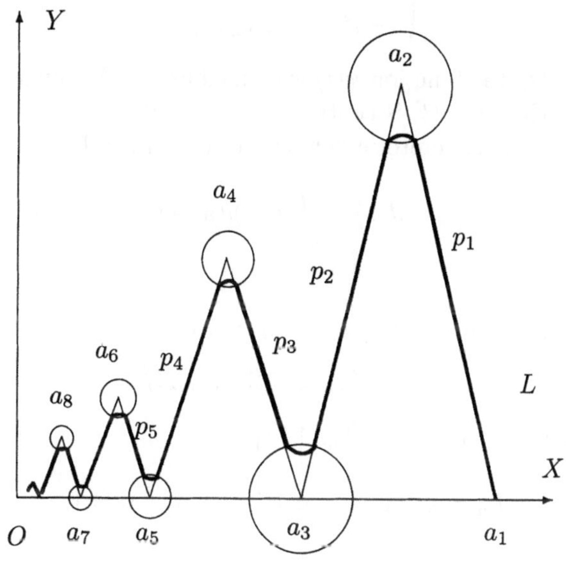

Fig. 3

Denote by Λ (see fig.3) the broken line with the vertexes at the points a_k, $k = 1, 2, \ldots$, where

$$a_{2n-1} = \frac{1}{2^n}, \quad a_{2n} = \frac{3}{2^{n+2}} + \frac{i}{2^n\sqrt{n}}, \quad n = 1, 2, \ldots. \tag{32.1}$$

By p_k, $k = 1, 2 \ldots$ denote the segment joining a_k and a_{k+1}. Let

$$q_k = \{z : \ |z - a_k| < \eta_k\}, \quad k = 2, 3, \ldots,$$

$$\eta_{2n} = \eta_{2n-1} = \frac{1}{n2^{n+3}}.$$

Denoting $\lambda_k = \Lambda \cap q_k$, replace each broken line λ_k by the arc of circle l_k chosen such that the curve

$$L = \Big(\Lambda \backslash \bigcup_{k=2}^{\infty} \lambda_k\Big) \cup \bigcup_{k=2}^{\infty} l_k$$

is smooth. It should be noted that the smoothness at the point $z = 0$ follows from (32.1). Direct the curves L, Λ, l_k, λ_k, and p_k in such a way that $\Re z$ increases when moving in positive direction.

Define the function $\psi(z)$ on L and Λ by setting

$$\psi(t) = \begin{cases} -4n^2, & t \in p_{2n}, \\ -n^2, & t \in p_{2n-1}, \end{cases} \qquad n = 1, 2, \ldots.$$

On the arcs l_k the function $\psi(t)$ is defined as a real-valued linear-fractional function such that $\psi(t)$ is continuous on $L \backslash \{0\}$.

Consider the asymptotic behavior of the integral

$$J(z) = \int_L \psi(\tau) d \arg(\tau - z)$$

as $z \to 0$.

Lemma 32.3 *For sufficiently small $|z|$ we have*

$$|J(z)| < C\big(\log(1/|z|)^3\big).$$

Proof. Let $|z| < 1/4$, $m = [\log(1/|z|)/\log 2]$ and

$$L_n = L \cap \Big(\frac{1}{2^{n+1}} < |\tau| < \frac{1}{2^n}\Big), \quad n = 1, 2, \ldots.$$

We have

$$J(z) = \Big(\sum_{n=m+2}^{\infty} \int_{L_n} + \sum_{n=m-1}^{m+1} \int_{L_n} + \sum_{n=1}^{m-2} \int_{L_n}\Big) \times$$
$$\times \psi(\tau) d \arg(\tau - z) = T_1 + T_2 + T_3.$$

The following inequalities hold:

$$\text{length}\,(L_n) < 2^{-n+2}, \quad \max_{\tau \in L_n} |\psi(\tau)| \leq 4n^2$$

$$\text{dist}(z, L_n) \geq 2^{-m-2} \quad \text{for } n \geq m+2 \text{ and}$$

$$\text{dist}(z, L_n) \geq 2^{-n-2} \quad \text{for } n \leq m-2.$$

Using this inequalities and applying Lemma 32.1, we obtain

$$|T_1| \le 16 \sum_{n=m+2}^{\infty} n^2 2^{m-n} =$$

$$= 4\sum_{n=0}^{\infty}(n+m+2)^2 2^{-n} < C_1\big(\log(1/|z|)\big)^2,$$

$$|T_2| \le C_2(m+1)^2 \le C_3\big(\log(1/|z|)\big)^2,$$

$$|T_3| \le 32\sum_{n=1}^{m-2} n^2 \le C_4\big(\log(1/|z|)\big)^3,$$

that prove the lemma.

Lemma 32.4 *For sufficiently small $|t| > 0$, $t \in L$, the following inequality holds*

$$J(t) < -C\big(\log(1/|t|)\big)^{5/2}.$$

Proof. Let $t \in L$, $|t| < 1/4$ and $m = [\log(1/|t|)/\log 2]$. Consider the curve

$$\sigma_n = \begin{cases} \lambda_{2n} \cup \lambda_{2n+1} \cup & \\ \cup(-l_{2n}) \cup (-l_{2n+1}), & 1 \le n \le m-3, \\ \lambda_{2m-4} \cup (-l_{2m-4}) \cup \{\lambda_{2m-3} \cup & \\ \cup(-l_{2m-3})\} \cap \{|\tau| > 2^{-m+1}\}, & n = m-2, \end{cases}$$

where the curve $-l_k$ differs by orientation from the curve l_k . We have

$$J(t) = \Big(\sum_{n=m+2}^{\infty}\int_{L_n} + \sum_{n=m-1}^{m+1}\int_{L_n} + \sum_{n=1}^{m-2}\int_{\sigma_n} + \Big)\psi(\tau)d\arg(\tau - t) -$$

$$-\sum_{n=1}^{m-2} n^2\{4\Delta_{p_{2n}}\arg(\tau - t) + \Delta_{p_{2n-1}}\arg(\tau - t)\} =$$

$$= P_1 + P_2 + P_3 + P_4.$$

By means ofthe estimates obtained when proving the previous lemma we get

$$|P_n| < C_1\big(\log(1/|t|)\big)^2, \quad n = 1, 2.$$

Now let us use the inequalities

$$\text{length } (\sigma_n) < \frac{1}{n2^n}, \quad \text{dist } (t, \sigma_n) > \frac{1}{2^{n-3}},$$

$$\max_{\tau\in\sigma_n} |\psi(\tau)| \le 4n^2.$$

Together with Lemma 32.1 they imply

$$|P_3| < 6\sum_{n=1}^{m-2} \frac{2^3}{n} n^2 < C_2 m^2 < C_3\big(\log(1/|z|)\big)^2.$$

It follows from the geometrical reasons (see fig. 3), that, for $1 \le n \le m-2$,

$$\Delta_{p_{2n}} \arg(\tau - t) > \arg a_{2n} = \arctan\frac{4}{3\sqrt{n}} > \frac{\pi}{3\sqrt{n}},$$

$$|\Delta_{p_{2n+1}} \arg(\tau - t)| < \Delta_{p_{2n}} \arg(\tau - t).$$

Therefore

$$P_4 < -\pi\sum_{n=1}^{m-2} n^{3/2} < -\int_0^{m-2} x^{3/2}\,dx < -C_4\big(\log(1/|z|)\big)^{5/2},$$

which gives the desired estimate.

Lemma 32.5 *Let*

$$\Gamma(z) = \frac{1}{2\pi i}\int_L \frac{\psi(\tau)}{\tau - z}\,d\tau.$$

Then, for sufficiently small z,

$$\Re\Gamma(z) < -C\big(\log(1/|z|)\big)^{5/2}. \tag{32.2}$$

Proof. Since the function $\psi(\tau)$ satisfies the Hölder condition on $L \cap \{z : |z| > \epsilon\}$, $\epsilon > 0$, then the Sohotski formula gives

$$\Re\Gamma^{\pm}(t) = \Re\Big\{\frac{1}{2\pi i}\int_L \frac{\psi(\tau)}{\tau - z}\,d\tau \pm \frac{\psi(t)}{2}\Big\} =$$

$$= \frac{1}{2\pi} J(t) \pm \frac{\psi(t)}{2}.$$

The inequality

$$|\psi(t)| < 4\{\log(1/|t|)/\log 2\}^2$$

allows us to apply Lemma 32.4. We obtain

$$\Re\Gamma^{\pm}(t) < -C_1\big(\log(1/|t|)\big)^{5/2}.$$

for sufficiently small $|t|$, $t \in L$. On the other hand, by Lemma 32.3, we have

$$|\Re\Gamma(z)| = \left|\frac{1}{2\pi}J(z)\right| < C_2\big(\log(1/|z|)\big)^3.$$

Applying Lemma 32.2 to the function $u(z) = \Re\Gamma(z)$, we obtain desired estimate (32.2).

Remark. As can be easily seen, Lemma 32.5 still remains true if one changes the function $\psi(t)$ in a neighborhood of the right-hand end $t = 1/2$ of the curve L in such a way that $\psi(1/2) = 0$ and $\psi(t)$ is linear in a neighborhood of the point $t = 1/2$ and continuous on L. In what follows we assume this change to be already done.

Now let D be the complex plane cutted along the curve L. Consider the two homogeneous boundary problems in D;

$$\Phi_1^+(t) = C_1(t)\Phi_1^-(t),$$

$$G_1(t) = \exp\{\psi(t) + 2\pi ik(1-2t)\}, \tag{32.3}$$

where k is an arbitrary positive number and

$$\Phi_2^+(t) = C_2(t)\Phi_2^-(t), \quad G_2(t) = \exp\{\psi(t) - i\log(2t)\}. \tag{32.4}$$

Clearly both the function $G_1(t)$ and $G_2(t)$ are continuous on L and $G_1(1/2) = 1$, $G_2(1/2) = 1$. The index of the both problems is negative, in addition, $\mathrm{Ind}_L G_1(t) = -k$ and $\mathrm{Ind}_L G_2(t) = -\infty$.

Theorem 32.1 *Each of the homogeneous boundary problems (32.3) and (32.4) admits infinitely many linearly independent, bounded (and continuous up to the curve L) solutions.*

Proof. Set

$$\Phi_1(z) = P\Big(\frac{1}{z}\Big)\exp\Big\{\frac{1}{2\pi i}\int_L \frac{\psi(\tau)}{\tau - z}d\tau + k\int_L \frac{1-2\tau}{\tau - z}d\tau\Big\}, \tag{32.5}$$

$$\Phi_2(z) = P\Big(\frac{1}{z}\Big)\exp\Big\{\frac{1}{2\pi i}\int_L \frac{\psi(\tau)}{\tau - z}d\tau - \frac{1}{2\pi}\int_L \frac{\log(2\tau)}{\tau - z}d\tau\Big\}, \tag{32.6}$$

where $P(\zeta)$ is an arbitrary polynomial. It follows from the Sohotski formulas and the equality $G_1(1/2) = G_2(1/2) = 1$ that the functions $\Phi_1(z)$ and $\Phi_2(z)$ satisfy conditions (32.3) and (32.4) respectively for $t \neq 0$. Also it is clear that the both functions are bounded in the domain $D \cap \{z : |z| > \epsilon\}$ for any $\epsilon > 0$ and continuous up to the boundary of this domain.

Let us verify that $\lim_{z\to 0}\Phi_k(z) = 0$, $k = 1, 2$. Using the known properties of the Cauchy integral near the points of discontinuity of its density (see [2, p. 73–75]) we have

$$\Re\Big\{k\int_L \frac{1-2\tau}{\tau - z}d\tau\Big\} = k\log\frac{1}{|z|} + O(1), \quad z \to 0,$$

$$\Re\Big\{-\frac{1}{2\pi}\int_L \frac{\log(2\tau)}{\tau - z}d\tau\Big\} = \frac{1}{4\pi}\Big(\log\frac{1}{|z|}\Big)^2 + O\Big(\log\frac{1}{|z|}\Big), \quad z \to 0.$$

Since $\log|P(1/z)| = O(\log(1/|z|))$ as $z \to 0$, then, with the help of Lemma 32.5, we obtain that

$$\log|\Phi_k(z)| < -C_k(\log(1/|z|))^{5/2}$$

for $k = 1, 2$ and sufficiently small $|z|$.

Remark 1. As can be easily proved, problems (32.3) and (32.4) admit bounded solutions that differ from the solutions described by (32.5) and (32.6).

Remark 2. In problems (32.3) and (32.4) we consider the open curve L. One can get rid of this restriction by completing L to a closed curve L_1 in an arbitrary way and then setting $G_k(t) = 1$ for $t \in L_1 \backslash L$. The functions

$$\Phi_k(z) = P\Big(\frac{1}{z}\Big)\exp\Big\{\frac{1}{2\pi i}\int_{L_1}\frac{\log G_k(\tau)}{\tau - z}d\tau\Big\}, \quad k = 1, 2,$$

still coincide with (32.5) and (32.6) and are bounded solutions of the corresponding problems.

33 Conditions of unsolvability of the homogeneous problem with a negative index

If considering the examples constructed in the previous section we see that the function $\log G(t)$ approaches $-\infty$ as $t \to 0$ rather non-monotonically. The curve L oscillates as well, and, moreover, its oscillations are coordinated with those of the function $\log|G(t)|$. This specific resonance is the reason for the considered boundary problem with negative index to be solvable. If one requires that either the function $\log|G(t)|$ or the curve L be regular at a neighborhood of the singular point, then the problem with negative index becomes unsolvable independently of whether the index is finite or infinite. The results of this section demonstrate this rule; in Theorem 33.1 we introduce some conditions on the function $\log|G(t)|$ and in Theorem 33.2 we introduce some conditions on the curve L.

Consider a simple finite smooth closed curve L which divides the plane into two parts D^+ and D^-. Assume that L is tangent to the real axis at the point $z = 0$ and in a neighborhood of this point the domain D^+ lies above L. Let $\tilde{L} = L\backslash\{0\}$, $L_{r,R} = L \cap \{z : r \leq |z| \leq R\}$ and $L_r = L_{r,\infty}$. We write $f(t) \in H(r \to 0)$ if the function $f(t)$ satisfies the Hölder condition on L_r for any $r > 0$.

Theorem 33.1 *Let L be a curve of the form described above and a function G(t) on L satisfies*

$$\log G(t) \in H(r \to 0), \quad \limsup_{r \to 0} \Delta_{L_r} \arg G(t) = -a < 0. \tag{33.1}$$

Assume that in a neighborhood of the point $t = 0$ the function $|G(t)|$ admits the representation $|G(t)| = \nu(t)\mu(t)$, where the functions ν and μ are positive for $t \neq 0$ and also $|\log \nu(t)| < C$ and the function $\mu(t)$ is monotonic both from the right and from the left of the point $t = 0$. Then the boundary problem

$$\Phi^+(t) = G(t)\Phi^-(t), \quad t \in L, \tag{33.2}$$

does not have bounded solutions.

First we shall prove three lemmas.

Lemma 33.1 *Let functions $u^\pm(z) = u^\pm(r, \alpha)$ $(z = re^{i\alpha})$ be harmonic in the domains $D^\pm$ respectively and both they and their conjugate functions are continuous up to $\tilde{L}$. Then, for sufficiently small $r_0 > 0$ and $0 < r < r_0$, the following equalities hold*

$$\int_r^{r_0} dR \int_{\alpha_1(R)}^{\alpha_2(R)} \frac{\partial u^+(R, \alpha)}{\partial R} d\alpha = \int_{\alpha_1(r_0)}^{\alpha_2(r_0)} u^+(r_0, \alpha) d\alpha -$$

$$-\int_{\alpha_1(r)}^{\alpha_2(r)} u^+(r,\alpha)d\alpha + \int_{L_{r,r_0}} u^+(t)d\arg(t), \tag{33.3}$$

$$\int_r^{r_0} dR \int_{\alpha_2(R)}^{\alpha_1(R)+2\pi} \frac{\partial u^-(R,\alpha)}{\partial R} d\alpha = \int_{\alpha_2(r_0)}^{\alpha_1(r_0)+2\pi} u^-(r_0,\alpha)d\alpha -$$

$$-\int_{\alpha_2(r)}^{\alpha_1(r)+2\pi} u^-(r,\alpha)d\alpha - \int_{L_{r,r_0}} u^-(t)d\arg(t). \tag{33.4}$$

Here $\alpha_1(r)$ and $\alpha_2(r)$ are the arguments of the intersection points of the curve L and the circle $|z| = r$ lying respectively on the left and from the right of the point $t = 0$.

Proof. We shall prove equality (33.3), the proof of (33.4) is similar. Let r_0 be sufficiently small. Put

$$L_{r,r_0,1} = L_{r,r_0} \cap \{z : \Re z > 0\}, \quad L_{r,r_0,2} = L_{r,r_0} \cap \{z : \Re z < 0\} \tag{33.5}$$

and denote by $D^+_{r,\epsilon}$ the part of the domain D^+ which is bounded by arcs of the circles $|z| = r$ and $|z| = r_0$ and the curves $L_{r,r_0,\epsilon,1}$ and $L_{r,r_0,\epsilon,2}$ that are obtained by the rotation of the curves $L_{r,r_0,1}$ and $L_{r,r_0,2}$ about the angles ϵ and $-\epsilon$ respectively. Since the function $u^+(z)$ is continuous in the closure of $D^+_{r,\epsilon}$, the Green formula yields

$$\int_r^{r_0} dR \int_{\alpha_1(R)+\epsilon}^{\alpha_2(R)-\epsilon} \frac{\partial u^+(R,\alpha)}{\partial R} d\alpha = \int_{\alpha_1(r_0)+\epsilon}^{\alpha_2(r_0)-\epsilon} u^+(r_0,\alpha)d\alpha -$$

$$-\int_{\alpha_1(r)+\epsilon}^{\alpha_2(r)-\epsilon} u^+(r,\alpha)d\alpha + \int_{L_{r,r_0,\epsilon,1}} u^+(t)d\alpha_1(t) -$$

$$-\int_{L_{r,r_0,\epsilon,2}} u^+(t)d\alpha_2(t). \tag{33.6}$$

By the hypothesis the function $v^+(z)$ conjugated to the function $u^+(z)$ is continuous in $D_{r,0}$. Therefore the integral

$$\int_{\alpha_1(R)}^{\alpha_2(R)} \frac{\partial u^+(R,\alpha)}{\partial R} d\alpha = \lim_{\epsilon\to 0} \int_{\alpha_1(R)+\epsilon}^{\alpha_2(R)-\epsilon} \frac{\partial u^+(R,\alpha)}{\partial R} d\alpha =$$

$$= \frac{1}{R} \lim_{\epsilon\to 0} \{v^+(R,\alpha_2(R)-\epsilon) - v^+(R,\alpha_1(R)+\epsilon)\}$$

converges uniformly with respect to R on the segment $[r, r_0]$. Sending $\epsilon \to 0$ in (33.6), we obtain (33.3).

Lemma 33.2 *Let the conditions of Theorem 33.1 be fulfilled and let $\Phi(z)$ be a solution of boundary problem (33.2) and the quantity a be defined by (33.1). Then, for any $b < a$, the following relation holds for all sufficiently small r_0 and $0 < r < r_0$:*

$$\int_0^{2\pi} \log|\Phi(re^{i\alpha})|d\alpha > C_{b,r_0} + b\log\frac{r_0}{r} + \int_{L_{r,r_0}} \log|G(t)|d\arg t. \quad (33.7)$$

Proof. First let us prove that $\Phi(z)$ admits the representation $\Phi(z) = X(z)F(1/z)$, where $X(z)$ is a solution of problem (33.2) which does not vanish for $z \neq 0$ and $F(z)$ is an entire function. Actually, the function

$$X_r(z) = \exp\Big\{\frac{1}{2\pi i}\int_{L_r} \frac{\log G(\tau)}{\tau - z}d\tau\Big\}$$

satisfies the relation $X_r^+(t) = G(t)X_r^-(t)$ for $t \in L_r$ and does not vanish in $\{z : |z| > r\}$. Therefore the function $\Phi(z)/X_r(z)$ is analytic in $\{z : |z| > r\}$. Thus the zeros $z_1, z_2, \ldots$ of the function $\Phi(z)$ may accumulate at the point $z = 0$ only. Now taking an arbitrary entire function $F(z)$ with the zero set $\{1/z_k : z_k \neq 0\}$ and setting $X(z) = \Phi(z)/F(1/z)$, we obtain the desired representation.

Let us return to inequality (33.7). We have

$$\int_0^{2\pi} \log|\Phi(re^{i\alpha})|d\alpha = \int_0^{2\pi} \log|X(re^{i\alpha})|d\alpha + \int_0^{2\pi} \log|F(r^{-1}e^{i\alpha})|d\alpha. \quad (33.8)$$

Fix an $r > 0$ such that for any $0 < r < r_0$ the circle $|z| = r$ intersects the curve L at exactly two points and, besides, $\Delta_{L_r} \arg G(t) = -b$. The functions $u^\pm(z) = \log|X^\pm(z)|$ and $v^\pm(z) = \arg X^\pm(z)$ are mutually conjugate harmonic functions in the domains D^+ and D^- respectively and, besides, they are continuous up to $\tilde{L}$. For $R \leq r_0$ we have

$$\int_{\alpha_1(R)}^{\alpha_2(R)} \frac{\partial v^+(Re^{i\alpha})}{\partial\alpha}d\alpha + \int_{\alpha_2(R)}^{\alpha_1(R)+2\pi} \frac{\partial v^-(Re^{i\alpha})}{\partial\alpha}d\alpha =$$

$$= \Delta_{L_R} \arg G(t) \leq -b.$$

From here by mean of the Cauchy-Riemann condition we find

$$\int_r^{r_0} dR \int_{\alpha_1(R)}^{\alpha_2(R)} \frac{\partial u^+(Re^{i\alpha})}{\partial R}d\alpha +$$

$$+ \int_r^{r_0} dR \int_{\alpha_2(R)}^{\alpha_1(R)+2\pi} \frac{\partial u^-(Re^{i\alpha})}{\partial R}d\alpha \leq -b\log\frac{r_0}{r}.$$

Using Lemma 33.1 and taking into account that $u^+(t)-u^-(t)=\log|G(t)|$, we obtain the inequality

$$\int_0^{2\pi}\log|X(r_0e^{i\alpha})|d\alpha-\int_0^{2\pi}\log|X(re^{i\alpha})|d\alpha+$$

$$+\int_{L_{r,r_0}}\log|G(t)|d\arg t<-b\log\frac{r_0}{r},$$

whence, by (33.8),

$$\int_0^{2\pi}\log|\Phi(r_0e^{i\alpha})|d\alpha-\int_0^{2\pi}\log|\Phi(re^{i\alpha})|d\alpha+$$

$$+\int_{L_{r,r_0}}\log|G(t)|d\arg t<-b\log\frac{r_0}{r}+$$

$$+\int_0^{2\pi}\log|F(r_0^{-1}e^{i\alpha})|d\alpha-\int_0^{2\pi}\log|F(r^{-1}e^{i\alpha})|d\alpha.$$

By virtue of the Jensen formula the integral

$$\int_0^{2\pi}\log|F(r^{-1}e^{i\alpha})|d\alpha$$

is bounded from below as $r\to\infty$. Since r_0 is fixed, we obtain estimate (33.7)

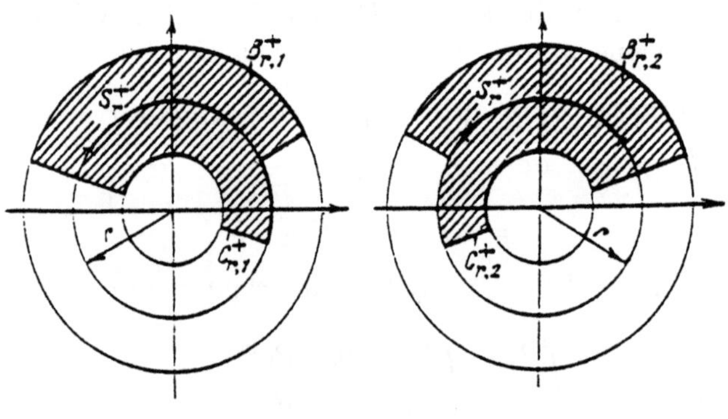

Fig. 4

Lemma 33.3 *Let the conditions of Theorem 33.1 be fulfilled, exept possibly for the second condition in (33.1). Let $\Phi(z)$ be a solution of problem (33.2) such that $|\Phi(z)| \le 1$. Then there exists a small $r_0 > 0$ such that, for all r, $0 < r < \frac{2}{3} r_0$, at least one of the following inequalities holds:*

$$\max_{\pi/4 \le \alpha \le 3\pi/4} \log |\Phi(re^{i\alpha})| \le C_1 - C_2 \max_{t \in L_{r,r_0}} |\log \mu(t)|,$$

$$\max_{5\pi/4 \le \alpha \le 7\pi/4} \log |\Phi(re^{i\alpha})| \le C_1 - C_2 \max_{t \in L_{r,r_0}} |\log \mu(t)|.$$

Proof. Put (see fig.4)

$$B^+_{r,1} = \Big\{\Big\{z:\ \frac{1}{2}r < |z| < \frac{3}{2}r\Big\} \cap \Big\{z:\ \frac{1}{8}\pi < \arg z < \frac{7}{8}\pi\Big\}\Big\} \cup$$

$$\cup \Big\{\Big\{z:\ \frac{1}{2}r < |z| < r\Big\} \cap \Big\{z:\ -\frac{1}{8}\pi < \arg z \le \frac{1}{8}\pi\Big\}\Big\},$$

$$B^+_{r,2} = \Big\{\Big\{z:\ \frac{1}{2}r < |z| < \frac{3}{2}r\Big\} \cap \Big\{z:\ \frac{1}{8}\pi < \arg z < \frac{7}{8}\pi\Big\}\Big\} \cup$$

$$\cup \Big\{\Big\{z:\ \frac{1}{2}r < |z| < r\Big\} \cap \Big\{z:\ \frac{7}{8}\pi \le \arg z < \frac{9}{8}\pi\Big\}\Big\},$$

$$C^+_{r,1} = \Big\{z:\ \arg z = -\frac{1}{8}\pi\Big\} \cap \Big\{z:\ \frac{1}{2}r \le |z| \le r\Big\},$$

$$C^+_{r,2} = \Big\{z:\ \arg z = \frac{9}{8}\pi\Big\} \cap \Big\{z:\ \frac{1}{2}r \le |z| \le r\Big\},$$

$$S^+_r = \{z:\ |z| = r\} \cap \Big\{z:\ \frac{1}{4}\pi \le \arg z \le \frac{3}{4}\pi\Big\}.$$

Denote by $B^-_{r,n}$, S^-_r and $C^-_{r,n}$ the reflections of $B^+_{r,n}$, S^+_r and $C^+_{r,n}$ with respect to the real axis respectively. Let $\omega(z, C^\pm_{r,n}, B^\pm_{r,n})$ be the harmonic measure of the segment $C^\pm_{r,n}$ with respect to the domain $B^\pm_{r,n}$. The value

$$N = \min_{z \in S^\pm_r;\, n=1,2} \omega(z, C^\pm_{r,n}, B^\pm_{r,n}) > 0$$

is clearly an absolute constant.

Let $\Phi(z)$ satisfy equation (33.2) and $|\Phi(z)| \le 1$. Since $|G(t)| = \nu(t)\mu(t)$ and $|\log \nu(t)| < C$, then, for sufficiently small $t \in L$, at least one of the following inequalities holds:

$$\log |\Phi^+(t)| < C_3 - |\log \mu(t)|,$$

$$\log |\Phi^-(t)| < C_3 - |\log \mu(t)|.$$

Assume that the function $\log \mu(t)$ is unbounded as $t \to 0$, otherwise the assertion of the lemma is trivial. Since the function $\log \mu(t)$ is monotonic, then, for sufficiently small $r > 0$, at least one of the following four inequalities holds:

$$\max_{t \in L_{r/2,r,n}} \log |\Phi^{\pm}(t)| < C_4 - \max_{r \le |t| \le 3r/2} |\log \mu(t)|, \quad n = 1, 2,$$

where $L_{r_1,r_2,n}$ is defined in (33.5). So much the more at least one of the following four inequalities holds:

$$\max_{t \in L_{r/2,r,n}} \log |\Phi^{\pm}(t)| < C_4 - \Lambda(r),$$

$$\Lambda(r) = \max_{t \in L_{r,r_0}} |\log \mu(t)|, \quad r < \frac{2}{3} r_0.$$

Suppose, for definiteness, that one of the following two inequalities holds:

$$\max_{t \in L_{r/2,r,n}} \log |\Phi^{+}(t)| < C_4 - \Lambda(r), \quad n = 1, 2.$$

Since the function $\log \mu(t)$ is unbounded, we obtain $C_4 - \Lambda(r) < 0$ for small r.

Let

$$D_r^+ = D^+ \cap \left\{ z : \frac{1}{2} r < |z| < \frac{3}{2} \right\}.$$

Since $\log |\Phi(z)| \le 0$ in D_r^+, we obtain

$$\log |\Phi(\zeta)| \le$$

$$\le (C_4 - \Lambda(r)) \omega(\zeta, L_{r/2,r,n}, D_r^+), \quad \zeta \in D_r^+, \quad n = 1, 2, \tag{33.9}$$

where $\omega(\zeta, L_{r/2,r,n}, D_n^+)$ is the harmonic measure of the curve $L_{r/2,r,n}$ with respect to the domain D_r^+. According to the principle of extension of domain, we have

$$\omega(\zeta, L_{r/2,r,n}, D_r^+) \ge \omega(\zeta, C_{r,n}^+, B_{r,n}^+)$$

for $\zeta \in D_r^+ \cap B_{r,n}^+$. Using (33.9) and the inequality $C_4 - \Lambda(r) < 0$, we obtain

$$\max_{\zeta \in S_r^+} \log |\Phi(\zeta)| \le$$

$$\le (C_4 - \Lambda(r)) \min_{\zeta \in S_r^+, n=1,2} \omega(\zeta, C_{r,n}^+, B_{r,n}^+) = (C_4 - \Lambda(r)) N,$$

which completes the proof.

Proof of Theorem 33.1 Suppose that $\Phi(z)$ is a bounded solution of problem (33.2). We may assume $|\Phi(z)| \leq 1$. Take r_0 sufficiently small such that estimate (33.7) is fulfilled. Since $|G(t)| = \nu(t)\mu(t)$, one can rewrite this estimate in the form

$$\int_0^{2\pi} \log|\Phi(re^{i\alpha})|d\alpha > C_{b,r_0} + b\log\frac{r_0}{r} + J_1(r) + J_2(r), \tag{33.10}$$

where

$$J_1(r) = \int_{L_{r,r_0}} \log\nu(t)d\arg t, \quad J_2(r) = \int_{L_{r,r_0}} \log\mu(t)d\arg t.$$

To estimate the first integral represent the curve $L_{r,r_0,1} = L_{r,r_0} \cap \{z : \Re z > 0\}$ in the form

$$L_{r,r_0,1} = \{t : \ t(R) = Re^{i\alpha(R)}, \quad r \leq R \leq r_0\}.$$

Since $|\log\nu(t)| \leq C$ and L is a smooth curve, we have

$$\Big|\int_{L_{r,r_0,1}} \log\nu(t)d\arg t\Big| \leq C\int_{L_{r,r_0,1}} \Big|\Im\frac{dt}{t}\Big| \leq$$

$$\leq C\int_r^{r_0} |t'(R)|\,|\sin(t'(R) - t(R))|\frac{dR}{R} = o\Big(\log\frac{r_0}{r}\Big), \quad r \to 0.$$

The integral

$$\int_{L_{r,r_0,2}} \log\nu(t)d\arg(t)$$

may be estimated in a similar way. We obtain

$$|J_1(r)| = o\Big(\log\frac{r_0}{r}\Big), \quad r \to 0.$$

To estimate the integral $J(r)$ let us apply the second mean value theorem. We have

$$|J_2(r)| \leq \max_{t\in L_{r,r_0}} |\log\mu(t)|\,\Big|\Big(\int_{L_1} + \int_{L_2} + \int_{L_3} + \int_{L_4}\Big)d\arg t\Big|,$$

where L_k are parts of the curve L_{r,r_0}. From here, using that L is a smooth curve, we get

$$|J_2(r)| = o(\max_{t\in L_{r,r_0}} |\log\mu(t)|), \quad r \to 0.$$

Substituting the estimates of the integrals $J_1(r)$ and $J_2(r)$ into (33.10), we get

$$\int_0^{2\pi} \log|\Phi(re^{i\alpha})|d\alpha >$$

$$> (b+o(1))\log\frac{r}{r_0} + o\big(\max_{t\in L_{r,r_0}} |\log\mu(t)|\big), \quad r\to 0.$$

On the other hand, Lemma 33.3 yields that

$$\int_0^{2\pi} |\log\Phi(re^{i\alpha})|d\alpha < \frac{\pi}{2}(C_1 - C_2 \max_{t\in L_{r,r_0}} |\log\mu(t)|)$$

for sufficiently small $r>0$. Thus we have

$$\big(b+o(1)\big)\log\frac{r_0}{r} + \Big(\frac{\pi}{2}C_2 + o(1)\Big)\max_{t\in L)r,r_0} |\log\mu(t)| < \frac{\pi}{2}C_1$$

for sufficiently small $r>0$, which is impossible. This completes the proof.

Corollary 33.1 *Consider the homogeneous boundary problem $\Phi^+(t) = G(t)\Phi^-(t)$, $t\in L$ for the curve L, satisfying the conditions of Sect. 18. Let all conditions (18.3)–(18.5) be fulfilled except the condition $\lambda>0$ which is replaced by $\lambda<0$. Then the boundary problem is unsolvable in the class B.*

Proof. By changing the variables and complementing the curve one may pass to the curve satisfying the conditions indicated in the beginning of this section. Then it suffices to use Theorem 33.1.

Remark. If L is the ray $1\le t<\infty$, then the corollary follows directly from Theorem 25.2, since $|\Phi(z)|\le C$ and $\lim_{x\to\infty}\arg G(x) = -\infty$ then the left-hand side of (25.4) is bounded as $r\to\infty$ and the right-hand side is unbounded.

Recall that a smooth curve L is called a Lyapunov curve if the angle between the tangent to the curve and the real axis considered as a function of the curve length satisfies the Hölder condition.

Theorem 33.2 *Let L be a closed Lyapunov curve satisfying the same conditions as in Theorem 33.1 and let a function $G(t)$ on L satisfy (33.1). Then the homogeneous problem (33.2) does not have bounded solutions.*

Let us fix a small $r_0>0$ and introduce the following notation:

$$r_n = r_0 2^{-n}, \quad l_n = L\cap\{z:\ r_n\le|z|\le r_{n-1}\},$$

$$\mu_n = \frac{1}{r_n}\int_{l_n} |\log|G(t)||ds.$$

First we need some auxiliary statements.

Lemma 33.4 *Let conditions of Theorem 33.2 be fulfilled and let* $0 < b < a$. *Then any solution* $\Phi(z)$ *of problem (33.2) satisfies the inequality*

$$\int_0^{2\pi} \log|\Phi(r_n e^{i\alpha})|d\alpha > -C + nb\log 2 - \epsilon_n \max_{1\le k\le n} \mu_k, \tag{33.11}$$

for all sufficiently large n, *here* $\lim_{n\to\infty}\epsilon_n = 0$.

Proof. According to Lemma 33.2, we have

$$\int_0^{2\pi} \log|\Phi(r_n e^{i\alpha})|d\alpha > C_1 + nb\log 2 -$$

$$-\sum_{k=1}^{n}\int_{l_k} |\log|G(t)||\;|d\arg t|. \tag{33.12}$$

Since L is a Lyapunov curve, then $\arg t'(s) \in H(\beta)$ for some $\beta \in (0,1]$. Therefore

$$|d\arg t| = \frac{1}{|t(s)|}|\sin(\arg t'(s) - \arg t(s))|ds \le$$

$$\le \frac{C_2 s^\beta ds}{|t(s)|} \le C_3|t|^{\beta-1}ds$$

for sufficiently large n, and hence,

$$\int_{l_k} |\log|G(t)||\;|d\arg t| \le$$

$$\le C_3 r_k^{\beta-1}\int_{l_k}|\log|G(t)||ds = C_3 r_k^\beta \mu_k. \tag{33.13}$$

If the sequence $\{\mu_k\}$ is bounded, then (33.13) yields

$$\sum_{k=1}^{n}\int_{l_k} |\log|G(t)||\;|d\arg t| = C_3 r_0^\beta \sum_{k=1}^{n}\mu_k 2^{-k\beta} \le C_4,$$

and taking into account (33.12) we obtain (33.11) (one may put $\epsilon_n = 0$). Now let $M_n = \max_{1\le k\le n}\mu_k \to \infty$ as $n\to\infty$. Choose numbers $p_n < n$, $p_n \to \infty$ such that $M_n/M_{p_n} \to \infty$ as $n \to \infty$. We obtain the estimate

$$\sum_{k=1}^{n}\int_{l_k} |\log|G(t)||\;|d\arg t| \le C_3\sum_{k=1}^{n} r_k^\beta \mu_k \le$$

$$\leq C_3 r_0^\beta \Big(M_{p_n} \sum_{k=1}^{p_n} 2^{-k\beta} + M_n \sum_{k=p_n+1}^{n} 2^{-k\beta} \Big) \leq$$

$$\leq C_3 r_0^\beta \Big(\frac{M_{p_n}}{2^\beta - 1} + \frac{M_n 2^{-p_n\beta}}{2^\beta - 1} \Big) = \epsilon_n M_n,$$

where $\lim_{n\to\infty} \epsilon_n = 0$.

Lemma 33.5 *Let the conditions of Theorem 33.2 be fulfilled, except for possibly the second of the conditions (33.1). Let $\Phi(z)$ be a solution toproblem (33.2) such that $|\Phi(z)| \leq 1$. Then we have*

$$\int_0^{2\pi} \log |\Phi(r_n e^{i\alpha})| d\alpha < C_1 - C_2 \mu_n. \tag{33.14}$$

for sufficiently large n.

Proof. First we obtain an auxiliary estimate. For any $z_0 \in D^+$ let the function $\zeta = f(z, z_0)$ map the domain D^+ conformally onto the disk $|\zeta| < 1$ such that $f(z_0, z_0) = 0$, $f(0, z_0) = 1$. Let

$$s_n = \{z : \ |z| = r_n\} \cap \Big\{ z : \ \frac{\pi}{3} \leq \arg z \leq \frac{2\pi}{3} \Big\}.$$

We shall prove that

$$\min_{t \in l_n, \ z_0 \in s_n} \Big| \frac{df}{dz}(t, z_0) \Big| > C_3 / r_n \tag{33.15}$$

for sufficiently large n.

Actually, let $w = \varphi(z)$ be an arbitrary function which maps D^+ conformally onto $|w| < 1$ such that $\varphi(0) = 1$. We have

$$f(z, z_0) = e^{i\alpha} \frac{\varphi(z) - \varphi(z_0)}{1 - \overline{\varphi(z_0)}\varphi(z)},$$

where α is a real number. Hence,

$$\Big| \frac{df}{dz}(t, z_0) \Big| = \frac{1 - |\varphi(z_0)|^2}{|1 - \varphi(t)\overline{\varphi(z_0)}|^2} |\varphi'(t)|. \tag{33.16}$$

Since the boundary ∂D^+ is a Lyapunov curve, both the functions $\varphi'(z)$ and $1/\varphi'(z)$ are continuous in $D^+ \cup \partial D^+$ (see[18, p. 411]). Therefore, for $z \in D^+ \cup \partial D^+$, we have

$$\varphi(z) = 1 + z\varphi'(0) + o(|z|), \quad z \to 0,$$

and also $\varphi'(0) \neq 0$. Using the equality $\arg \varphi'(0) = \pi/2$, we see that, for $z_0 = r_n e^{i\theta} \in s_n$ and $t \in l_n$,

$$\varphi(t)\overline{\varphi(z_0)} - 1 = \varphi'(0)(t - \bar{z}_0) + o(|z_0|) + o(|t_0|),$$

$$|z_0| \to 0, \quad |t| \to 0,$$

$$|\varphi(z_0)|^2 - 1 = 2i\varphi'(0)rn \sin\theta + o(|z_0|), \quad |z_0| \to 0.$$

Since $\sin\theta > \sin\frac{\pi}{3}$ and $\min\{|t - \bar{z}_0| : t \in l_n, z_0 \in s_n\} \geq r_n/2$ for sufficiently large n, relation (33.16) yields (33.15).

Now let $\Phi(z)$ be a solution of problem (33.2) and $|\Phi(z)| \leq 1$. Using the well-known formula for the solution of the Dirichlet problem, we see that, for $z_0 \in s_n$,

$$\log|\Phi(z_0)| \leq \int_{l_n} \log|\Phi^+(t)|\left|\frac{df}{dz}(t, z_0)\right|ds.$$

By virtue of (33.15) this yields

$$\max_{\pi/3 \leq \alpha \leq 2\pi/3} \log|\Phi(r_n e^{i\alpha})| \leq \frac{C_4}{r_n}\int_{l_n} \log|\Phi^+(t)|ds. \tag{33.17}$$

One can similarly obtain the relation

$$\max_{4\pi/3 \leq \alpha \leq 5\pi/3} \log|\Phi(r_n e^{i\alpha})| \leq \frac{C_5}{r_n}\int_{l_n} \log|\Phi^-(t)|ds. \tag{33.18}$$

Since $\log|\Phi^{\pm}(t)| \leq 0$ and

$$\log|\Phi^+(t)| = \log|\Phi^-(t)| + \log|G(t)|,$$

we get

$$\log|\Phi^+(t)| + \log|\Phi^-(t)| \leq -|\log|G(t)||, \quad t \in L.$$

Now we conclude, from (33.17) and (33.18), that

$$\int_0^{2\pi} \log|\Phi(r_n e^{i\alpha})|d\alpha \leq -\frac{C_6}{r_n}\int_{L_n} |\log|G(t)|\,|ds,$$

which proves the lemma.

Proof of Theorem 33.2 Suppose that $\Phi(z)$ is a bounded solution of problem (33.2). We may assume $|\Phi(z)| \leq 1$. Applying Lemmas 33.4 and 33.5, we obtain from relations (33.11) and (33.14) that

$$nb\log 2 < C + C_1 - C_2\mu_n + \epsilon_n \max_{1 \leq k \leq n} \mu_k. \tag{33.19}$$

If the sequence $\{\mu_k\}$ is unbounded, then letting $n \to \infty$ over the subsequence for which $\mu_n = \max_{1 \leq k \leq n} \mu_k$, we obtain a contradiction.

If the sequence $\{\mu_k\}$ is bounded, then inequality (33.19) is also impossible.

Corollary 33.2 *If L is a simple closed rectifiable curve containing the point $z = 0$ and a function $G(t)$ satisfies conditions (33.1), then problem (33.2) does not have stable bounded solutions.*

Recall (see [2, p.38]) that a solution is called a stable solution if it continuously depends on the function $G(t)$ and the curve L. The assertion follows from Theorem 33.2, since, by changing the curve L in a neighborhood of zero by a Lyapunov arc and then varying the function $G(t)$ to satisfy conditions (33.1), we obtain a problem without bounded solutions.

In the same way we obtain the following statement

Corollary 33.3 *Bounded solutions of problems (32.3) and (32.4) are unstable.*

34 Conditions of solvability of the non-homogeneous problem with an index $-\infty$

It had been proved by Gakhov (see [2, p.113]) the non-homogeneous Riemann boundary problem with the negative index $\kappa < -1$ possesses a unique bounded solution if and only if the function $g(t)$ satisfies $-\kappa - 1$ additional conditions. So it is natural to expect that if $\operatorname{Ind} G(t) = -\infty$, then there are infinitely many conditions of solvability. We prove that this is really the case and also find these conditions.

Consider the non-homogeneous problem on the curve $L = \{t : \ 1 \le t < \infty\}$:

$$\Phi^+(t) = G(t)\Phi^-(t) + g(t). \tag{34.1}$$

Assume all conditions (29.2)–(29.6) are satisfied, except the condition $\lambda > 0$ in (29.3) which is changed by $\lambda < 0$. Consider the homogeneous problem which corresponds to (34.1)

$$\Psi^+(t) = G(t)\Psi^-(t). \tag{34.2}$$

By the Corollary to Theorem 33.1 this problem has no solution in the class B. However for our purposes it suffices to solve this problem in a wider class.

Construct the function $\Psi_*(z)$ in the same way as the solution of the homogeneous problem in Sect. 31 (see 31.5) with the only difference being that we change the zeros to poles. Namely we put

$$\Psi_*(z) = \exp\Big\{\frac{z}{2\pi i}\int_1^\infty \frac{\log G(x) + 2\pi i p(x)}{x(x-z)}dx\Big\}\prod_{n=1}^\infty \frac{1 - z/q_n}{1 - z/z_n}, \tag{34.3}$$

where $z_n = q_n \exp(iq_n^{-\rho})$, $q_n \ge 1$, and the counting function $p(x)$ of the sequence $\{q_n\}$ is defined by the relation

$$p(x) = \Big[\max_{1\le t\le x}\Big\{\frac{1}{2} - \varphi(t)t^\rho\Big\}^+\Big].$$

Lemma 34.1 *The function $\Psi_*(z)$ defined by (34.3) possesses the following properties*

1) $\Psi_*(z)$ is meromorphic and non-vanishing in D and satisfies boundary condition (34.2) on $\tilde{L}$.

2)

$$g_*(t) = g(t)/\Psi_*^+(t) \in H_{[1+\epsilon,\infty)}, \quad g_*(\infty) = 0,$$

for any $\epsilon > 0$.

3) For sufficiently small $\epsilon > 0$ the following representation holds on the segment $[1, 1+\epsilon]$

$$g_*(t) = \psi(t)(t-1)^{-\alpha - i\beta}, \quad 0 \le \alpha < 1,$$

here $\psi(t) \in H_{[1,1+\epsilon]}$ and if $\alpha + i\beta = 0$, then $\psi(1) = 0$.

4) The estimate

$$C_2|z-1|^\alpha \le |\Psi_*(z)| \le C_1|z-1|^\alpha,$$

holds in a neighborhood of the point $z = 1$.

5) The boundary values of the function $\Psi_*(z)$ are bounded:

$$\sup_{1 \le t < \infty} |\Psi_*^{\pm}(t)| < \infty,$$

6) The function $\Psi_*(z)$ is uniformly bounded both from above and from below on a sequence of circles $|z| = r_n$, $r_n \to \infty$:

$$\max_{|z|=r_n} \{|\Psi_*(z)| + |\Psi_*(z)|^{-1}\} < C_3. \tag{34.4}$$

Proof. Consider the auxiliary problem $\Psi_1^+(t) = \tilde{G}(t)\Psi_1^-(t)$, where $\tilde{G}(t) = 1/G(t)$. The function $\Psi_1(z) = 1/\Psi_*(z)$ may be represented in form (31.5) and, by Remarks 2 and 3 to Lemma 31.1, is a solution of this problem bounded in any neighborhood of the point $z = 1$. Near the point $z = 1$ the function $\Psi_1(z)$ admits the representation

$$\Psi_1(z) = (z-1)^{\varphi(1)}\psi(z), \quad \psi(z) = O(1), \quad z \to 1,$$

which implies property 1). One can easily obtain properties 2)–5) by applying Remarks 2 and 3 to Lemma 31.1 to the function $\Psi_1(z)$.

To prove property 6) we verify relation (34.4) with $r_n = \frac{1}{2}(q_n + q_{n+1})$. From (34.3) we have

$$\log \Psi_*(z) = z \int_1^\infty \frac{\varphi(x)x^\rho + p(x)}{x(x-z)} dx +$$

$$+ \int_1^\infty \log \frac{x-z}{x - z\exp(-ix^\rho)} dp(x) + \frac{z}{2\pi i} \int_1^\infty \frac{\log|G(x)|}{x(x-z)} dx =$$

$$= J(z) + V(z) + U(z). \tag{34.5}$$

Let us estimate the three last integrals. First we mention that the arguments, similar to those of Lemmas 30.2 and 30.4, provide

$$C_1 q_n^{1-\rho} \le q_{n+1} - q_n \le C_2 q_n^{1-\rho}, \quad C_1 \le q_{n+1}^\rho - q_n^\rho \le C_2, \tag{34.6}$$

and also

$$\varphi(t)t^\rho + p(t) = \beta(t) + \gamma(t), \tag{34.7}$$

where

$$\beta(t) = \frac{q_{n+1}^\rho + q_n^\rho - 2t^\rho}{2(q_{n+1}^\rho - q_n^\rho)}, \quad q_n \le t \le q_{n+1}, \quad q_0 = 1. \tag{34.8}$$

In addition,

$$|\beta(t)| \le C_3, \quad \Big|\int_1^\infty \beta(x)x^{\rho-1}dx\Big| \le C_4, \tag{34.9}$$

$$\gamma(t) \in H_{[1,\infty)}, \quad \gamma(1) = \gamma(\infty) = 0. \tag{34.10}$$

It should be also mentioned that

$$\beta(q_n) = \varphi(q_n)q_n^p + p(q_n) = \frac{1}{2}, \quad \gamma(q_n) = 0.$$

By virtue of (34.7) we have

$$J(z) = z\int_1^\infty \frac{\gamma(x)dx}{x(x-z)} + z\int_1^\infty \frac{\beta(x)dx}{x(x-z)} = J_1(z) + J_2(z).$$

It follows from (34.10), that $|J_1(z)| \le C_5$, $z \in D$. Let $z = r_m e^{i\theta}$. Then

$$\Re J_2(z) = \int_{\omega_m} \frac{r_m(x\cos\theta - r_m)}{x^\rho(x^2 - 2xr_m\cos\theta + r_m^2)}\beta(x)x^{\rho-1}dx +$$

$$+\Re\int_{q_m}^{q_{m+1}} \frac{z(\beta(x) - \beta(r_m))}{x(x-z)}dx + \beta(r_m)\log\Big|\frac{(q_{m+1}-z)q_m}{(q_m - z)q_{m+1}}\Big| =$$

$$= J_2^{(1)}(z) + J_2^{(2)}(z) + J_2^{(3)}(z),$$

where $\omega_m = L\backslash[q_m, q_{m+1}]$. For any θ the function

$$\varphi(x, r_m) = \frac{r_m(x\cos\theta - r_m)}{x^\rho(x^2 - 2xr_m\cos\theta + r_m^2)}$$

has at most three extremal points on the axis $1 < x < \infty$ and also

$$\max_{x\in\omega_m} |\varphi(x, r_m)| < C_6.$$

Applying Lemma 22.2 and the second property from (34.9), we obtain $|J_2^{(1)}(z)| < C_7$. Since relations (34.8) and (34.6) yield

$$|\beta(x) - \beta(r_m)| < C_8|x - r_m|r_m^{\rho-1}$$

for $x \in [q_m, q_{m+1}]$, we have $|J_2^{(2)}(z)| < C_9$. Finally, using the inequality $|\beta(t)| \leq C_3$ and relation (34.6) we conclude $|J_2^{(3)}(z)| < C_{10}$. So we have proved that $|\Re J_2(z)| < C_{11}$, and hence, $|\Re J(z)| < C_{12}$ for $|z| = r_m = \frac{1}{2}(q_m + q_{m+1})$.

Setting

$$\lambda_m = \left(\frac{1}{2}r_m, \frac{1}{2}r_{m+1}\right), \quad \Lambda_m = L \backslash \lambda_m,$$

and integrating by parts we obtain (in the same way as in (30.25))

$$V(z) = z\int_{\Lambda_m} \frac{(\exp(ix^{-\rho}) - 1)p(x)dx}{(x\exp(ix^{-\rho}) - z)(x-z)} + i\rho z\int_{\Lambda_m} \frac{x^{-\rho-1}p(x)dx}{x\exp(ix^{-\rho}) - z} +$$

$$+z\int_{\lambda_m} \frac{(\exp(ix^{-\rho}) - 1)(p(x) - p(r_m))}{(x\exp(ix^{-\rho}) - z)(x-z)}dx +$$

$$+i\rho z\int_{\lambda_m} \frac{x^{-\rho-1}(p(x) - p(r_m))}{x\exp(ix^{-\rho}) - z}dx - \{p(x) - p(r_m)\}\times$$

$$\times \log\left\{1 + \frac{z(1 - \exp(-ix^{-\rho}))}{x - z}\right\}\Bigg|_{x=\frac{1}{2}r_m}^{x=\frac{3}{2}r_{m+1}} = \sum_{j=1}^{5} V_j(z).$$

Since $|\exp(\pm ix^{-\rho}) - 1|p(x) = O(1)$ as $x \to \infty$, we have

$$\max_{|z|=r_m} |V_1(z)| \leq C_{13}r_m \int_{\Lambda_m} \frac{dx}{(x - r_m)^2} \leq C_{14},$$

$$\max_{|z|=r_m} |\Re V_2(z)| \leq C_{15}r_m \leq \max_{0\leq\theta\leq 2\pi} \int_{\Lambda_m} \frac{|\sin(\theta - x^{-\rho})|dx}{|x\exp(ix^{-\rho}) - z|^2} \leq C_{16}.$$

Since $p(x) - p(r_m) = 0$ for $q_m < x < q_{m+1}$, we get

$$\Re V_3(z) =$$

$$= \left(\int_{\frac{1}{2}r_m}^{q_m} + \int_{q_{m+1}}^{\frac{3}{2}r_m}\right) \frac{2r_m(r_m^2 - x^2)\sin\left(\theta - \frac{1}{2}x^{-\rho}\right)\sin\left(\frac{1}{2}x^{-\rho}\right)}{|x\exp(ix^{-\rho}) - z|^2|x - z|^2}\times$$

$$\times (p(x) - p(r_m))dx = V_3^{(1)}(z) + V_3^{(2)}(z).$$

The following inequality is similar to (30.26)

$$|p(x) - p(r_m)| \leq C_{17}|x - r_m|r_m^{\rho-1}, \quad x \in [r_m/2, 3r_{m+1}/2]. \tag{34.11}$$

Using it we obtain

$$|V_3^{(1)}(z)| \leq$$

$$\leq C_{17} r_m^{\rho+1} \int_{r_m/2}^{q_m} \frac{x^{-\rho}|2\theta - x^{-\rho}|dx}{(x-r_m)^2 + 4xr_m \sin^2 \frac{1}{2}(\theta - x^{-\rho})}.$$

Consider the two possible cases: $|\theta| < 2^{\rho+1} r_m^{-\rho}$ and $|\theta| \geq 2^{\rho+1} r_m^{-\rho}$. In the first case taking into account (34.11) we obtain

$$|V_3^{(1)}(z)| \leq C_{18} r_m \int_{r_m/2}^{q_m} \frac{x^{-\rho}dx}{(x-r_m)^2} \leq$$

$$\leq \frac{C_{19}}{r_m^{\rho-1}(q_{m+1} - q_m)} \leq C_{20}.$$

In the second case we have $|2\theta - x^{-\rho}| < 3|\theta|$ for $x \in [r_m/2, q_m]$ and there exists a constant $a > 0$ such that

$$|V_3^{(1)}(z)| \leq$$

$$\leq C_{21} r_m |\theta| \int_{-\infty}^{\infty} \frac{dx}{(x-r_m)^2 + axr_m\theta^2} = \frac{\pi}{\sqrt{a}} C_{21}.$$

Likewise,

$$\max_{|z|=r_m} |\Re V_3(z)| \leq C_{22}$$

Now, using again (34.11), we obtain

$$\max_{|z|=r_m} |V_4(z)| \leq r_m^{\rho} \int_{\lambda_m} x^{-\rho-1}dx \leq C_{23}.$$

and, clearly,

$$\max_{|z|=r_m} |V_5(z)| \leq C_{24}.$$

Thus we have obtained the bound

$$\max_{|z|=r_m} |V(z)| \leq C_{25}$$

of the second integral in the right-hand side of (34.5).

The estimate of the last integral $U(z)$ in (34.5) is obtained from the relation

$$\Re U(z) = \arg z \cdot \log G(\infty) + o(1), \quad z \to \infty,$$

which is proved similarly as done for (19.11). We have

$$\max_{|z|=r_m} |\Re U(z)| \leq C_{26}.$$

The obtained estimates of the integrals $J(z)$, $V(z)$, $U(z)$ together with equality (34.5) provide

$$\max_{|r|=r_m} |\log|\Psi_*(z)|| < C_{27},$$

which is equivalent to (34.4). We have completed the proof.

Lemma 34.2 *Let $\Psi(z)$ be an arbitrary function possessing properties 1)–6) from Lemma 34.1. Then the integral*

$$\Gamma(z) = \frac{1}{2\pi i}\int_1^{\infty} \frac{g(x)dx}{\Psi^+(x)(x-z)} \tag{34.12}$$

converges and is bounded inside any domain of the form $|z-1| \geq \epsilon > 0$. In addition,

$$\Gamma(z) = o(1), \quad z \to \infty,$$

$$\Gamma(z) = O(|z-1|^{-\alpha}), \quad z \to 1.$$

The lemma is a direct consequence of the wellknown properties of the Cauchy-type integral (see [2, p.74]).

Theorem 34.1 *Let $\Psi(z)$ be an arbitrary function possessing properties 1)–6) from Lemma 34.1*[22]. *Then*

1) The set of poles $\{z_n\}$ of the function $\Psi(z)$ if infinite and $\lim_{n\to\infty} z_n = \infty$.

2) In order that the non-homogeneous problem (34.1) with the index $-\infty$ have a solution it is necessary and sufficient that the function $\Gamma(z)$ defined by (34.12) has zeros at all points z_n with the multiplicities not exceeding multiplicities of the corresponding poles.

Under this condition the solution of problem (34.1) is unique and admits the representation

$$\Phi(z) = \frac{\Psi(z)}{2\pi i}\int_1^{\infty} \frac{g(x)dx}{\Psi^+(x)(x-z)}. \tag{34.13}$$

Proof. Suppose that the set of the poles $\{z_1, \dots, z_m\}$ of the function $\Psi(z)$ is finite. Then the function $\Psi_1(z) = \Psi(z)\prod_{k=1}^{m}(z-z_k)$ is a solution of the boundary problem (34.1) in the class B. By Theorem 25.2 we have

$$\int_0^{2\pi} \log|\Psi_1(re^{i\theta})|d\theta = -\int_1^r \arg G(x)\frac{dx}{x} + O(\log r), \quad r \to \infty.$$

[22]The existence of such a function is provided by Lemma 34.1. One can verify that there are infinitely many such functions.

Therefore

$$\int_0^{2\pi} \log|\Psi(re^{i\theta})|d\theta = -\int_1^r \arg G(x)\frac{dx}{x} + O(\log r), \quad r \to \infty. \tag{34.14}$$

Since $\arg G(x) = \lambda x^\rho + o(x^\rho)$, $x \to \infty$ and $\lambda < 0$, the right-hand side of equality (34.14) approaches $+\infty$ as $z \to +\infty$. On the other hand, by the hypothesis, the function $\Psi(z)$ possesses the property 6) (see Lemma 34.2), and hence, the left hand-side of (34.14) is bounded from above on a sequence $\{r_n\}$ producing a contradiction. Thus we have proved that the set $\{z_n\}$ of the poles of the function $\Psi(z)$ is infinite an;, clearly, it does not have finite condensation points.

Let the zeros of the function $\Gamma(z)$ satisfy the conditions of Theorem 34.1. Then the function $\Phi(z)$ defined by (34.13) is analytic in D and satisfies boundary condition (34.1). It follows from Lemma 34.3 and the properties of the function $\Psi(z)$ that $\Phi(z) \in \tilde{B}$. Let $\{r_n\}$ be the sequence from property 6) of the function $\Psi(z)$. Then the function $\Phi(z)$ is uniformly bounded on the boundaries of the domains $D \cap \{z : \ |z| < r_n\}$ and, by the maximum principle, $\Phi(z) \in B$. So, we have proved the sufficiency of the condition of Theorem 34.1. Now let us prove the necessity.

Let problem (34.1) have a solution $\Phi_0(z) \in B$ and $\Phi(z)$ be defined by (34.13). Using the properties of $\Psi(z)$, we see that the function

$$F(z) = \{\Phi(z) - \Phi_0(z)\}/\Psi(z) = \Gamma(z) - \big(\Phi_0(z)/\Psi(z)\big) \tag{34.15}$$

is analytic in D, satisfies the boundary condition $F^+(t) = F^-(t)$, $1 < t < \infty$, and also $|F(z)| = O(|z-1|^{-\alpha})$ as $z \to 1$ for some α, $0 \le \alpha < 1$. Therefore $F(z)$ is an entire function. Since it is uniformly bounded on the sequence of the circles $\{|z_n| = r_n\}$, applying the maximum modulus principle and the Liouville theorem, we see that $F(z) \equiv \text{const}$. Since $\Gamma(z) \to 0$ as $z \to \infty$ and $\Psi(z_n) = \infty$, then, refering to (34.15) $F(z_n) \to 0$ as $z_n \to \infty$, and hence, $F(z) \equiv 0$. Thus $\Phi_0(z) = \Phi(z) = \Psi(z)\Gamma(z) \in B$, whence $\Gamma(z)$ should have zeros of the appropriate multiplicities at the points z_n. We have proved the necessity of the condition of Theorem 34.1 and simultaneously the uniqueness of the solution and also the fact that it may be represented by (34.13). This complete the proof.

In the case $0 < \rho < 1/2$, the solution of the non-homogeneous problem (34.1) admits a representation wich is simpler than (34.13). To this end, introduce the function $X(z)$ by relation (24.1) as in Sect. 24 and call by it the canonical function of problem (34.2). Clearly, properties 1)–5) of this function indicated in Sect. 19 remain true as well as Theorem 24.1. It should be mentioned, that, since $\lambda < 0$ and $0 < \rho < 1/2$, relation (24.2) yields

$$\min_{0\leq\theta\leq 2\pi} |X(re^{i\theta})| > \exp(Cr^{\rho}), \quad r > r_0. \tag{34.16}$$

One can prove the following statement in the same way as Theorem 24.2 (here again $0 < \rho < \infty$).

Lemma 34.3 *The general solution of the homogeneous problem (34.2) in the class $\tilde{B}$ has the form $\Psi(z) = X(z)F(z)$, where $F(z)$ is an arbitrary entire function.*

Theorem 34.2 *If the non-homogeneous problem (34.1) has a bounded solution $\Phi(z) \in B$ and $0 < \rho < 1/2$, then*

$$\Phi(z) = \frac{X(z)}{2\pi i} \int_1^{\infty} \frac{g(x)dx}{X^+(x)(x-z)}, \tag{34.17}$$

where $X(z)$ is the canonical function of the problem defined by (24.1) with $q = 0$.

Proof. By (34.16) the integral in the right-hand side of (34.17) converges. Denote it by $\Gamma_0(z)$ and consider an arbitrary solution $\Phi(z) \in B$ of problem (34.1). The function

$$\Psi_0(z) = \Phi(z) - \frac{1}{2\pi i} X(z)\Gamma_0(z)$$

is a solution of the homogeneous problem (34.2) in the class $\tilde{B}$. By Lemma 34.3 we have $\Psi_0(z) = X(z)F(z)$, where $F(z)$ is an entire function. Therefore

$$\Phi(z) = X(z)\Big\{\frac{1}{2\pi i}\Gamma_0(z) + F(z)\Big\}.$$

If $F(z) \not\equiv 0$, then there exists a sequence $a_n \to \infty$ such that $\inf_n |F(a_n)| > 0$. Since $\Gamma_0(z) = o(1)$ as $z \to \infty$, then, taking into account (34.16), we obtain $\lim_{n\to\infty} |\Phi(a_n)| = \infty$, which contradicts $\Phi(z) \in B$. So we have proved (34.17).

Corollary 34.1 *Let $0 < \rho < 1/2$. For the right-hand side of (34.17) to be bounded it is necessary and sufficient that for any function $\Psi(z)$ possessing properties 1)–6) from Lemma 34.1, the function $\Gamma(z)$ defined by (34.12) has zeros at all poles of $\Psi(z)$ and the multiplicities of these zeros are at least such as those of the corresponding poles.*

In conclusion we give an example of a solvable non-homogeneous problem with the index $-\infty$.

Example. Set

$$\Phi(z) = \exp\{-(z-1)^{1/4} e^{i\pi/4}\}, \quad z \in D,$$

where the branch of the function $(z-1)^{1/4}$ is chosen to be positive on the upper side of the cut $1 < t < \infty$. Let $G(t) = \exp(-i(t-1)^{\rho})$, $1 \le t < \infty$, where $\rho \in (0, \infty)$ and

$$g(t) = \Phi^+(t) - G(t)\Phi^-(t), \quad 1 \le t < \infty. \tag{34.18}$$

Clearly, $g(1) = g(\infty) = 0$ and also $g(t) \in H_{[1,\infty)}(1/4)$. Thus the function $\Phi(z)$ is a bounded solution of non-homogeneous problem (34.1) and both $G(t)$ and $g(t)$ satisfy all the conditions indicated at the beginning of this section.

By Theorem 34.1 the boundary problem mentioned above satisfies the conditions of solvability. One can verify it directly. Indeed, let $\Psi(z)$ be an arbitrary function possessing properties 1)–6) from Lemma 34.2. Since $\Psi^+(t) = G(t)\Psi^-(t)$, taking into account (34.18) one may represent the function $\Gamma(z)$ defined by (34.12) in the form

$$\Gamma(z) = \frac{1}{2\pi i}\int_1^\infty \frac{\Phi^+(x)dx}{\Psi^+(x)(x-z)} - \frac{1}{2\pi i}\int_1^\infty \frac{\Phi^-(x)dx}{\Psi^-(x)(x-z)}.$$

By the properties of the function $\Psi(z)$, the function $\Phi(z)/\Psi(z)$ is analytic in D and continuous up to the point $z = 1$. Its behavior in the neiborhood of this point is described by property 4). Let $\epsilon > 0$ and r_n be defined by property 6) of the function $\Psi(z)$. Using the Cauchy theorem for the domain

$$D \cap \{z : |z-1| > \epsilon\} \cap \{z : |z| < r_r\},$$

we obtain

$$\int_{|\zeta|=r_n} \frac{\Phi(\zeta)d\zeta}{\Psi(\zeta)(\zeta-z)} - \int_{|\zeta-1|=\epsilon} \frac{\Phi(\zeta)d\zeta}{\Psi(\zeta)(\zeta-z)} +$$

$$\int_\epsilon^{r_n} \frac{\Phi^+(x)dx}{\Psi^+(x)(x-z)} - \int_\epsilon^{r_n} \frac{\Phi^-(x)dx}{\Psi^-(x)(x-z)} = 2\pi i\frac{\Phi(z)}{\Psi(z)},$$

$$z \in D \cap \{z : |z-1| > \epsilon\} \cap \{z : |z| < r_n\}.$$

Sending $\epsilon \to 0$ and $r_n \to \infty$, we see that $\Gamma(z) = \Phi(z)/\Psi(z)$, $z \in D$, and hence, at the poles of the function $\Psi(z)$, the function $\Gamma(z)$ has zeros of at least the same multiplicity.

APPENDIX

I.V.Ostrovskii

ON THE PALEY PROBLEM

A.1 Formulation of the problem and proof of the main inequality

Assume the standard notation and the main properties of the Nevanlinna characteristics of meromorphic functions to be known (see, for example, [19, Ch. 1]).

In 1932 Paley had formulated the following conjecture. For every entire function $f(z)$ the inequality

$$\liminf_{r\to\infty} \frac{\log M(r,f)}{T(r,f)} \le \begin{cases} \pi\rho/\sin\pi\rho, & 0\le\rho\le 1/2, \qquad \text{(A.1.1)}\\ \pi\rho, & \rho>1/2 \qquad \text{(A.1.2)}\end{cases}$$

holds. Earlier Valiron [40] and Walhund [41] had proved the first inequality, so the open problem (which is called *the Paley problem*) is to prove inequality (A.1.2). It should be mentioned that one can easily give examples of an entire function for which the equality sign is attained both in (A.1.1) and (A.1.2). One of such functions is the well-known Mittag-Leffler function $E_\rho(z)$, $0<\rho<\infty$.

In his lecture at the Kharkov function theory seminar in 1967 N.V. Govorov presented a complete solution of the Paley problem. This solution was published later [14]. The solution was based on the following statement:

For every entire function $f(z)$, $f(0)=1$, of finite order ρ the following inequality

$$\log M(r,f) \le x^2 r^x \int_0^\infty \frac{T(t,f)t^{x-1}dt}{(t^x+r^x)^2}, \qquad r>r_0. \qquad \text{(A.1.3)}$$

holds for any $x>\max(2\rho,1)$.

The proof of this inequality used the approach which was created to study the Riemann boundary problem with infinite index.

In 1968 V.P. Petrenko found a principally new approach which allowed him to obtain a more general result (see [26]).

Theorem A.1.1 *Let $f(z)$, $f(0)=1$, be a meromorphic function and $\{b_k\}$ be its poles. For any $x>1/2$ the following inequality holds*

$$\log M(r,f) \le x^2 r^x \int_0^R \frac{m(t,f)t^{x-1}dt}{(t^x+r^x)^2} +$$

$$+\sum_{|b_k|<R}\log\frac{1+(r/|b_k|)^x}{|1-(r/|b_k|)^x|}+Ax\left(\frac{r}{R}\right)^x\{T_1(8R,f)+B\}, \tag{A.1.4}$$

here $4\le r\le R/4$, *A is an absolute constant and the constant B is independent of r, R and x, and*

$$T_1(r,f)=\int_0^r\frac{T(t,f)}{t}dt. \tag{A.1.5}$$

If $f(z)$, $f(0)=1$, is an entire function of order ρ, then $m(t,f)=T(t,f)$ and the set $\{b_k\}$ is empty. Therefore setting $x>\max(2\rho,1)$ in (A.1.4) and sending $R\to\infty$ we obtain (A.1.3).

Theorem A.1.1. became the basis of a new theory developed by Petrenko, i.e. the theory of growth of meromorphic functions. This theory is presented in monograph [27]. In particular, this theorem yields the following result (also due to Petrenko) which is stronger than the Paley conjecture.

Theorem A.1.2 *For a meromorphic function $f(z)$ of finite lower order λ the following inequality holds*

$$\liminf_{r\to\infty}\frac{\log M(r,f)}{T(r,f)}\le\begin{cases}\pi\lambda/\sin\pi\lambda, & 0\le\lambda\le 1/2, \qquad (A.1.6)\\ \pi\lambda, & \lambda>1/2. \qquad (A.1.7)\end{cases}$$

In this section we show that one can obtain Theorem A.1.1 by the same approach as was used in [14] for the proof of inequality (A.1.3). In the next section we prove Theorem A.1.2. In what follows the letter A denotes various absolute constants. We need the following lemma.

Lemma A.1.1 *Let $f(z)$, $f(0)=1$, be a meromorphic function free of poles and zeros on the ray $\arg z=\alpha$. Then for $\beta>1/2$, $4\le r_2\le r_1\le R_1/2$, the following relation holds*

$$\int_1^{r_1}\frac{|\log f(te^{i\alpha})|}{t}\eta\left(\left(\frac{t}{r_2}\right)^\beta\right)dt\le AT_1(4R_1,f),$$

here $\log f(te^{i\alpha})$,$0\le t$ *is the branch of logarithm, wich vanishes at* $t=0$, $T_1(r,f)$ *is defined by (A.1.5) and*

$$\eta(x)=\frac{1}{2\pi}\int_0^{2\pi}|1-xe^{i\theta}|^{-1}d\theta,\quad 0\le x<\infty. \tag{A.1.8}$$

Proof. We have

$$\int_1^{r_1}\frac{|\log f(te^{i\alpha})|}{t}\eta\left(\left(\frac{t}{r_2}\right)^\beta\right)dt=\int_1^{r_2/2}+\int_{r_2/2}^{r_1}=I_1+I_2.$$

Since $\eta(u) \le A$ for $0 \le u \le (1/2)^{\beta}$, we get

$$I_1 \le A \int_1^{r_2/2} |\log|f(te^{i\alpha})|| \frac{dt}{t} + A \int_1^{r_2/2} |\arg f(te^{i\alpha})| \frac{dt}{t} = I_{11} + I_{12}.$$

By the well-known Nevanlinna theorem (see [19, p. 55])

$$\frac{1}{r} \int_1^r \{\log^+ M(t,f) + \log^+ M(t,1/f)\} dt \le AT(2r,f),$$

and using the inequality

$$|\log|f(te^{i\alpha})|| \le \log^+ M(t,f) + \log^+ M(t,1/f),$$

we have

$$I_{11} \le A \int_1^{r_2/2} \{\log^+ M(t,f) + \log^+ M(t,1/f)\} \frac{dt}{t} =$$

$$= \frac{1}{t} \int_1^t \{\log^+ M(u,f) + \log^+ M(u,1/f)\} du \Big|_1^{r_2/2} +$$

$$+ \int_1^{r_2/2} \frac{dt}{t^2} \int_1^t \{\log^+ M(u,f) + \log^+ M(u,1/f)\} du \le$$

$$\le AT(r_2,f) + AT_1(r_2,f).$$

Now the Nevanlinna-Schwartz formula (see [19, p. 16]) gives

$$\log f(te^{i\alpha}) = \frac{1}{2\pi} \int_0^{2\pi} \log|f(Re^{i\theta})| \frac{Re^{i\theta} + te^{i\alpha}}{Re^{i\theta} - te^{i\alpha}} d\theta +$$

$$+ \sum_{|c_m|<R} \delta_m \log\Big\{ \frac{|c_m|}{\bar{c}_m} \frac{R^2 - \bar{c}_m te^{i\alpha}}{R(c_m - te^{i\alpha})} \Big\}, \quad 0 \le t < R.$$

Here $\{c_m\}$ is the union of the zeros and the poles of the function $f(z)$ and $\delta_m = -1$ (respectively $+1$) if the corresponding c_m is a zero (pole) of $f(z)$. We choose all branches of logarithm to be real at $t = 0$. Since

$$\Big| \Im \log\Big\{ \frac{|c_m|}{\bar{c}_m} \frac{R^2 - \bar{c}_m te^{i\alpha}}{R(c_m - te^{i\alpha})} \Big\} \Big| \le 2\pi, \quad 0 \le t < R,$$

we obtain by setting $R = 2t$,

$$|\arg f(te^{i\alpha})| \le 3\{m(2t,f) + m(2t,1/f)|\} +$$

$$+2\pi\{n(2t,f)+n(2t,1/f)\}\le AT(4t,f).$$

Hence, $I_{12}\le AT_1(2r_2,f)$ and

$$I_1\le A\{T(2r_2,f)+T_1(2r_2,f)\}.$$

Consider the integral I_2. Setting $R=R_1$ and $0\le t\le R_1/2$ in the Nevanlinna-Schwartz formula and using the inequality

$$\left|\log\left\{\frac{|c_m|}{\bar{c}_m}\frac{R^2-\bar{c}_m te^{i\alpha}}{R(c_m-te^{i\alpha})}\right\}\right|\le\log\frac{2R_1}{|t-|c_m||}+2\pi,$$

we obtain

$$|\log f(te^{i\alpha})|\le 3\{m(R_1,f)+m(R_1,1/f)\}+$$
$$+2\pi\{n(R_1,f)+n(R_1,1/f)\}+\sum_{|c_m|<R_1}\log\frac{2R_1}{|t-|c_m||}.$$

Hence

$$I_2\le A\{T(R_1,f)+n(R_1,f)+n(R_1,1/f)\}\times$$
$$\times\int_{r_2/2}^{r_1}\eta\Big(\Big(\frac{t}{r_2}\Big)^\beta\Big)\frac{dt}{t}+$$
$$+\sum_{|c_m|<R_1}\int_{r_2/2}^{r_1}\Big\{\log\frac{2R_1}{|t-|c_m||}\Big\}\eta\Big(\Big(\frac{t}{r_2}\Big)^\beta\Big)\frac{dt}{t}\qquad\text{(A.1.9)}$$

We need the two estimates

$$\int_{r_2/2}^{r_1}\eta\Big(\Big(\frac{t}{r_2}\Big)^\beta\Big)\frac{dt}{t}\le A,\quad\int_{r_2/2}^{r_1}\eta^2\Big(\Big(\frac{t}{r_2}\Big)^\beta\Big)\frac{dt}{t}\le A.\qquad\text{(A.1.10)}$$

Since they are similar, we shall prove only the first one. We have

$$\int_{r_2/2}^{r_1}\eta\Big(\Big(\frac{t}{r_2}\Big)^\beta\Big)\frac{dt}{t}\le\frac{1}{\beta}\int_{(1/2)^\beta}^{\infty}\eta(u)\frac{du}{u}=$$

$$=\frac{1}{\beta}\Big(\int_{(1/2)^\beta}^{1/\sqrt{2}}+\int_{1/\sqrt{2}}^{\infty}\Big)\eta(u)\frac{du}{u}\le$$

$$\le\frac{1}{\beta}\Big(A\int_{(1/2)^\beta}^{1/\sqrt{2}}\frac{du}{u}+\int_{1/\sqrt{2}}^{\infty}\eta(u)\frac{du}{u}\Big)=$$

$$=\frac{1}{\beta}\Big(A\Big(\beta-\frac{1}{2}\Big)\log 2+A\Big)\le A$$

(We have used the trivial inequalities $\eta(u) \le A$, $0 \le u \le 1/\sqrt{2}$, and $\eta(u) = O(1/u)$, $u \to \infty$, and also the fact that $\eta(u)$ has a logarithmic singularity at $u = 1$.)

Using inequality (A.1.10), we have

$$\int_{r_2/2}^{r_1} \Big\{\log \frac{2R_1}{|t-|c_m||}\Big\}\eta\Big(\Big(\frac{t}{r_2}\Big)^\beta\Big)\frac{dt}{t} \le \Big(\log\frac{2R_1}{|c_m|}\Big)\int_{r_2/2}^{r_1}\eta\Big(\Big(\frac{t}{r_2}\Big)^\beta\Big)\frac{dt}{t}+$$

$$+\int_{r_2/2}^{r_1}\Big(\log^+\frac{1}{|1-t/|c_m||}\Big)\eta\Big(\Big(\frac{t}{r_2}\Big)^\beta\Big)\frac{dt}{t} \le A\log\frac{2R_1}{|c_m|}+$$

$$+\Big\{\int_{r_2/2}^{r_1}\Big(\log^+\frac{1}{|1-t/|c_m||}\Big)^2\frac{dt}{t}\cdot\int_{r_2/2}^{r_1}\eta^2\Big(\Big(\frac{t}{r_2}\Big)^\beta\Big)\frac{dt}{t}\Big\}^{1/2} \le$$

$$\le A\log\frac{2R_1}{|c_m|}+\Big\{\int_0^\infty\Big(\log^+\frac{1}{|1-u|}\Big)^2\frac{du}{u}\cdot A\Big\}^{1/2} \le$$

$$\le A\log\frac{R_1}{|c_m|}+A.$$

Since

$$\sum_{|c_m|<R_1}\log\frac{R_1}{|c_m|} = N(R_1,f)+N(R_1,1/f) \le 2T(R_1,f),$$

relation (A.1.9) implies $I_2 \le AT(2R_1,f)$.

Now, combining the estimates of the integrals I_1, I_2 we have

$$\int_1^{r_1}\frac{|\log f(te^{i\alpha})|}{t}\eta\Big(\Big(\frac{t}{r_2}\Big)^\beta\Big)dt \le A\{T(2R_1,f)+T_1(2R_1,f)\}.$$

and since

$$T(r,f) \le \frac{1}{\log 2}\int_r^{2r}\frac{T(t,f)}{t}dt \le \frac{1}{\log 2}T_1(2r,f), \tag{A.1.11}$$

we have obtained the proof of the lemma.

Proof of Theorem A.1.1 Take an arbitrary σ such that $0<\sigma<1/x<1$ and also the function $f(z)$ has neither zeros nor poles on the rays $\arg z = \pm\pi\sigma$. Consider the function

$$\Phi(\zeta) = f(\zeta^\sigma e^{-i\pi\sigma}), \quad 0<\arg\zeta<2\pi. \tag{A.1.12}$$

which is meromorphic in the domain D obtained by cutting the complex plane along the ray $L=\{t:\ 0\le t<\infty\}$. The boundary values on L

$$\Phi^+(t) = f(t^\sigma e^{-i\pi\sigma}),\quad \Phi^-(t) = f(t^\sigma e^{i\pi\sigma}),\quad 0\le t<\infty, \tag{A.1.13}$$

are finite and differ from zero. Denote

$$G(t) = \Phi^+(t)/\Phi^-(t), \quad t \geq 0, \tag{A.1.14}$$

$$X_R(\zeta) = \exp\left\{\frac{\zeta}{2\pi i}\int_0^R \frac{\log G(t)}{t(t-\zeta)}dt\right\}, \tag{A.1.15}$$

where the branch of logarithm is chosen such that $\log G(0) = 0$. The function $\log G(t)$ satisfies the Hölder condition on $[0, R]$. Therefore arguing in the same way as in Sect. 19 one can prove that the function $X_R(\zeta)$ possesses the properties similar to those of the canonical function. Namely,

1) $X_R(\zeta)$ is analytic in D, continuous up to L except at the points 0 and R and also non-vanishing in $(D \cup L)\backslash\{0, R\}$.

2) $X_R(\zeta)$ satisfies the boundary condition

$$X_R^+(t) = G(t)X_R^-(t), \quad 0 < t < R.$$

3) $X_R(\zeta)$ is normalized by the relation $X_R(0) = 1$.

4) In a neighborhood of the point $\zeta = 0$ we have

$$C_1 \leq |X_R(\zeta)| \leq C_2.$$

Since the function $\Phi(\zeta)$ satisfies the boundary condition

$$\Phi^+(t) = G(t)\Phi^-(t), \quad 0 < t < \infty, \tag{A.1.16}$$

the function

$$\Omega_R(\zeta) = \Phi(\zeta)/X_R(\zeta) \tag{A.1.17}$$

is meromorphic in the disk $|\zeta| < R$, has the same zeros and poles as the function $\Phi(\zeta)$ in the disk and also $\Omega_R(0) = 1$.

Let $\{u_k\}$ and $\{v_n\}$ be the sets of poles and zeros of respectively the function $\Phi(\zeta)$ in the disk $|\zeta| < R$. Denote

$$\pi_R^{(1)}(\zeta) = \prod_k (1 - \zeta/u_k), \quad \pi_R^{(2)}(\zeta) = \prod_n (1 - \zeta/v_n),$$

$$\Lambda_R(\zeta) = \Omega_R(\zeta)\pi_R^{(1)}(\zeta)/\pi_R^{(2)}(\zeta). \tag{A.1.18}$$

The function $\Lambda_R(\zeta)$ is analytic and nonvanishing for $|\zeta| < R$ and also $\Lambda_R(0) = 1$. Let us verify the inequality

$$\log|\Lambda_R(\zeta)| \leq \frac{A}{\sigma}\frac{|\zeta|}{R}\{T_1(8R^\sigma, f) + B\} \tag{A.1.19}$$

for $|\zeta| \leq R/2$. First estimate the quantity $T(r, \Lambda_R)$ for $2 \leq r \leq R$. We have

$$T(r, \Lambda_R) = m(r, \Lambda_R) \leq$$

$$\leq m(r, \Omega_R) + m(r, \pi_R^{(1)}) + m(r, 1/\pi_R^{(2)}). \qquad \text{(A.1.20)}$$

Consider the sum of the the two last summands in (A.1.20). We have

$$m(r, \pi_R^{(1)}) \leq \log M(r, \pi_R^{(1)}) \leq \sum_k \log(1 + r/|u_k|) \leq$$

$$\leq \sum_k \{\log 2 + \log^+(r/|u_k|)\} =$$

$$= \log 2 \cdot n(R, 1/\pi_R^{(1)}) + N(R, 1/\pi_R^{(1)}).$$

The similar estimate holds for $m(r, \pi_R^{(2)})$. Therefore

$$m(r, 1/\pi_R^{(2)}) = m(r, \pi_R^{(2)}) - N(r, 1/\pi_R^{(2)}) \leq$$

$$\leq \log 2 \cdot n(R, 1/\pi_R^{(2)}),$$

$$m(r, \pi_R^{(1)}) + m(r, 1/\pi_R^{(2)}) \leq$$

$$\leq \log 2 \cdot \big(n(R, 1/\pi_R^{(1)}) + n(R, 1/\pi_R^{(2)})\big) + N(r, 1/\pi_R^{(1)}).$$

Since $n(R, 1/\pi_R^{(1)}) \leq n(R^\sigma, f)$ and $n(R, 1/\pi_R^{(2)}) \leq n(R^\sigma, 1/f)$, we obtain

$$m(r, \pi_R^{(1)}) + m(r, 1/\pi_R^{(2)}) \leq$$

$$\leq \log 2\big(n(R^\sigma, f) + n(R^\sigma, 1/f)\big) + \frac{1}{\sigma} N(R^\sigma, f) \leq$$

$$\leq N(2R^\sigma, f) + N(2R^\sigma, 1/f) + \frac{1}{\sigma} N(R^\sigma, f) \leq$$

$$\leq \Big(2 + \frac{1}{\sigma}\Big) T(2R^\sigma, f) \leq \frac{1}{\log 2}\Big(2 + \frac{1}{\sigma}\Big) T_1(4R^\sigma, f) \qquad \text{(A.1.21)}$$

(We have also used inequality (A.1.11)). Now consider $m(r, \Omega_R)$. We have

$$m(r, \Omega_R) \leq \frac{1}{2\pi} \int_0^{2\pi} \log^+ |\Phi(re^{i\theta})| d\theta +$$

$$+ \frac{1}{2\pi} \int |_0^{2\pi} \log |X_R(re^{i\theta})| \, |d\theta. \qquad \text{(A.1.22)}$$

Clearly,

$$\frac{1}{2\pi}\int_0^{2\pi}\log^+|\Phi(re^{i\theta})|d\theta \le \frac{1}{\sigma}m(r^\sigma, f) \tag{A.1.23}$$

Using equalities (A.1.15), (A.1.8), (A.1.14), (A.1.13) and applying the Foubini theorem, we obtain

$$\frac{1}{2\pi}\int |\log|X_R(re^{i\theta})||d\theta \le \frac{1}{2\pi}\int_0^{2\pi}\Big\{\frac{r}{2\pi}\int_0^R \frac{|\log G(t)|}{t|t-re^{i\theta}|}dt\Big\}d\theta =$$

$$= \frac{1}{2\pi}\int_0^R \frac{|\log G(t)|}{t}\eta\Big(\frac{t}{r}\Big)dt \le$$

$$\le \frac{1}{2\pi}\int_0^R \frac{|\log f(t^\sigma e^{-i\pi\sigma})| + |\log f(t^\sigma e^{i\pi\sigma})|}{t}\eta\Big(\frac{t}{r}\Big)dt =$$

$$= \frac{1}{2\pi\sigma}\int_0^{R^\sigma} \frac{|\log f(te^{-i\pi\sigma})| + |\log f(te^{i\pi\sigma})|}{t}\eta\Big(\frac{t^{1/\sigma}}{r}\Big)dt.$$

Let us split the last integral into two ones over the segments $[0,1]$ and $[1,R^\sigma]$ respectively. One can easily estimate the integral over $[0,1]$ by a constant B independent of r, R and σ. For estimating the second integral apply Lemma A.1.1 with $\alpha = \pm\pi\sigma$, $\beta = 1/\sigma$, $r_1 = R^\sigma$, $r_2 = r^\sigma$, $R_1 = 2R^\sigma$. We obtain

$$\frac{1}{2\pi}\int_0^{2\pi} |\log|X_R(re^{i\theta})||d\theta \le \frac{A}{\sigma}T_1(8R^\sigma, f) + \frac{B}{\sigma}. \tag{A.1.24}$$

Now (A.1.20)–(A.1.24) imply

$$T(r,\Lambda_R) \le \frac{A}{\sigma}T_1(8R^\sigma, f) + \frac{B}{\sigma}, \quad 2 \le r \le R. \tag{A.1.25}$$

One may use the last relation to obtain (A.1.19). Since the function $\Lambda_R(\zeta)$ is analytic and non-vanishing in $|\zeta| < R$ and $\Lambda_R(0) = 1$, the Poisson formula yields

$$\log|\Lambda_R(\zeta)| =$$

$$= \frac{1}{2\pi}\int_0^{2\pi}\Big\{\Re\frac{R_2e^{i\theta}+\zeta}{R_2e^{i\theta}-\zeta} - 1\Big\}\log|\Lambda_R(R_2e^{i\theta})|d\theta =$$

$$= \frac{1}{2\pi}\int_0^{2\pi}\Big\{\Re\frac{2\zeta}{R_2e^{i\theta}-\zeta}\Big\}\log|\Lambda_R(R_2e^{i\theta})|d\theta, \quad |\zeta| < R_2 < R,$$

whence

$$\log|\Lambda_R(\zeta)| \le \frac{2|\zeta|}{R_2-|\zeta|}\cdot\frac{1}{2\pi}\int_0^{2\pi}|\log|\Lambda_R(R_2e^{i\theta})||d\theta \le$$

$$\leq \frac{4|\zeta|}{R_2 - |\zeta|} T(R_2, \Lambda_R).$$

Setting $R_2 = \frac{3}{4}R$ and assuming $|\zeta| < R/2$, we obtain

$$\log |\Lambda_R(\zeta)| \leq \frac{16|\zeta|}{R} T\Big(\frac{3}{4}R, \Lambda_R\Big).$$

which in view of (A.1.25) implies (A.1.19).

In what follows we need the equality

$$\frac{1}{2\pi} \int_0^{2\pi} \log |\Phi(re^{i\theta})| d\theta =$$

$$= \int_0^r \Big\{ \hat{n}_\Phi(t) - \frac{1}{2\pi} \arg G(t) \Big\} \frac{dt}{t}, \quad 0 < t < \infty, \tag{A.1.26}$$

where $\hat{n}_\Phi(t)$ is the difference between the numbers of zeros and poles of the function $\Phi(\zeta)$ in the disk $|\zeta| \leq t$. One may prove this relation in the same way as Theorem 25.2 (since $\Phi(0) = 1$, relation (A.1.26) does not contain the summand similar to $c_\Phi(r)$ in (25.4).) We also need the following representation of $\Phi(\zeta)$ which is similar to relation (26.1)

$$\Phi(\zeta) = \exp\Big\{ \frac{\zeta}{2\pi i} \int_0^R \frac{\log G(x) - 2\pi i \hat{n}_\Phi(x)}{x(x-\zeta)} dx - \zeta \int_R^\infty \frac{\hat{n}_\Phi(R) dx}{x(x-\zeta)} \Big\} \times$$

$$\times \prod_n \frac{1 - \zeta/v_n}{1 - \zeta/|v_n|} \cdot \prod_k \frac{1 - \zeta/|u_k|}{1 - \zeta/u_k} \cdot \Lambda_R(\zeta). \tag{A.1.27}$$

To obtain this representation it suffices to mention that relations (A.1.17), (A.1.18) and (A.1.15) imply

$$\Phi(\zeta) = \exp\Big\{ \frac{\zeta}{2\pi i} \int_0^R \frac{\log G(x) dx}{x(x-\zeta)} \times$$

$$\times \prod_n \Big(1 - \frac{\zeta}{v_n}\Big) \prod_k \Big(1 - \frac{\zeta}{u_k}\Big)^{-1} \cdot \Lambda_R(\zeta),$$

and then use the identity

$$\prod_n \Big(1 - \frac{\zeta}{|v_n|}\Big) \prod_k \Big(1 - \frac{\zeta}{|u_k|}\Big)^{-1} =$$

$$= \exp\Big\{ -\zeta \int_0^R \frac{\hat{n}_\Phi(x) dx}{x(x-\zeta)} - \zeta \int_R^\infty \frac{\hat{n}_\Phi(R) dx}{x(x-\zeta)} \Big\}$$

(this identity is similar to (26.9) and admits the same proof.)

Setting $\zeta = -r$, $2 \le r < R$ in (A.1.27), we have

$$|\Phi(-r)| \le \exp\left\{ r\int_0^R \frac{\hat{n}_\Phi(x) - \frac{1}{2\pi}\arg G(x)}{x(x+r)}dx + r\int_R^\infty \frac{\hat{n}_\Phi(R)dx}{x(x+r)}\right\}\times$$

$$\times \prod_k \frac{1+r/|u_k|}{|1-r/|u_k||}\cdot|\Lambda_R(-r)|.$$

Now using (A.1.19) and the simple estimate

$$r\int_R^\infty \frac{\hat{n}_\Phi(R)dx}{x(x+r)} = \hat{n}_\Phi(R)\log\left(1+\frac{r}{R}\right) \le \frac{r}{R}\hat{n}_\Phi(R) \le$$

$$\le \frac{r}{R}\{n(R^\sigma, f) + n(R^\sigma, 1/f) \le A\frac{r}{R}T(2R^\sigma, f),$$

we obtain

$$\log|\Phi(-r)| \le r\int_0^R \frac{\hat{n}_\Phi(x) - \frac{1}{2\pi}\arg G(x)}{x(x+r)}dx +$$

$$+\sum_k \log\frac{1+r/|u_k|}{|1-r/|u_k||} +$$

$$+\frac{A}{\sigma}\cdot\frac{r}{R}\{T_1(8R^\sigma, f) + B\}, \quad 2 \le r \le \frac{1}{2}R. \tag{A.1.28}$$

To transform the integral in the left-hand side denote the expression in the left-hand side of (A.1.26) by $m_\Phi(r)$ and notice that by (A.1.12) we have

$$m_\Phi(r) = \frac{1}{2\pi\sigma}\int_{-\pi\sigma}^{\pi\sigma} \log|f(r^\sigma e^{i\theta})|d\theta \le \frac{1}{\sigma}m(r^\sigma, f),$$

$$m_\Phi(r) \ge -\frac{1}{\sigma}m(r^\sigma, 1/f) \ge -\frac{1}{\sigma}T(r^\sigma, f).$$

Using (A.1.26) and integrating by parts, we obtain

$$r\int_0^R \frac{\hat{n}_\Phi(x) - \frac{1}{2\pi}\arg G(x)}{x(x+r)}dx = r\int_0^R \frac{dm_\Phi(x)}{x+r} = \frac{rm_\Phi(z)}{R+r} +$$

$$+r\int_0^R \frac{m_\Phi(x)dx}{(x+r)^2} \le$$

$$\le \frac{1}{\sigma}T(R^\sigma, f) + \frac{r}{\sigma^2}\int_0^{R^\sigma} \frac{dm(u,f)u^{1/\sigma-1}du}{(u^{1/\sigma}+r)^2}.$$

Substitute this estimate into (A.1.28). Noting that $\Phi(-r) = f(r^\sigma)$ and putting $r^\sigma = q$, $R^\sigma = Q$, we obtain

$$\log|f(q)| \leq \frac{q^{1/\sigma}}{\sigma^2}\int_0^Q \frac{m(u,f)u^{1/\sigma-1}du}{(u^{1/\sigma}+q^{1/\sigma})^2}+$$

$$+\sum_{|b_k|\leq Q}\log\frac{1+(q/|b_k|)^{1/\sigma}}{|1-(q/|b_k|)^{1/\sigma}|}+A\Big(\frac{q}{Q}\Big)^{1/\sigma}\frac{1}{\sigma}\{T_1(8Q,f)+B\},$$

$$2^\sigma \leq q \leq 2^{-\sigma}Q. \tag{A.1.29}$$

Here, as earlier, b_k denote the poles of the function $f(z)$. The quantity σ satisfies the following conditions:

a) $1/\sigma > x > 1/2$,

b) the function $f(z)$ has neither poles nor zeros on the rays $\arg z = \pm\pi\sigma$.

Sending $\sigma \to 1/x$ over the appropriate sequence, we reduce (A.1.29) to the following

$$\log|f(q)| \leq x^2q^x\int_0^Q \frac{m(u,f)u^{x-1}du}{(u^x+q^x)^2}+$$

$$+\sum_{|b_k|\leq Q}\log\frac{1+(q/|b_k|)^x}{|1-(q/|b_k|)^x|}+A\Big(\frac{q}{Q}\Big)^x x\{T_1(8Q,f)+B\},$$

$$4 \leq q \leq Q/4. \tag{A.1.30}$$

Apply this relation to the function $f_\theta(z) = f(ze^{i\theta})$, $0 \leq \theta < 2\pi$. All the summands of the right-hand side, except possibly B are independent of θ. The constant B has appeared when estimating the expression

$$\frac{1}{2\pi\sigma}\int_0^1 \frac{|\log f(te^{-i\pi\sigma})|+|\log f(te^{i\pi\sigma})|}{t}\eta\Big(\frac{t^{1/\sigma}}{r}\Big)dt,$$

and one can easily verify that this constant may be taken independent of θ. Therefore one may put $\log|f(qe^{i\theta})|$ instead of $\log|f(q)|$ in the left-hand side of (A.1.30). Now taking the maximum over θ, we obtain the inequality which differs from (A.1.4) by notation only.

A.2 Solution of the Paley problem

Now proceed to the proof of Theorem A.1.2. Without loss of generality we may assume $f(0) = 1$, otherwise one may consider the function $f(z)/c_k z^k$, where $c_k z^k$ is the first non-zero term in the Laurent expansion at the point $z = 0$.

First obtain inequality (A.1.6). We need not Theorem A.1.1 for this aim. Instead of this we use the following result (see [19, p. 267]).

Lemma A.2.1 *Let $f(z)$, $f(0) = 1$, be a meromorphic function of lower order $\lambda < 1$. For any $\sigma \in (\lambda, 1)$ the following inequality holds*

$$\limsup_{r\to\infty}\Big\{\log \mu(r,f) + \frac{\pi\sigma}{\sin \pi\sigma}(N(r,f) - \cos \pi\sigma N(r,1/f))\Big\} \geq 0,$$

here

$$\mu(r,f) = \min_{|z|=r} |f(z)|.$$

Apply the lemma to the function $1/f(z)$. Since $\log \mu(r,1/f) = -\log M(r,f)$, we obtain

$$\limsup_{r\to\infty}\Big\{-\log M(r,f) + \frac{\pi\sigma}{\sin \pi\sigma}(N(r,1/f) - \cos \pi\sigma N(r,f))\Big\} \geq 0 \quad \text{(A.1.1)}$$

for any $\sigma \in (\lambda, 1)$.

Let $\lambda < 1/2$. Taking $\sigma \in (\lambda, 1/2)$ we obtain $\cos \pi\sigma > 0$ and

$$\limsup_{r\to\infty}\Big\{-\log M(r,f) + \frac{\pi\sigma}{\sin \pi\sigma} N(r,1/f)\Big\} \geq 0.$$

Since $N(r,1/f) \leq T(r,f)$, we have

$$\liminf_{r\to\infty} \frac{\log M(r,f)}{T(r,f)} \leq \frac{\pi\sigma}{\sin \pi\sigma},$$

and letting $\sigma \to \lambda$ we obtain (A.1.6) for $\lambda < 1/2$. If $\lambda = 1/2$, then applying (A.2.1) with $\sigma \in (1/2, 1)$ and taking into accoun that $\cos \pi\sigma < 0$ and $N(r,f) \leq T(r,f)$, we have

$$\liminf_{r\to\infty} \frac{\log M(r,f)}{T(r,f)} \leq \frac{\pi\sigma}{\sin \pi\sigma}(1 - \cos \pi\sigma).$$

Letting $\sigma \to \lambda = 1/2$, we complete the proof of (A.1.6).

Let us proceed to the proof of (A.1.7).

First verify that, if y belongs to a fixed finite segment $[\alpha, \beta] \in (1/2, \infty)$, then

$$\int_4^{R/4} r^{-y-1} \log M(r, f) dr \le$$

$$\le \pi y \int_4^{R/4} r^{-y-1} T(r, f) dr + C R^{-y} T_1(8R, f) + C \tag{A.2.2}$$

for all sufficiently large R. Here and below C denotes positive constants independent of R, r and y. We use Theorem A.1.1. Since

$$\int_0^4 \frac{m(t, f) t^{x-1} dt}{(t^x + r^x)^2} = O(r^{-2x}), \quad r \to +\infty$$

uniformly with respect to x belonging to any finite segment $[\delta, \gamma] \subset (1/2, \infty)$, then (A.1.4) implies

$$\log M(r, f) \le$$

$$\le x^2 r^x \int_4^R \frac{m(t, f) t^{x-1} dt}{(t^x + r^x)^2} + \sum_{|b_k| < R} \log \frac{1 + (r/|b_k|)^x}{|1 - (r/|b_k|)^x|} +$$

$$+ Ax \left(\frac{r}{R}\right)^x \{T_1(8R, f) + B\} + D r^{-x}, \quad 4 \le r \le R/4, \tag{A.2.3}$$

where D is independent of r, R, x.

Set $x = 2y$ in (A.2.3), multiply the both sides by r^{-y-1} and then integrate over r from 4 to $R/4$. We have

$$\int_4^{R/4} r^{-y-1} \left\{ x^2 r^x \int_4^R \frac{m(t, f) t^{x-1} dt}{(t^x + r^x)^2} \right\} dr =$$

$$= (2y)^2 \int_4^R m(t, f) t^{2y-1} \left\{ \int_4^{R/4} \frac{r^{y-1} dr}{(t^{2y} + r^{2y})^2} \right\} dt \le$$

$$\le (2y)^2 \int_4^R m(t, f) t^{2y-1} \left\{ \int_0^\infty \frac{r^{y-1} dr}{(t^{2y} + r^{2y})^2} \right\} dt =$$

$$= \pi y \int_4^R m(t, f) t^{-y-1} dt \le$$

$$\le \pi y \int_4^R m(t, f) t^{-y-1} dt + C R^{-y} T(R, f),$$

$$\int_4^{R/4} r^{-y-1} \left\{ \sum_{|b_k| < R} \log \frac{1 + (r/|b_k|)^x}{|1 - (r/|b_k|)^x|} \right\} dr \le$$

$$\leq \sum_{|b_k|<R} \int_0^\infty r^{-y-1} \log \frac{1+(r/|b_k|)^{2y}}{|1-(r/|b_k|)^{2y}|} dr =$$

$$= \frac{\pi}{y} \sum_{|b_k|<R} |b_k|^{-y} = \frac{\pi}{y} \int_0^R t^{-y} dn(t,f) =$$

$$= \frac{\pi}{y}\{R^{-y}n(R,f) + yR^{-y}N(R,f) + y^2 \int_0^R t^{-y-1}N(t,f)dt\} \leq$$

$$\leq \pi y \int_4^{R/4} t^{-y-1}N(t,f)dt + CR^{-y}T(2R,f) + C.$$

Therefore, taking into account (A.1.11), we obtain (A.2.2).

Now assume inequality (A.1.7) to be false. Then there exists $\tau > \pi\lambda$ such that $\log M(r,f) \geq \tau T(r,f)$ for all sufficiently large r, say, $r > r_0$. Applying (A.2.2), we see that if y belongs to a fixed finite segment $[\alpha,\beta] \subset (1/2,\infty)$, then the following relation holds for all sufficiently large R:

$$(\tau - \pi y) \int_{r_0}^{R/4} r^{-y-1}T(r,f)dr \leq CR^{-y}T_1(8R,f) + C. \qquad \text{(A.2.4).}$$

Since

$$\int_{r_0}^{R/4} r^{-y-1}T_1(r,f)dr = -\frac{1}{y}T_1(r,f)r^{-y}\Big|_{r_0}^{R/4} +$$

$$+\frac{1}{y}\int_{r_0}^{R/4} r^{-y-1}T(r,f)dr \leq C + C\int_{r_0}^{R/4} r^{-y-1}T(r,f)dr,$$

(A.2.4) yields

$$(\tau - \pi y) \int_{r_0}^{R/4} r^{-y-1}T_1(r,f)dr \leq CR^{-y}T_1(8R,f) + C \qquad \text{(A.2.5)}$$

for $y \in [\alpha,\beta]$ and sufficiently large R.

We need the notion of the lower proximate order. Recall that by the *lower proximate order of a positive non-decreasing function* $\alpha(r)$ we mean the proximate order $\lambda(r)$ such that

$$0 < \liminf_{r\to\infty} r^{-\lambda(r)}\alpha(r) < \infty.$$

Definition of proximate order see [19, p. 69]. It is known (see [19, p. 70]), that for any positive non-decreasing function $\alpha(r)$ of finite lower order λ, there exists a lower proximate order $\lambda(r)$ and for each such order $\lim_{r\to\infty}\lambda(r) = \lambda$. We also need the following lemma ([19, p. 74]).

Lemma A.2.2 *Let* $\lambda(r)$ *be an arbitrary proximate order. For any fixed* $k > 0$ *the following relation holds*

$$\lim_{R\to\infty} \int_1^{kR} r^{\lambda(r)-\lambda(R)-1} dr = \infty.$$

Denote by $\lambda(r)$ the lower proximate order of the function $T_1(r,f)$. Clearly, the lower order of $T_1(r,f)$ is the same as the lower order λ of the function $T(r,f)$.By the definition of $\lambda(r)$ there are positive constants C_1 and C_2 such that

$$T_1(r,f) \geq C_1 r^{\lambda(r)}, \tag{A.2.6}$$

for sufficiently large r, say $r > r_1$ and also

$$T_1(r_k,f) \leq C_2 r_k^{\lambda(r_k)}. \tag{A.2.7}$$

for a sequence $r_k \to \infty$.

Let the quantity y from (A.2.5) vary in the segment $[\alpha,\beta]$ such that $\lambda \in (\alpha,\beta)$ and $[\alpha,\beta] \subset (1/2, \tau/\pi)$. Suppose R to be sufficiently large such that $\lambda(R) \in [\alpha,\beta]$ and set $y = \lambda(R)$ in (A.2.5). Let R approaches ∞ over the sequence $R_k = r_k/8$ where r_k are defined in (A.2.7). By (A.2.7) the expression in the right-hand side of (A.2.7) is bounded and, besides, $\tau - \pi\lambda(R) \to \tau - \pi\lambda > 0$. Therefore

$$(\tau - \pi\lambda) \limsup_{k\to\infty} \int_{r_0}^{R_k/4} r^{-\lambda(R_k)-1} T_1(r,f) dr < \infty.$$

Now, using (A.2.6), we obtain

$$\lim_{k\to\infty} \int_{r_0}^{R_k/4} r^{\lambda(r)-\lambda(R_k)-1} dr < \infty$$

in contrast to Lemma A.2.2. This completes the proof of Theorem A.1.2.

At the end of this section we will present a result from [14]. This result is not contained in Theorem A.1.2 and strengthens inequalities (A.1.1), (A.1.2) in a somewhat other direction.

Let $f(z)$ be an entire function of finite order ρ and $\rho(r)$ be its proximate order. Denote

$$\Delta_M = \limsup_{r\to\infty} r^{-\rho(r)} \log M(r,f),$$

$$\Delta_T = \limsup_{r\to\infty} r^{-\rho(r)} \log T(r,f).$$

By virtue of the definition of proximate order the both quantities Δ_M and Δ_T are finite and positive.

Theorem A.2.1 *The following inequalities hold:*

$$\Delta_M \le \begin{cases} (\pi\rho/\sin\pi\rho)\Delta_T, & 0 \le \rho \le 1/2, \quad \text{(A.2.8)} \\ \pi\rho\Delta_T, & 1/2 < \rho < \infty \quad \text{(A.2.9)} \end{cases}$$

Proof. One can easily see that

$$\liminf_{r\to\infty} \frac{\log M(r,f)}{T(r,f)} \le \frac{\Delta_M}{\Delta_T}.$$

Therefore both (A.1.1) and (A.1.2) are contained in Theorem A.2.1. Without loss of generality we may assume $f(0) = 1$. Let us prove (A.2.8). We have

$$\log M(r,f) \le \int_0^\infty \log\Big(1+\frac{r}{t}\Big)dn(t,1/f) =$$

$$= r\int_0^\infty \frac{n(t,1/f)}{t}\frac{dt}{r+t} = r\int_0^\infty N(t,1/f)\frac{dt}{(r+t)^2} \le$$

$$\le r\int_0^\infty T(t,f)\frac{dt}{(r+t)^2} = \int_0^\infty T(sr,f)\frac{ds}{(1+s)^2}.$$

Suppose that for $r \ge r_0$ we have $T(r,f) \le (\Delta_T+\epsilon)r^{\rho(r)}$. Then

$$\log M(r,f) \le T(r_0,f)\int_0^{r_0/r}\frac{ds}{(1+s)^2} + (\Delta_T+\epsilon)\int_{r_0/r}^\infty (rs)^{\rho(rs)}\frac{ds}{(1+s)^2}.$$

Whencc

$$\Delta_M = \limsup_{r\to\infty} r^{-\rho(r)}\log M(r,f) \le$$

$$\le (\Delta_T+\epsilon)\lim_{r\to\infty}\int_{r_0/r}^\infty \frac{(rs)^{\rho(rs)}}{r^{\rho(r)}}\frac{ds}{(1+s)^2} =$$

$$= (\Delta_T+\epsilon)\int_0^\infty \frac{s^\rho ds}{(1+s)^2} = (\Delta_T+\epsilon)\frac{\pi\rho}{\sin\pi\rho}.$$

That we can change the passage to the limit and integration follows from the definition of the proximate order, (see [19, p. 74]). Since ϵ may be chosen arbitrarily small, we obtain (A.2.8).

To prove (A.2.9) use inequality (A.1.3) with $x = 2\rho$. Suppose that for $r \ge r_0$ we have $T(r,f) \le (\Delta_T+\epsilon)r^{\rho(r)}$. Then

$$\log M(r,f) \le (2\rho)^2 r^{2\rho}\int_0^\infty \frac{T(t,f)t^{2\rho-1}dt}{(t^{2\rho}+r^{2\rho})^2} =$$

$$= (2\rho)^2\int_0^\infty \frac{T(sr,f)s^{2\rho-1}ds}{(s^{2\rho}+1)^2} \le (2\rho)^2 T(r_0,f)\int_0^{r_0/r}\frac{s^{2\rho-1}ds}{(s^{2\rho}+1)^2} +$$

$$+(2\rho)^2(\Delta_T+\epsilon)\int_{r_0/r}^{\infty}\frac{(sr)^{\rho(sr)}s^{2\rho-1}ds}{(s^{2\rho}+1)^2},$$

whence

$$\Delta_M=\limsup_{r\to\infty} r^{-\rho(r)}\log M(r,f)\leq$$

$$\leq(2\rho)^2(\Delta_T+\epsilon)\lim_{r\to\infty}\int_{r_0/r}^{\infty}\frac{(sr)^{\rho(sr)}s^{2\rho-1}ds}{r^{\rho(r)}(s^{2\rho}+1)^2}=$$

$$=(2\rho)^2(\Delta_T+\epsilon)\int_0^{\infty}\frac{s^{3\rho-1}ds}{(s^{2\rho}+1)^2}=(\Delta_T+\epsilon)\pi\rho.$$

Since ϵ may be chosen arbitrarily small we obtain (A.2.9).

Dahlberg [35] have proved the inequalities which generalize (A.1.6) and (A.1.7) for subharmonic functions in R_n, $n\geq 2$. A generalization for function analytic within an angle was obtained in [15].

Bibliography

[1] Gakhov F.D. *On the Riemann problem.* —Mat. Sb., 1937(**2**), 673–683.

[2] Gakhov F.D. *Boundary problems.* —M.: Nauka, 1977.

[3] Govorov N.V. *On conditions of solvability of the homogeneous Riemann problem with a negative index of its coefficient.* Uchen. zap. Kabardino-Balkarskogo un-ta, 1963(**19**), 137–141.

[4] Govorov N.V. *An example of unsolvable homogeneous Riemann problem with a negative index of its coefficient.* —Uchen. zap. Kabardino-Balkarskogo un-ta, 1963(**19**), 141–144.

[5] Govorov N.V. *On the Riemann problem with infinite index.* —Izv. AN BSSR. Ser. fiz.-tekh. nauk, 1964(**1**), 12–17.

[6] Govorov N.V. *Non-homogeneous Riemann problem with infinite index.* —DAN SSSR, 1964(**159**), 961–964.

[7] Govorov N.V. *—On the Riemann problem with infinite index.* —DAN SSSR, 1964(**154**), 1247–1249.

[8] Govorov N.V. *On indicator of functions of non-integer order and analytic and of completely regular growth in half-plane.* —DAN SSSR, 1965(**162**), 495–498.

[9] Govorov N.V. *On indicator of functions analytic and of completely regular growth in half-plane.* —In: Communications of IMC, Sec. 4, —M., 1964, 45–46.

[10] Govorov N.V. *On indicator of functions of integer order and analytic and of completely regular growth in half-plane.* —DAN SSSR, 1967(**172**), 763–766.

[11] Govorov N.V. *On bounded solutions of the Riemann problem with infinite index of power type.* —DAN SSSR, 1968(**182**), 750–753.

[12] Govorov N.V. *On an estimate from below of functions subharmonic inside a disk.* —Kharkov: KhGU, 1968(**6**), 130–150.

[13] Govorov N.V. *The Riemann problem with an infinite index power type less than 1/2.* —In: Theory of functions, functional analysis and their applications. —Kharkov: KhGU, 1968(**6**), 151–176.

[14] Govorov N.V. *On the Paley conjecture.* Functional Analysis and its Applications, 1969(**3**), No.2, 41–45.

[15] Govorov N.V. *On the Paley inequality for functions analytic within an angle.* —In: Materialy vsesouznoi konferentsii po kraevym zadacham. —Kazan: Kazan.GU, 1970, 78–82.

[16] Govorov N.V. *On bounded solutions of the homogeneous Riemann problem with infinite index of power type.* —In: Theory of functions, functional analysis and their applications. —Kharkov: KhGU, 1970(**11**), 3–34.

[17] Govorov N.V. *On solution of a homogeneous edge problem with infinite index in the class of function of completely regular growth.* —Theory of functions, functional analysis and their applications, —Kharkov: KhGU, 1972(**15**), 3–34.

[18] Goluzin G.M. *Geometrical theory of functions of complex variable.* —M.: Nauka, 1966.

[19] Goldberg A.A. and Ostrovskii I.V. *Distribution of values of meromorphic functions.* —M.: Nauka, 1970.

[20] Hoffman K. *Banach spaces of analytic functions.* —M.: IL, 1963 (*Russian translation*).

[21] Grishin A.F *On regularity of growth of subharmonic functions.* —In: Theory of functions, functional analysis and their applications. —Kharkov: KhGU, 1968(**11**), 59–84.

[22] Kudina L.S. *Estimates for functions representable as difference of subharmonic ones in a ball.* —In: Theory of functions, functional analysis and their applications. —Kharkov: KhGU, 1971(**14**), 58–67.

[23] Landkof N.S. *Fundamentals of the modern potential theory.* —M.: Gostekhizdat, 1956.

[24] Levin B.Ya. *Distribution of zeros of entire functions.* —AMS, Providence, 1964.

[25] Muskhelishvily N.I. *Singular integral equations.* —M.: Nauka, 1968.

[26] Petrenko V.P. *Growth of meromorphic functions of integer lower order.* —Izv. AN SSSR. ser. mat., 1969(**33**), 414–454.

[27] Petrenko V.P. *Growth of meromorphic functions.* —Kharkov: Vischa shkola, 1978.

[28] Privalov I.I. *Boundary properties of analytic functions.* —M.: Gostekhizdat, 1950.

[29] Titchmarsh E. *The theory of functions.* —Oxford University Press, 1939

[30] Fainberg E.D. *Estimates of indicators of functions analytic and of a noninteger finite order in the halfplane.* —In: Theory of functions, functional analysis and their applications. —Kharkov: KhGU, 1974(**21**), 13–30.

[31] Fainberg E.D. *Estimates of indicators of special classes of functions.* —dep. VINITI, 1981, N4167.

[32] Heifiz A.I. *Indicators of analytical functions in the open half-plane with order less than one, having a completely regular growth within the inner angles.* —Izv. AN SSSR. ser. mat., 1975(**39**), 899–910.

[33] Heifiz A.I. *On subharmonic functions of completely regular growth in the half-plane.* —DAN SSSR, 1978(**239**), 282–285.

[34] Carleman T. *Über die Approximation analytischer Funktionen durch lineare Aggregate van vorgegeben Potenzen.* —Arkiv fór Mat., Astr. och Fysik, 1923(**17**), N9, 1–30.

[35] Dahlberg B. *Mean values of subharmonic functions.* —Arkiv för mat., 1972(**10**), 293–309.

[36] Hayman W.K. *Questions of regularity connected with the Phragmen -Lindelöf principle.* —J. Math. pures et appl., 1956(**35**), 115–126.

[37] Nevanlinna R. *Über die Eigenschaften meromorpher Funktionen in einem Winkelraum.* —Acta Soc. Sci. Fenn., 1925(**50**), N12, 1–45.

[38] Paley R.E.A.C. *A note on integral functions.* —Proc. Cambr. Phil. Soc., 1932(**28**), 261–265.

[39] Pfluger A. *Über gewisse ganze Funktionen vom Exponentialtypus.* —Comment. Math. Helv., 1944(**16**), 1–18.

[40] Valiron G. *Sur un thèoremè de M. Wiman.* —Opuscula math. A. Wiman dedicata, 1930, 1–12.

[41] Wahlund A. *Über einen Zusammenhang zwischen dem Jensensche Theoreme.* —Arkiv för Mat., 1929(**21**), N23, 1–34.

MIX
Papier aus verantwortungsvollen Quellen
Paper from responsible sources
FSC® C105338

If you have any concerns about our products,
you can contact us on
ProductSafety@springernature.com

In case Publisher is established outside the EU,
the EU authorized representative is:
Springer Nature Customer Service Center GmbH
Europaplatz 3, 69115 Heidelberg, Germany

Printed by Libri Plureos GmbH
in Hamburg, Germany